PROGRAMMING IN C#

A Primer

Second Edition

About the Author

E Balagurusamy, former Vice Chancellor, Anna University, Chennai, is currently Member, Union Public Service Commission, New Delhi. He is a teacher, trainer, and consultant in the fields of Information Technology and Management. He holds an ME (Hons) in Electrical Engineering and Ph.D. in Systems Engineering from the Indian Institute of Technology, Roorkee. His areas of interest include Object-Oriented Software Engineering, Electronic Business, Technology Management, Business Process Re-engineering, and Total Quality Management.

A prolific writer, he has authored a large number of research papers and several books. His best selling books, among others include:

- Programming in ANSI C, 4/e
- Programming in Java, 3/e
- Object-Oriented Programming with C++, 3/e
- Programming in BASIC, 3/e
- Numerical Methods
- Reliability Engineering

A recipient of numerous honours and awards, he has been listed in the Directory of Who's Who of Intellectuals and in the Directory of Distinguished Leaders in Education.

PROGRAMMING IN C#

A Primer

Second Edition

E BALAGURUSAMY
Member, UPSC
New Delhi

Tata McGraw-Hill Publishing Company Limited
NEW DELHI

McGraw-Hill Offices
New Delhi New York St Louis San Francisco Auckland Bogotá Caracas
Kuala Lumpur Lisbon London Madrid Mexico City Milan Montreal
San Juan Santiago Singapore Sydney Tokyo Toronto

Tata McGraw-Hill

Published by Tata McGraw-Hill Publishing Company Limited,
7 West Patel Nagar, New Delhi 110 008.

ISBN (13): 978-0-07-067813-2
ISBN (10): 0-07-067813-8

Managing Director: *Ajay Shukla*
General Manager: Publishing—SEM & Tech Ed: *Vibha Mahajan*
Asst. Sponsoring Editor: *Shalini Jha*
Executive—Editorial Services: *Sohini Mukherjee*
Senior Proof Reader: *Suneeta S Bohra*

General Manager: Marketing—Higher Education & School: *Michael J Cruz*
Product Manager: SEM & Tech Ed: *Biju Ganesan*

Controller—Production: *Rajender P Ghansela*
Asst. General Manager—Production: *B L Dogra*

Typeset at Bukprint India, B-180A, Guru Nanak Pura, Laxmi Nagar, Delhi 110 092,

DLLQCDDYRCCCC

The McGraw-Hill Companies

Preface

It gives me great pleasure to present the second edition of *Programming in C#*. During the years that the first edition has been in circulation, I have received a lot of favourable responses, comments and suggestions, and I have tried to keep them in mind while preparing the script of the second edition. The new edition has been updated with C# version 2005 and now contains updated topics and enhanced pedagogical features.

As stated earlier, this book has not been written keeping any specific syllabus in mind, rather it can be used by any one who is desirous of developing C# programs and will be very useful for the first course in C# taken by undergraduate students in Computers and Information Technology. The revised edition maintains the lucid flow and continuity which has been the strength of the book. The topics and the script in itself are well-graded and it takes the student through a step-by-step process, starting from simple programming problems to more complex and difficult ones. The specific improvements that have been added in the edition are the following:

- The book has been updated with C# version 2005. Validated HTML coding (part of Web 2.0) is used in the examples, which has been generated by the Visual Studio 2005 IDE
- Very important inclusion would be new chapters on 'Multithreading' and 'Windows and Web-based Application Development on .NET'
- Two sample programs in each chapter using latest version of C#
- One case study per chapter
- Additional new problems/debugging exercises/programs and review questions
- Major and Mini projects
- Walkthrough of the major project

The book has been divided into twenty chapters. Chapter 1 introduces the C# language, its development, characteristics and evolution. Chapter 2 discusses the .NET Framework, its benefits and the relationship between C# and .NET. Chapter 3 introduces the reader to the first steps of building a C# program. Chapters 4 to 7 define literals, variables, data types, operators, expressions, decision making, branching and looping statements and how to use these in C# programs. Chapter 8 discusses declaring, invoking, nesting of methods and method overloading. Chapter 9 explains creation of arrays and the different types of arrays. Chapter 10 discusses string manipulation and Chapter 11 is on structures and enumerations.

The basic principles of object-oriented programming, classes, objects, constructors and destructors, are covered in Chapter 12. Chapter 13 explains inheritance and polymorphism in C# and Chapter 14 goes a bit further to introduce interfaces and multiple inheritance. Operator overloading is covered in Chapter 15 while delegates and events are discussed in Chapter 16. Chapter 17 deals with managing console I/O operations and their standard numeric format and custom numeric format. Chapter 18 discusses the various types of errors and exceptions occurring in C# and explains how to manage them. Chapter 19 extensively covers the concept of multithreading and finally, Chapter 20 discusses web-based applications and their development in .NET. Two major and two minor projects have been introduced in this edition, with one each being given in the book and the website respectively. The projects are elaborate case studies in themselves and offer ample hands-on practice to students at developing real-life C# applications.

The online learning center can be accessed at the http://www.mhhe.com/balagurusamy/csharp2e and has an exhaustive content which will prove very useful for both instructors and students alike.

For the Instructor

- Solutions to the debugging exercises.

For the Student

- Downloadable programs for students.
- Exclusive major (Employee Management System, Basic IO Example) and mini projects (Calculate Example, Random Number Application) for implementation with code, step-by-step description and user manual.
- Code for all major and mini projects. This would include code to the new major and mini projects as well.
- Web links to more information on C#.

I would like to thank all those who provided me with valuable feedback and inputs during the preparation of this book, and especially those at Tata McGraw-Hill, without whose help and cooperation, this book would not have had a timely release. The support, patience and inspiration that I got from my wife, Dr Sushila, is something which I cherish above all. And finally, a special thanks is also due to all my teacher friends and students for their encouragement.

Further suggestions for improvement will always be welcome.

E BALAGURUSAMY

Preface to the First Edition

C# (pronounced 'C Sharp') is a new programming language. It has been developed by Microsoft Corporation as part of their .NET strategy to provide web-based services. C# is a pure object-oriented language which supports the component-based approach for software development. According to Microsoft Chairman Bill Gates, the next revolutionary step in software development will be to make the Internet intelligent. C# as a programming tool fits in very well here. C# promises to help us ride the next wave of computing, namely, "software as a service".

C# is a simple but powerful language. It combines the concept of C, power of C++, elegance of Java, and productivity of Visual BASIC, besides having new features to support component-based programming. Since Microsoft introduced C# as a *de facto* language of their .NET platform, it supports all the key features of .NET natively. C# is, therefore, ideally suited for component-based distributed Web applications.

This book is written for inexperienced as well as experienced programmers. It does not consider any significant programming knowledge as a prerequisite but assumes that the primary objective of the reader is to develop C# programs. While experienced programmers in C, C++, or Java may find the material in some of the initial chapters simple and familiar the beginners will find them new and useful and therefore, would need to understand and master them carefully.

The book comprehensively covers all major aspects of C# language. Since C# is a part of .NET revolution, it is closely linked to the .NET Framework. The book therefore begins with the necessary introduction to the language and its relationship with .NET strategy and then moves on to explore the object-oriented features such as classes, inheritance, interfaces and polymorphism. It also introduces several new features that are unique to C# such as properties, indexers, delegates, events, and namespaces.

Large number of illustrations and example programs are provided to reinforce learning. Wherever necessary, concepts are explained pictorially to facilitate easy grasping and better understanding. Each chapter includes a set of review questions and programming exercises which can be used by the readers to test their understanding of the concepts discussed in the chapter.

I hope everyone who desires to be a part of the next generation of computing will find this book interesting and useful.

E BALAGURUSAMY

Contents

Chapter

1

Introducing C#

1.1 WHAT IS C#?

'C#' (pronounced as 'C sharp') is a new computer-programming language developed by Microsoft Corporation, USA. C# is a fully *object-oriented* language like Java and is the first *Component-oriented* language. It has been designed to support the key features of *.NET Framework*, the new development platform of Microsoft for building component-based software solutions. It is a simple, efficient, productive and type-safe language derived from the popular C and C++ languages. Although it belongs to the family of C / C++, it is a purely objected-oriented, modern language suitable for developing *Web-based* applications.

C# is designed for building robust, reliable and durable components to handle real-world applications. Major highlights of C# are:

- It is a brand new language derived from the C / C++ family
- It simplifies and modernizes C++
- It is the only component-oriented language available today
- It is the only language designed for the .NET Framework
- It is a concise, lean and modern language
- It combines the best features of many commonly used languages: the productivity of Visual Basic, the power of C++ and the elegance of Java
- It is intrinsically object-oriented and web-enabled
- It has a lean and consistent syntax
- It embodies today's concern for simplicity, productivity and robustness
- It will become the language of choice for .NET programming
- Major parts of .NET Framework are actually coded in C#

Today's World Wide Web consists of a large number of individual web sites that do not co-operate. The next generation of the Web is expected to have dynamic co-operating web sites on its network. The .NET platform and its technologies developed by Microsoft will enable such co-operation among web sites. C# is expected to play a major role in developing applications on co-operating networks of web sites.

1.2 WHY C#?

A large number of computer languages, starting from FORTRAN developed in 1957 to the object-oriented language Java introduced in 1995, are being used for various applications. The choice of a language depends upon many factors such as hardware environment, business environment, user requirements and so on. The primary motivation while developing each of these languages has been the concern that it be able to handle the increasing complexity of programs that are robust, durable and maintainable. The history of major languages developed during the last three decades is given in Fig. 1.1:

C and C++ have been the two most popular and most widely used languages in the software industry for the past two decades. They provide programmers with a tremendous amount of power and control for developing scientific, commercial and business applications. However, these languages suffer from a number of shortcomings in meeting the emerging World Wide Web requirements and standards. Of concern are the following:

- The high complexity of the language
- Their long cycle-time
- They are not truly object-oriented
- They are not suitable for working with new web technologies
- They have poor type-safety
- They are prone to costly programming errors
- They do not support versioning
- They are prone to memory leakages
- Their low productivity
- Their poor interoperability with the existing systems
- They are weak in consistency
- Their poor support for component programming

Visual Basic (VB), a language promoted by Microsoft for overcoming these problems, also could not meet some of the requirements of the www. Since VB is not truly an object-oriented programming language, it becomes increasingly difficult to use when systems become large.

Java, a language derived from C / C++ family, however, is truly object-oriented and has been widely used for web applications for the past five

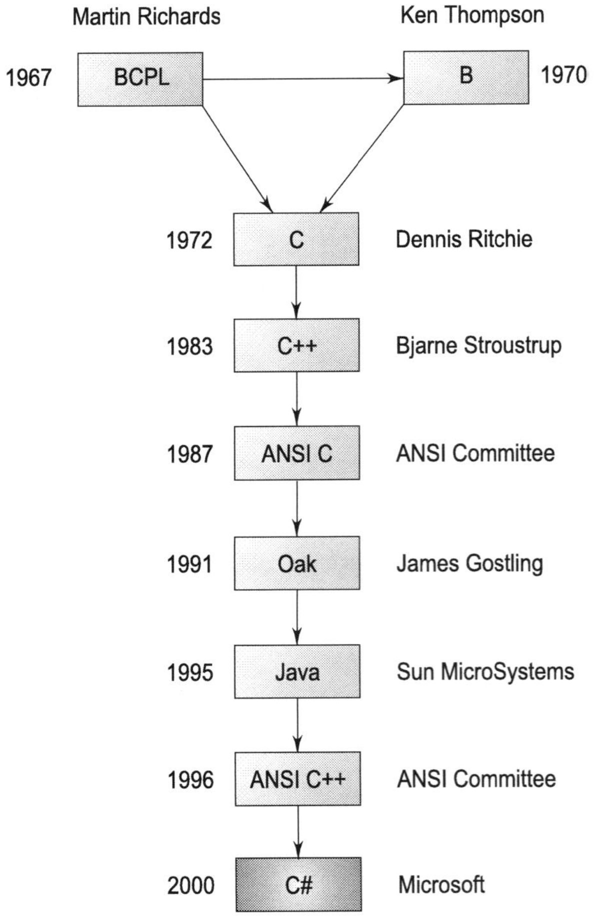

Figure 1.1 *Looking back*

years. Unfortunately, Java has not retained some powerful C++ features such as operator overloading. It also lacks inter-operability with code developed in other languages. Although Microsoft attempted to transform Java and make it a language of choice for their proposed new generation .NET platform, the law suit of Sun Microsystems against Microsoft prevented this from happening.

Microsoft wanted an environment that is completely in tune with current and emerging Web-programming practices and one that easily integrates with existing systems. Microsoft therefore decided to design a new language starting with a clean slate. The result is C#, a simple and modern language that directly addresses the needs of component-based software development.

1.3 EVOLUTION OF C#

The World Wide Web is growing everyday not only in terms of the number of web sites but also in terms of volume and variety of information it contains. Doing business across those web sites is very exciting and useful. The number of individuals and organizations using the Internet, which supports the World Wide Web is increasing exponentially. The Internet is therefore in the mainstream of many business organizations today.

But what is the Internet today? We have a number of limitations in using the WWW over the Internet.

- We can see only one site at a time
- The site has to be authored to our hardware environment
- The information we get is basically read-only
- We cannot compare dynamically similar information stored in different sites
- The Internet is a collection of many information islands that do not co-operate with each other. It continues to be a browsing and presentation network rather than an intelligent knowledge management network.

Microsoft Chairman Bill Gates, the architect of many innovative and path-breaking software products during the past two decades, wanted to develop a *software platform* which will overcome these limitations and enable users to get information anytime and anywhere, using a natural interface. The platform should be a collection of readily available Web services that can be distributed and accessed via standard Internet protocols. He wanted to make the Web both programmable and intelligent. The outcome is a new generation platform called *.NET*. .NET is simply the Microsoft's vision of 'software as a service.'

Although the research and development work of .NET platform began in the mid-90s, only during the Microsoft Professional Developers Conference in September 2000, was .NET officially announced to the developer community. At the same conference, Microsoft introduced C# as a de facto language of the .NET platform. In fact, they had already used C# to code key modules of the .NET platform. C# has been particularly designed to build software components for .NET and it supports key features of .NET natively. They are fairly tightly tied together. In fact, C# compiler is embedded into .NET as shown in Fig. 1.2.

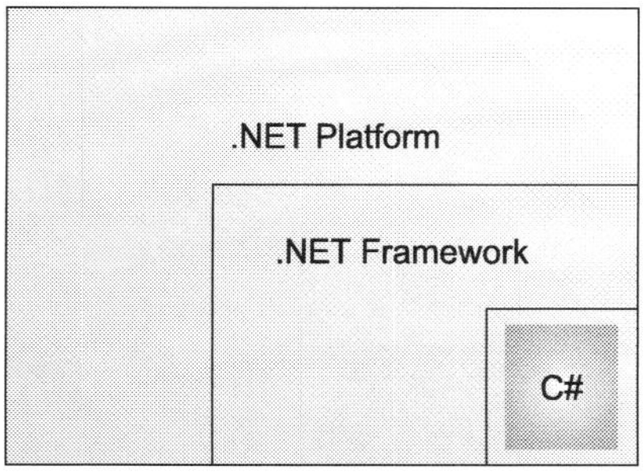

Figure 1.2 *C# inside the .NET*

Like Java, C# is a descendant of C++ which in turn is a descendant of C as illustrated in Fig. 1.3. C is the mother of all the three modern languages. C# modernizes C++ by enhancing some of its features and adding a few new features so as to help developers do more with fewer lines of code and fewer opportunities for error.

C# borrows Java's features such as grouping of classes, interfaces and implementation together in one file so that programmers can edit the code more easily. C# also handles objects using references, the same way as Java. C# uses VB's approach to form design, namely, dragging controls from a tool box, dropping them onto forms, and writing event handlers for them. C# applies this process not only to the development of user interface but also to the building of business objects.

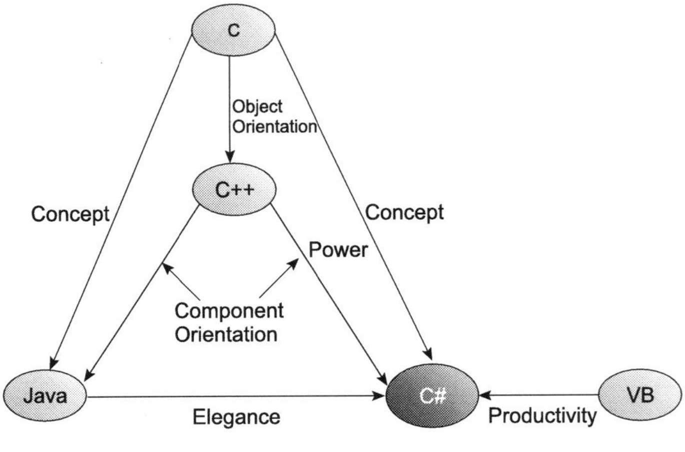

Figure 1.3 *Evolution of C# language*

1.4 CHARACTERISTICS OF C#

As pointed out earlier, the main design goal of C# was simplicity rather than pure power. C# fulfills the need for a language that is easy to write, read and maintain and also provides the power and flexibility of C++. The language that is designed for both computing and communications is characterized by several key features. It is

- Simple
- Consistent
- Modern
- Object-oriented
- Type-safe
- Versionable
- Compatible
- Interoperable and
- Flexible

1.4.1 Simple

C# simplifies C++ by eliminating irksome operators such as ->,:: and pointers. C# treats integer and Boolean data types as two entirely different types. This means that the use of = in place of = = in **if** statements will be caught by the compiler.

1.4.2 Consistent

C# supports an unified type system which eliminates the problem of varying ranges of integer types. All types are treated as objects and developers can extend the type system simply and easily.

1.4.3 Modern

C# is called a modern language due to a number of features it supports. It supports

- Automatic garbage collection
- Rich intrinsic model for error handling
- Decimal data type for financial applications
- Modern approach to debugging and
- Robust security model

1.4.4 Object-Oriented

C# is truly object-oriented. It supports all the three tenets of object-oriented systems, namely,

- Encapsulation
- Inheritance
- Polymorphism

The entire C# class model is built on top of the Virtual Object System (VOS) of the .NET Framework In C#, everything is an object. There are no more global functions, variables and constants.

1.4.5 Type-safe

Type-safety promotes robust programs. C# incorporates a number of type-safe measures.

- All dynamically allocated objects and arrays are initialized to zero
- Use of any uninitialized variables produces an error message by the compiler
- Access to arrays are range-checked and warned if it goes out-of-bounds
- C# does not permit unsafe casts
- C# enforces overflow checking in arithmetic operations
- Reference parameters that are passed are type-safe
- C# supports automatic garbage collection

1.4.6 Versionable

Making new versions of software modules work with the existing applications is known as *versioning*. C# provides support for versioning with the help of **new** and **override** keywords. With this support, a programmer can guarantee that his new class library will maintain binary compatibility with the existing client applications.

1.4.7 Compatible

C# enforces the .NET common language specifications and therefore allows inter-operation with other .NET languages.

C# provides support for transparent access to standard COM and OLE Automation.

C# also permits interoperation with C-style APIs.

1.4.8 Flexible

Although C# does not support pointers, we may declare certain classes and methods as 'unsafe' and then use pointers to manipulate them. However, these codes will not be type-safe.

1.4.9 Inter-operability

C# provides support for using COM objects, no matter what language was used to author them. C# also supports a special feature that enables a program to call out any native API.

1.5 APPLICATIONS OF C#

As pointed out earlier, C# is a new language developed exclusively to suit the features of .NET platform. It can be used for a variety of applications that are supported by the .NET platform:

- Console applications
- Windows applications
- Developing Windows controls
- Developing ASP.NET projects
- Creating Web controls
- Providing Web services
- Developing .NET component library

1.6 HOW DOES C# DIFFER FROM C++ ?

As stated earlier, C# was derived from C++ to make it the language of choice for C and C++ programmers. C# therefore shares major parts of syntax with C++. However, the C# designers introduced a few changes in the syntax of C++ and removed a few features primarily to reduce the common pitfalls that occurred in C++ program development. They also added a number of additional features to make C# a type-safe and web-enabled language.

Changes Introduced

01. C# compiles straight from source code to executable code, with no object files.
02. C# does not separate class definition from implementation. Classes are defined and implemented in the same place and therefore there is no need for header files.
03. In C#, class definition does not use a semicolon at the end.
04. The first character of the **Main( )** function is capitalized. The **Main** must return either **int** or **void** type value.
05. C# does not support **#include** statement. (Note that **using** is not the same as **#include).**
06. All data types in C# are inherited from the **object** super class and therefore they are objects.
07. All the basic value types will have the same size on any system. This is not the case in C or C++. Thus C# is more suitable for writing distributed applications.
08. In C#, data types belong to either **value types** (which are created in a stack) or **reference types** (which are created in a heap).
09. C# checks for uninitialized variables and gives error messages at compile time. In C++, an uninitialized variable goes undetected thus resulting in unpredictable output.
10. In C#, structs are value types.
11. C# supports a native string type. Manipulation of strings are easy.
12. C# supports a native Boolean data type and bool-type data cannot be implicitly or explicitly cast to any data type except object.
13. C# declares **null** as a keyword and considers it as an intrinsic value.
14. C# does not support pointer types for manipulating data. However, they are used in what is known as 'unsafe' code.
15. Variable scope rules in C# are more restrictive. In C#, duplicating the same name within a routine is illegal, even if it is in a separate code block.
16. C# permits declaration of variables between **goto** and label.
17. We can only create objects in C# using the **new** keyword.
18. Arrays are classes in C# and therefore they have built-in functionality for operations such as sorting, searching and reversing.
19. Arrays in C# are declared differently and behave very differently compared to C++ arrays.

(Contd.)

20. C# provides special syntax to initialize arrays efficiently.
21. Arrays in C# are always reference types rather than value types, as they are in C++ and therefore stored in a heap.
22. In C#, expressions in **if** and **while** statements must resolve to a **bool** value. Accidental use of the assignment operator (=) instead of equality operator = = will be caught by the compiler.
23. C# supports four iteration statements rather than three in C++ . The fourth one is the **foreach** statement.
24. C# does not allow silent fall-through in switch statements. It requires an explicit jump statement at the end of each case statement.
25. In C#, **switch** can also be used on string values.
26. C# does not support the labeled break and therefore jumping out of nested loops can be messy.
27. The set of operators that can be overloaded in C# is smaller compared to C++.
28. C# can check overflow of arithmetic operations and conversions using **checked** and **unchecked** keywords.
29. C# does not support default arguments.
30. Variable method parameters are handled differently in C#.
31. In exception-handling, unlike in C++, we cannot throw any type in C#. The thrown value has to be a reference to a derived class or **System.Exception** object.
32. C# requires ordering of **catch** blocks correctly.
33. General catch statement **catch (...)** in C++ is replaced by simple **catch** in C#.
34. C# does not provide any defaults for constructors.
35. Destructors in C# behave differently than in C++.
36. In C#, we cannot access static members via an object, as we can in C++.
37. C# does not support multiple code inheritance.
38. Casting in C# is much safer than in C++.
39. When overriding a virtual method, we must use the **override** keyword.
40. Abstract methods in C# are similar to virtual functions in C++, but C# abstract methods cannot have implementations.
41. Command-line parameters array behave differently in C# as compared to C++.

C++ features dropped

The following C++ features are missing from C#:

1. Macros
2. Multiple inheritance
3. Templates
4. Pointers
5. Global variables
6. **Typedef** statement
7. Default arguments
8. Constant member functions or parameters
9. Forward declaration of classes

Enhancements to C++

C# modernizes C++ by adding the following new features:

1. Automatic garbage collection
2. Versioning support
3. Strict type-safety
4. Properties to access data members
5. Delegates and events
6. Boxing and unboxing
7. Web services

1.7 HOW DOES C# DIFFER FROM JAVA ?

Like C#, Java was also derived from C++ and therefore they have similar roots. Moreover, C# was developed by Microsoft as an alternative to Java for web programming. C# has borrowed many good features from Java, which has already become a popular Internet language. However, there exist a number of differences between C# and Java:

01. Although C# uses .NET runtime that is similar to Java runtime, the C# compiler produces an executable code.
02. C# has more primitive data types.
03. Unlike Java, all C# data types are objects.
04. Arrays are declared differently in C#.
05. Although C# classes are quite similar to Java classes, there are a few important differences relating to constants, base classes and constructors, static constructors, versioning, accessibility of members etc.
06. Java uses **static final** to declare a class constant while C# uses **const.**
07. The convention for Java is to put one public class in each file and in fact, some compilers require this. C# allows any source file arrangement.
08. C# supports the **struct** type and Java does not.
09. Java does not provide for operator overloading.
10. In Java, class members are virtual by default and a method having the same name in a derived class overrides the base class member. In C#, the base member is required to have the **virtual** keyword and the derived member is required to use the **override** keyword.
11. The **new** modifier used for class members has no complement in Java.
12. C# provides better versioning support than Java.
13. C# provides **static** constructors for initialization.
14. C# provides built-in delegates and events. Java uses interfaces and inner classes to achieve a similar result.
15. In Java, parameters are always passed by value. C# allows parameters to be passed by reference by using the **ref** keyword.
16. C# adds **internal**, a new accessibility modifier. Members with internal accessibility can be accessed from other classes within the same project, but not from outside the project.
17. C# includes native support for properties, Java does not.
18. Java does not directly support enumerations.
19. Java does not have any equivalent to C# indexers.
20. Both Java and C# support interfaces. But, C# does not allow type definitions in interfaces, while Java interfaces can have **const** type data.
21. In Java, the **switch** statement can have only integer expression, while C# supports either an integer or string expressions.
22. C# does not allow free fall_through from case to case.
23. C# provides a fourth type of iteration statement, **foreach** for quick and easy iterations over collections and array type data.
24. Catch blocks should be ordered correctly in C#.
25. C# checks overflows using **checked** statements.
26. C# uses **is** operator instead of **instanceof** operator in Java.
27. C# allows a variable number of parameters using the **params** keyword.
28. There is no labeled **break** statement in C#. The **goto** is used to achieve this.

Case Study

Problem Statement

WebDev Enterprises is a medium sized company that develops Web applications using J2EE technology for its overseas clients such as BlueSoft Software Solutions and InterSolutions Private Limited. The management at WebDev recently conducted an analysis and discovered that it has not been able to use the code of its existing applications developed in J2EE when developing new J2EE applications. As a result, the deadlines for the new projects were not being met causing financial loss to WebDev Enterprises. Because of not being able to meet deadlines, WebDev Enterprises was losing its clients. In addition, the management of WebDev Enterprises found that it was a complex task to develop applications in J2EE. How can the management of WebDev solve the problem?

Solution

In order to be able to reuse the code of existing applications, WebDev must start creating Web based applications using C# and ASP.NET. C# provides features such as **inheritance** and **interface** that allow the reusability of code. This means that code written for a specific application in C# can be easily used for another C# application. The **inheritance** feature of C# allows a class to inherit properties, which contain data and methods of another class. For example, if we create a class **A** and define certain properties such as name and address in it, and a **display** method to display the values contained in the properties, then we can use the properties and methods of class **A** in another class **B**.

Interfaces in C# allow us to declare abstract methods. These interfaces can be implemented in multiple classes. Each class in which the interface is implemented can contain a specific code for the abstract methods defined in the interface. Thus, with the use of C# features such as **inheritance** and **interface**, code of one C# application can be reused in another C# application. As a result, development time for creating an application in C# is reduced.

With the use of C#, programmers working at WebDev Enterprises could develop applications efficiently and in less time.

C# programming language is supported by Microsoft Visual Studio 2005, which is a rapid development tool for developing Windows based and Web based applications. The support for .NET Framework in C# allows us to create interactive user interfaces for applications in less time. The .NET Framework provides the **Control** class that helps use controls, their properties and methods to develop interactive user interfaces for an application in less time.

Because of all these reasons, which include C# support in Microsoft Visual Studio 2005 and inheritance, the programmers working in WebDev were able to meet the deadlines for the new projects and satisfy the requirements of the overseas clients.

ASP.NET is a technology supported by Microsoft Visual Studio 2005. It helps in rapid development of Web based applications. ASP.NET also supports C# programming language for development of Web based applications. In ASP.NET, programmers at WebDev Enterprises can use the controls available in Microsoft Visual Studio 2005 through the Microsoft .NET Framework. The controls available in Visual Studio 2005 and supported by ASP.NET can be used to create a graphical user interface for Web based applications in less amount of time.

Thus, with the use of C# and ASP.NET, WebDev has been able to solve its problem and rapidly develop applications for its overseas clients.

Review Questions

1.1 State whether the following statements are true or false.
   (a) C# is the first component-oriented language.
   (b) C# is not suitable for developing scientific applications.
   (c) C# compiler is embedded into the .NET Framework.
   (d) C# is a simple extension of Visual Basic language.
   (e) C# supports a feature known as versioning.
   (f) In C#, all data types are treated as objects.
   (g) Class definition of C# is identical to the class definition of C++.
   (h) Like C++, C# supports multiple code inheritance.
   (i) Array declarations in C# are identical to Java array declarations.
   (j) C# was first officially announced by Microsoft in 2000.

1.2 State atleast five most important highlights of C# language.

1.3 What are the shortcomings of C++ language? List atleast five of them.

1.4 What are the features of Java which have been implemented in C#?

1.5 How is C# better than Java?

1.6 What are the limitations of using the Internet today?

1.7 What is the primary motivation of Microsoft Corporation in developing C# language?

1.8 C# is a modem language. Comment.

1.9 Why is C# called type-safe language?

1.10 State ten significant differences between C# and C++.

1.11 State ten significant differences between C# and Java.

1.12 State five features of C++ that have not been incorporated into C#.

1.13 State some of the new features that are unique to C# language.

1.14 Explain the need for Virtual Object System (VOS) in C#.

Chapter

2

Understanding .NET: The C# Environment

2.1 THE .NET STRATEGY

We briefly discussed in the previous chapter how C# was created as a language for developing systems for the .NET platform. As pointed out, Microsoft wanted to make the World Wide Web (WWW) more vibrant by enabling individual devices, computers and web services to work together intelligently to provide richer solutions to users. By the intelligent integration of web sites on the Internet, users can create a wide variety of value-based applications such as unified banking services, electronic bill payment, stock trading, insurance services and comprehensive supply chain management. Microsoft calls this 'Web services' and the software strategy for implementing and delivering these services is '.NET'.

.NET is a software framework that includes everything required for developing software for web services. It integrates presentation technologies, component tech-nologies and data technologies on a single platform so as to enable users to develop Internet applications as easily as they do on desktop systems. Microsoft took many of the best ideas in the industry, added their own creativity and innovations and produced a coherent systems solution popularly known as *Microsoft .NET*. The *Microsoft .NET* software solution strategy includes three key approaches as shown in Fig. 2.1.

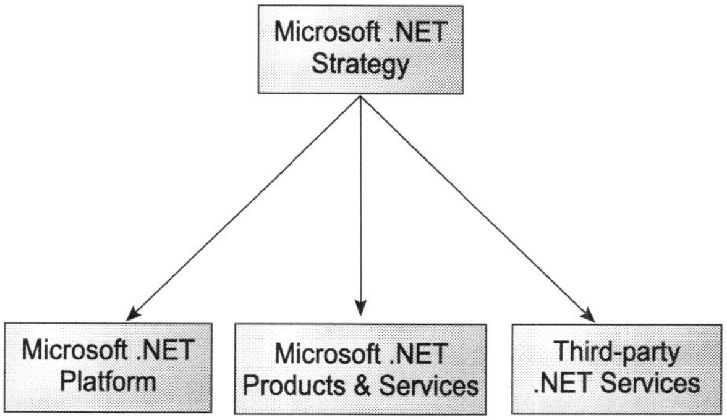

Figure 2.1 *Microsoft .NET strategy*

Microsoft .NET platform includes the following components that would help develop a new generation of smart Internet services:

- .NET infrastructure and tools
- .NET user experience
- .NET building block
- .NET device software

Microsoft .NET products and services consist of the following:

- Windows .NET
- MSN .NET
- Office .NET
- Visual Studio .NET
- Personal subscription services
- bCentral for .NET

Third-party .NET services will provide opportunities to a vast range of developers and users to produce corporate and vertical services using .NET platform.

2.2 THE ORIGINS OF .NET TECHNOLOGY

Before going into the details of .NET features further, we shall see how the concept of .NET was evolved over the past ten years. The current technology of .NET has gone through three significant phases of development:

- OLE technology
- COM technology
- .NET technology

These developments that took place during the 1990s are illustrated in Fig. 2.2.

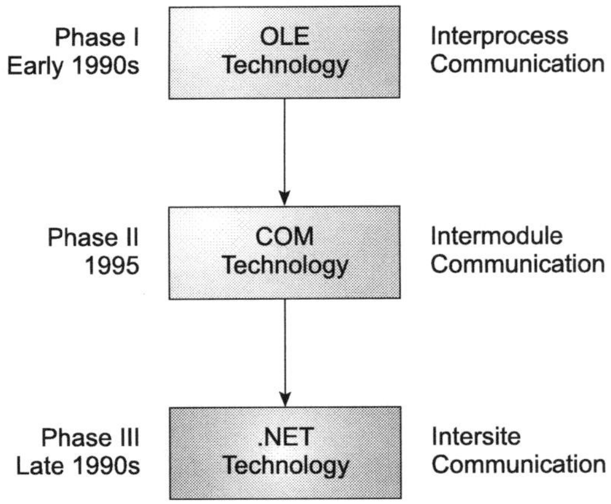

Figure 2.2 *Three generations of component model*

2.2.1 OLE Technology

OLE (Object Linking and Embedding) technology was developed by Microsoft in the early 1990s to enable easy interprocess communications. OLE provided support to achieve the following:

- To embed documents from one application into another application
- To enable one application to manipulate objects located in another application

This enabled users to develop applications which required inter-operability between various products such as MS Word and MS Excel.

2.2.2 COM Technology

Till the advent of COM technology, the *monolithic* approach had been used for developing software. But when programs become too large and complex, the monolithic approach leads to a number of problems in terms of maintainability and testing of software. To overcome these problems, Microsoft introduced the component-based model for developing software programs. In the component-based approach, a program is broken into a number of independent components where each one offers a particular service. Each component can be developed and tested independently and then integrated it into the main system. This technology is known as the *Component Object Model* (COM) and the software built using COM is referred to as *componentware.*

COM technology offers a number of benefits to developers and users. It

- reduces the overall complexity of software
- enables distributed development across multiple organizations or departments and
- enhances software maintainability

2.2.3 .NET Technology

.NET technology is a third-generation component model. This provides a new level of inter-operability compared to COM technology. COM provides a standard binary mechanism for inter-module communication. This mechanism is replaced by an intermediate language called *Microsoft Intermedia Language* (MSIL) or simply IL in the .NET technology. Various .NET-language compilers enforce inter-operability by compiling code into IL, which is automatically compatible with other IL modules. An inherent characteristic of IL code is *metadata*. Metadata is data about data and describes its characteristics, including data types and locations. IL allows for true cross-language integration. In addition to IL, .NET includes a host of other technologies and tools that will enable us develop and implement Web-based applications easily.

2.3 THE .NET FRAMEWORK

The .NET Framework is one of the tools provided by the .NET infrastructure and tools component of the .NET platform as shown in Fig. 2.3. As pointed out earlier, the .NET platform provides a new environment for creating and running robust, scalable and distributed applications over the Web. A detailed description of either the .NET strategy or the .NET platform is beyond the scope of this book. Since C# derives much of its power from the .NET Framework on which it runs, we shall briefly discuss the key features of the .NET Framework.

The .NET Framework provides an environment for building, deploying and running web services and other applications. It consists of three distinct technologies as shown in Fig. 2.4.

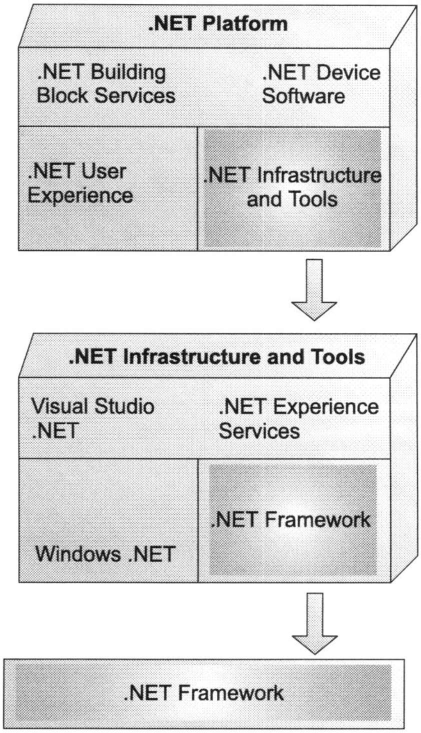

Figure 2.3 *Various Components of .NET platform*

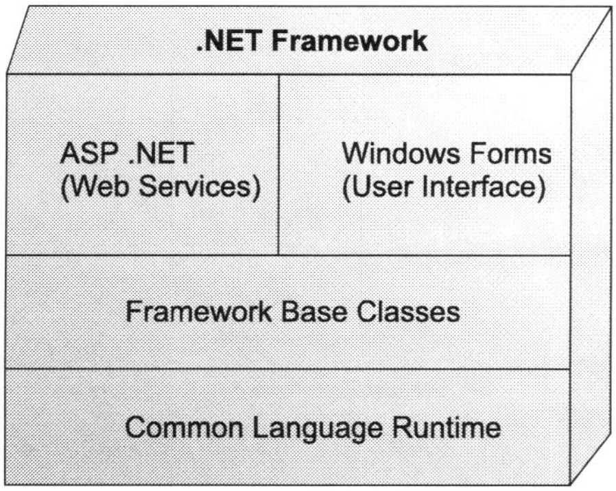

Figure 2.4 *Architecture of .NET Framework*

- Common Language Runtime (CLR)
- Framework Base Classes
- User and program interfaces (ASP .NET and Winforms)

The CLR is the core of the .NET Framework and is responsible for loading and running C# programs. Base classes provide basic data types, collection classes and other general classes for use by C# and other .NET languages. The top layer contains a set of classes for developing web services and to deal with the user interface.

2.4 THE COMMON LANGUAGE RUNTIME

The Common Language Runtime, popularly known as CLR is the heart and soul of the .NET Framework. As the name suggests, CLR is a runtime environment in which programs written in C# and other .NET languages are executed. It also supports *cross-language interoperability*. Figure 2.5 gives a diagrammatic summary of the major components of the CLR.

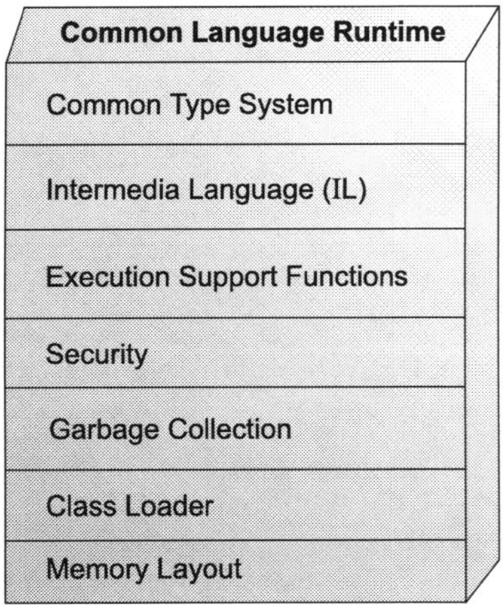

Figure 2.5 *Components of CLR*

The CLR provides a number of services that include:

- Loading and execution of programs
- Memory isolation for applications
- Verification of type safety
- Compilation of IL into native executable code
- Providing metadata
- Memory management (automatic garbage collection)
- Enforcement of security
- Interoperability with other systems
- Managing exceptions and errors
- Support for tasks such as debugging and profiling

Figure 2.6 shows a flow chart of CLR activities that go on when an application is executed. The source code is compiled to IL while the metadata engine creates metadata information. IL and metadata are linked with other native code if required and the resultant IL code is saved. During execution, the IL code and any requirement from the base class library are brought together by the class loader. The combined code is tested for type-safety and then compiled by the JIT compiler to produce native machine code, which is sent to the runtime manager for execution.

2.4.1 Common Type System (CTS)

The .NET Framework provides multiple language support using the feature known as *Common Type System* that is built into the CLR. The CTS supports a variety of types and operations found in most programming languages and therefore calling one language from another does not require type conversions. Although C# is specially designed for the .NET platform, we can build .NET programs in a number of other languages including C++ and Visual Basic.

2.4.2 Common Language Specification (CLS)

The Common Language Specification defines a set of rules that enables interoperability on the .NET platform. These rules serve as a guide to third-party compiler designers and library builders. The CLS is a subset of CTS and therefore the languages supporting the CLS can use each others' class libraries as if they are their own. Application Program Interfaces (APIs) that are designed following the rules of CLS can easily be used by all the .NET languages.

2.4.3 Microsoft Intermediate Language (MSIL)

MSIL, or simply IL, is an instruction set into which all the .NET programs are compiled. It is akin to assembly language and contains instructions for loading, storing, initializing and calling methods. When we compile a C# program or any program written in a CLS-compliant language, the source code is compiled into MSIL.

2.4.4 Managed Code

As we know, the CLR is responsible for managing the execution of code compiled for the .NET platform. The code that satisfies the CLR at runtime in order to execute is referred to as *managed code*. Compilers that are compatible to the .NET platform generate managed code. For example, the C# compiler generates managed code. The managed code generated by C# (and other compilers capable of generating managed code) is IL code. The IL code is then converted to *native machine code* by the JIT compilers.

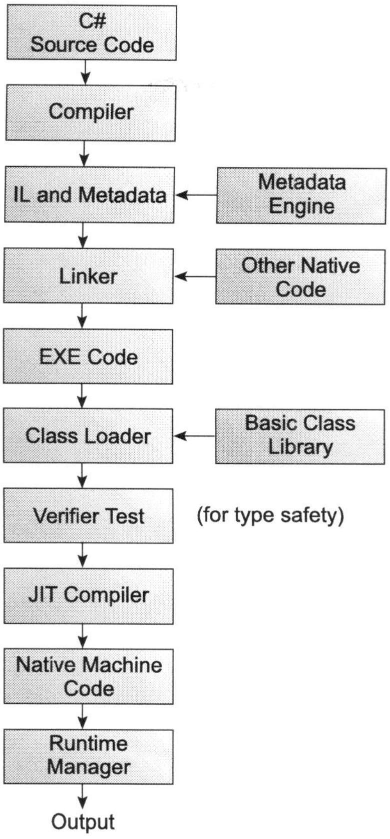

Figure 2.6 *Flowchart of CLR activities for executing a program*

2.5 FRAMEWORK BASE CLASSES

.NET supplies a library of base classes that we can use to implement applications quickly. We can use them by simply instantiating them and invoking their methods or by inheriting them through derived classes, thus extending their functionality.

Much of the functionality in the base framework classes resides in the vast namespace called **System**. We can use the base classes in the system namespace for many different tasks including:

- Input/Output Operations
- String handling
- Managing arrays, lists, maps, etc
- Accessing files and file systems
- Accessing the registry
- Security
- Windowing
- Windows messages
- Database management
- Evaluation of mathematical functions
- Drawing
- Managing errors and exceptions
- Connecting to the Internet
- And many more

2.6 USER AND PROGRAM INTERFACES

The .NET Framework provides the following tools for managing user- and application interfaces:

- Windows forms
- Web forms
- Console Applications
- Web services

These tools enable users to develop user-friendly desktop-based as well as web-based applications using a wide variety of languages on the .NET platform.

2.7 VISUAL STUDIO .NET

Visual Studio .NET (VS .NET) supports an Integrated Development Environment (IDE) with a rich set of features and productivity tools. These features and tools allow developers to build web applications faster and easier. Using web services and XML regardless of the language chosen for development, there is now one environment to learn, configure and use. We need not have to switch back and forth between environments to build, debug and deploy our code. VS .NET provides tools that extends support to the development lifecycle. As shown in Fig. 2.7, it acts as a foundation for the lifecycle platform. A detailed discussion on these tools is beyond the scope of this book.

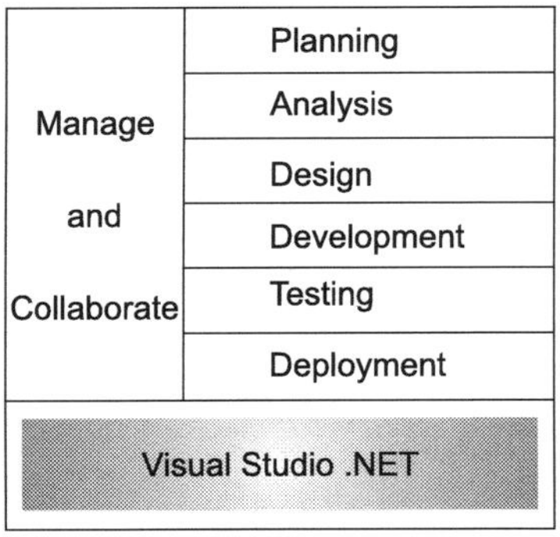

Figure 2.7 *Development lifecycle*

2.8 .NET LANGUAGES

The .NET Framework is language neutral. Currently, we can use a number of languages for developing .NET applications. They include:

2.8.1 Native to .NET

- C# (Specially created for .NET)
- C++
- Visual Basic
- JScript

2.8.2 Third-party Languages

- COBOL
- Eiffel
- Perl
- Python
- SmallTalk
- Mercury
- Scheme

All .NET languages are not created equal. Some can use the components of other languages, some can use the classes produced in other languages to create objects, and some languages can extend the classes of other languages using the inheritance features of .NET.

2.9 BENEFITS OF THE .NET APPROACH

Microsoft has advanced the .NET strategy in order to provide a number of benefits to developers and users. Some of the major benefits envisaged are:

- Simple and faster systems development
- Rich object model
- Enhanced built-in functionality
- Many different ways to communicate with the outside world
- Integration of different languages into one platform
- Easy deployment and execution
- Wide range of scalability
- Interoperability with existing applications
- Simple and easy-to-build sophisticated development tools
- Fewer bugs
- Potentially better performance

2.10 C# AND THE .NET

C# is a new programming language introduced with .NET. It is a concise, elegant .NET language. In many respects it is a version of the .NET object model. With C#, developers can quickly implement applications and components using the built-in capabilities of the .NET Framework. Since the C# code is managed by the CLR (of the .NET Framework) it becomes leaner and safer than C++.

The CLR extends a number of benefits to C# when it is implemented on the .NET platform. These include:

- Interoperability with other languages
- Enhanced security
- Versioning support
- Debugging support
- Automatic garbage collection
- XML support for web-based applications

Case Study

Problem Statement

ABC Finance Solutions is a finance company involved in providing financial solutions to different individuals. The main task of ABC Finance Solutions is to provide suggestions to individuals on how to invest money. The IT Research and Development department in ABC Finance Solutions is currently involved in developing a software for financial planning to automate the task of creating financial plans, which can be provided to individuals so that they can invest money according to that plan. The development of software for financial planning is a difficult task as it involves doing a lot of complex calculations. How can the programmers involved in developing the financial planning software make their task easy?

Solution

Since the development of financial planning software involves performing a lot of complex calculations, the team of programmers at ABC Finance Solutions must use C# to create the financial planning software. In the financial planning software, the functionality for Monte Carlo modelling has to be provided, which is a very complex task. A large number of objects need to be created for different classes in the financial planning software. The objects in the financial planning software need to interact with each other through sending and receiving of messages. If the programmers at ABC Finance Solutions use C++ programming language and create Component Object Model (COM) and ALT objects in the financial planning software then the financial planning software will become complex and time consuming. With the use of COM and ALT objects in a C++ application, it will take the programmers of ABC Finance Solutions a large amount of time to develop the financial planning software. Debugging and writing code for financial planning software will become easy with the use of C# as most of the functionality required for financial planning software is already available in C#. As the development time for creating the financial planning software reduces with the use of C# programming language, the cost of developing the financial planning software also decreases. This results in financial benefits to the ABC Finance Solutions company.

Review Questions

2.1 What is .NET?
2.2 What is .NET Framework?
2.3 What is .NET technology? Describe briefly its origins.
2.4 List the tools provided by .NET Framework for managing the user- and application interfaces.
2.5 What is Web service? How is it achieved using the .NET strategy?
2.6 What is the Microsoft Intermediate Language?
2.7 What is the Common Language Runtime (CLR)?
2.8 Enumerate major services provided by the CLR.
2.9 What are the additional benefits provided by CLR to programs developed using C#?
2.10 How does the CLR implements a C# program?
2.11 What is the Common Type System?
2.12 What is the Common Language Specification?
2.13 What is managed code?
2.14 List some of the important services the Framework Base Classes can offer to the users.
2.15 Describe the application of Visual Studio .NET?
2.16 List the languages supported by the .NET Framework.
2.17 What are the benefits of .NET strategy advanced by Microsoft?

Chapter

3

Overview of C#

3.1 INTRODUCTION

'C#' seems a strange name for a modern programming language. Perhaps Microsoft named their new language 'C sharp' because they wanted it to be better, smarter and 'sharper' than its ancestors C and C++. C# is designed to bring rapid development to C++ programmers without sacrificing the power and control that have been the hallmarks of C and C++. C# is the only language designed specially for the .NET platform which provides tools and services that fully exploit both computing and communications.

Since one of the designers of C# (Anders Hejsberg) was a Java expert, it is natural that many Java features have been incorporated into the design of C#. In fact, in many cases, the C# code may bear a striking resemblance to the functionally equivalent Java code. Unlike C++, both Java and C# promote a *one-stop coding* approach to code maintenance. They group classes, interfaces and implementations together in one file so that programmers can edit the code more easily.

C# can be used to develop two categories of programs, namely,

- Executable application programs and
- Component libraries

as illustrated in Fig. 3.1. Executable programs are written to carry out certain tasks and require the method **Main** in one of the classes. In contrast, component libraries do not require a **Main** declaration because they are not standalone application programs. They are written for use by other applications. This concept is something similar to applets and application programs in Java.

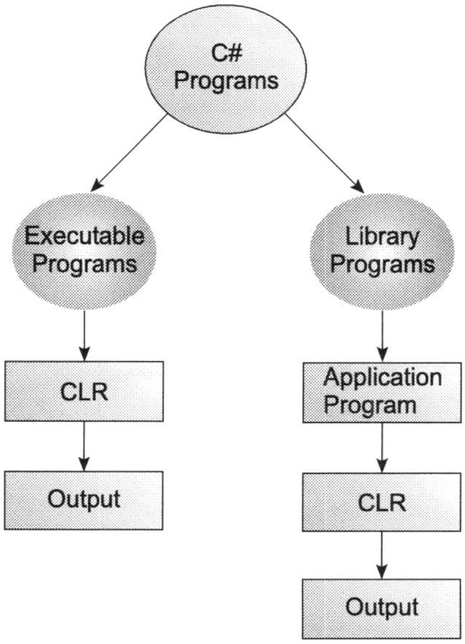

Figure 3.1 *Two kinds of C# programs*

3.2 A SIMPLE C# PROGRAM

The best way to learn a new language like C# is to write a few simple programs and execute them. This will enable us to analyze and understand how C# programs work. We begin with a very simple program that displays a line of text. Consider the code given in Program 3.1.

Program 3.1 A simple C# program

```
class SampleOne
{
    public static void Main( )
    {
    System.Console.WriteLine("C# is sharper than C++.");
    }
}
```

Program 3.1 is the simplest of all C# programs. Nevertheless, it brings out many salient features of the language. Let us therefore discuss the program line by line and understand the unique features that constitute a C# program.

3.2.1 Class Declaration

The first line

```
class SampleOne
```

declares a class, which is an object-oriented construct. As stated earlier, C# is a true object-oriented language and therefore, 'everything' must be placed inside a class. **class** is a keyword and declares that a new class definition follows. **SampleOne** is a C# *identifier* that specifies the name of the class to be defined.

A class is a template for what an object looks like and how it behaves. (More about classes are discussed in Chapter 12).

3.2.2 The Braces

C# is a *block-structured* language, meaning code blocks are always enclosed by braces { and }. Therefore, every class definition in C# begins with an opening brace '{'and ends with a corresponding closing brace '}' that appears in the last line of the program. This is similar to class constructs of Java and C++. Note that there is no semicolon after the closing brace.

3.2.3 The Main Method

The third line

```
public static void Main( )
```

defines a method named **Main**. Every C# executable program must include the **Main( )** method in one of the classes. This is the 'starting point' for executing the program. A C# application can have any number of classes but 'only one' class can have the **Main** method to initiate the execution. Note that C# component libraries will not use the **Main** method at all.

This line contains a number of keywords: **public**, **static** and **void.** This is very similar to the **main** of C++ and Java. In contrast to Java and C++, **Main** has a capital, not lowercase M. The meaning and purpose of these keywords are give below:

public	The keyword **public** is an access modifier that tells the C# compiler that the **Main** method is accessible by anyone
static	The keyword **static** declares that the **Main** method is a global one and can be called without creating an instance of the class. The compiler stores the address of the method as the entry point and uses this information to begin execution before any objects are created.
void	The keyword **void** is a type modifier that states that the **Main** method does not return any value (but simply prints some text to the screen).

Note: It is not essential to declare **Main** as **public**. However, this is how Visual Studio 2005 declares it. Also, many authors prefer this approach. We also therefore declare **Main** as **public** in our examples. It is likely that some authors declare it without the **public** specifier.

3.2.4 The Output Line

The only executable statement in the program is

```
System.Console.WriteLine("C# is sharper than C++.");
```

This has a striking resemblance to the output statement of Java and similar to the **printf ( )** of C or **cout<<** of C++. Since C# is a pure object-oriented language, every method should be part of an object. The **WriteLine** method is a static method of the **Console** class, which is located in the namespace **System**. This line prints the string

```
C# is sharper than C++.
```

to the screen. The method **WriteLine** always appends a new-line character to the end of the string. This means, any subsequent output will start on a new line.

Note the semicolon at the end of the statement. *Every C# statement must end with a semicolon.* And also note that there is no semicolon at the end of the class.

3.2.5 Executing the Program

After creating the program (source code), save it with **.cs** file extension in a folder in your system. We may use 'any suitable name' for the file. In this case, we may use the class name itself. Example:

```
sampleone.cs
```

For *compiling* the program, go to the folder where you have stored the file **sampleone.cs** and then type the following command:

```
csc sampleone.cs
```

The C# Compiler (CSC) compile your code and create an executable file (IL code) by name

```
sampleone.exe
```

in the same directory (if the code is error-free). If there are errors, the compiler will produce appropriate error messages. Errors should be corrected and the program should be compiled again.

For *executing* the program, simply type in the name of the executable file at the command prompt. That is,

```
sampleone
```

Note that no file extension is used. This will display the output

```
C# is sharper than C++.
```

on the screen. Figure 3.2 shows how the code is transformed from source to output.

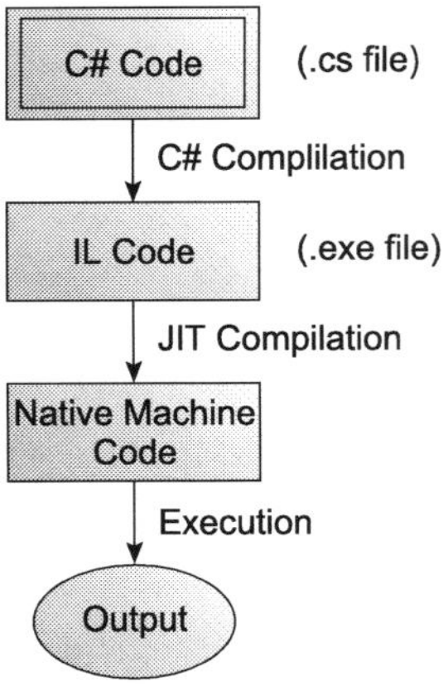

Figure 3.2 *Three stages of code transformation*

3.3 NAMESPACES

Let us consider the output statement of the previous sample program again:

```
System.Console.WriteLine ( );
```

In this, we noted that **System** is the namespace (scope) in which the **Console** class is located. A class in a namespace can be accessed using the dot operator as illustrated in the statement above. Namespaces are the way C# segregates the .NET library classes into reasonable groupings.

C# supports a feature known as **using** directive that can be used to import the namespace **System** into the program. Once a namespace is imported, we can use the elements of that namespace without using the namespace as prefix. This is illustrated in Program 3.2.

Program 3.2 A program using namespaces and comments

```
/*
This program uses namespaces and comment lines
*/
using System; // System is a namespace
class SampleTwo
{
    // Main method begins
    public static void Main( )
    {
        Console.WriteLine("Hello!");
    }
    // Main method ends
}
```

Note that the first statement in the program is

```
using System;
```

This tells the compiler to look in the **System** library for unresolved class names. Note that we have not used the **System** prefix to the **Console** class in the output line.

When the compiler parses the **Console.WriteLine** method, it will understand that the method is undefined. However, it will then search through the namespaces specified in **using** directives and, upon finding the method in the **System** namespace, will compile the code without any complaint.

3.4 ADDING COMMENTS

Comments play a very important role in the maintenance of programs. They are used to enhance readability and understanding of code. All programs should have information such as implementation details, change history and tasks performed. C# permits two types of comments, namely,

- Single-line comments
- Multiline comments

Program 3.2 illustrates the use of these two approaches.

Single-line comments begin with a double backslash (//) symbol and terminate at the end of the line. We can use // on a line of its own or after a code statement. This can also be used to comment out an entire line or part of a line of source code. Everything after the // on a line is considered a comment.

If we want to use multiple lines for a comment, we must use the second type known as multi-line comment. This comment starts with the /* characters and terminates with */ as shown in the beginning of the program. These comments can also occur within a line. For example:

```
using /* namespace */ System; //o.k
```

Remember, we cannot insert a //-style comment within a line of code. However, we can use the // style for commenting multi-lines.

```
// This is an example of
// Multiline comments
// in C# language
```

The /*...*/ is a C-type comment containing a caveat. Consider the following code segment:

```
/*
.................................................
/* using System;*/
.................................................
*/
```

This will cause an error because of a mismatch between the starting and ending symbols. That is, we cannot nest the /* */ comments. Every /* symbol should have a corresponding */ symbol.

3.5 MAIN RETURNING A VALUE

Another important aspect is the return type of **Main( ).** We have used **void** as the return type in earlier programs. **Main( )** can also return a value if it is declared as **int** type instead of **void**. When the return type is **int**, we must include a **return** statement at the end of the method as shown in Program 3.3.

Program 3.3 Main returning a value

```
// Main returning a value
using System;
class SampleThree
{
    public static int Main( )
    {
        Console.WriteLine ("Hello!");
        return 0; // Return statement
    }
}
```

Program 3.3 returns an integer-type value to the system. The value returned serves as the program's *termination status code*. The purpose of this code is to allow communication of success or failure to the execution environment.

3.6 USING ALIASES FOR NAMESPACE CLASSES

We have seen how we can avoid the prefix **System** to the **Console** class by implementing the **using System**; statement. Can we use this approach to avoid prefixing of **System.Console** to the method **WriteLine ( )**? The answer is "no."

System is a namespace and **Console** is a class. The **using** directive can be applied only to namespaces and cannot be applied to classes. Therefore the statement

```
using System.Console;
```

is illegal. However, we can overcome this problem by using aliases for namespace classes. This takes the form:

```
using alias-name = class-name;
```

Program 3.4 illustrates the use of aliases for classes.

Program 3.4 Use of aliases for classes

```
using A = System.Console; //A is alias for System.Console
class SampleFour
{
    public static void Main( )
    {
        A.WriteLine("Hello!");
    }
}
```

3.7 PASSING STRING OBJECTS TO WRITELINE METHOD

So far, we have seen only constant string output to the **Console**. We can store string values in string objects and use these objects as parameters to the **WriteLine** method. We can use **string** data type to create a string variable and assign a string constant to it as below:

```
string s = "abc" ;
```

The content of **s** may be printed out using the **WriteLine** method.

```
System.Console.WriteLine(s);  //s is string object
```

Program 3.5 illustrates the use of a string object as a parameter to the output method **WriteLine.**

Program 3.5 Using string objects in output

```
using System;
class SampleFive
{
    public static void Main( )
    {
        string name = "C sharp";  // name is assigned a string
        Console.WriteLine(name);
    }
}
```

3.8 COMMAND LINE ARGUMENTS

There may be occasions when we may like our program to behave in a particular way depending on the input provided at the time of execution. This is achieved in C# by using what are known as *command line arguments.* Command line arguments are parameters supplied to the **Main** method at the time of invoking it for execution. Program 3.6 accepts a name from the command line and writes it to the console.

Program 3.6 Command line arguments

```
/*
This program uses command line arguments as input
*/
using System;
class SampleSix
{
    public static void Main (string [ ] args)
    {

        Console.Write ("welcome to");
        Console.Write (" "+args[0]);
        Console.WriteLine (" "+args[1]);
    }
}
```

In earlier examples, we have used the **Main** method with no parameters. Notice how the **Main** is declared in this example.

```
public static void Main (string [ ] args)
```

Main is declared with a parameter **args**. The parameter **args** is declared as an array of strings (known as string objects). Any arguments provided in the command line (at the time of execution) are passed to the array **args** as its elements. We can access the array elements by using a subscript like **args[0], args[1]** and so on. For example, consider the command line

```
SampleSix C sharp
```

This command line contains two arguments which are assigned to the array **args** as follows:

C	⟶	args [0]
sharp	⟶	args [1]

Note that C# subscripts begin with 0 and not 1. (Arrays and strings are discussed in detail later). If we execute Program 3.6 with the above command line, we will get the following output:

```
Welcome to C sharp
```

Notice the use of a new output method

```
Console.Write( )
```

Its only difference from the **WriteLine** ( ) method is that **Write ( )** does not cause a line break and therefore the next output statement is printed on the same line. Thus, all the three output statements combine to produce one line of output. See Chapter 17 for more details on output formats.

Note how the output statements concatenate strings using the + operator while printing. Here, empty strings are used to provide blank space between strings. (The operator + is overloaded to perform the task of adding two strings. See Chapter 15 for more details on operator overloading.)

Program 3.7 shows another example of use of command-line arguments in C#. The program calculates the square root of a number passed as a command-line argument and displays the output.

Program 3.7 Command line example

```
using System;
class CommandLineExample
{
    static void Main(string[] args)
    {
        double argValue=0.0;
        double sqrtValue=0.0;
        if (args.Length == 0)
        {
            Console.WriteLine(“No argument has been specified.”);
            Console.ReadLine();
            return;
        }
        argValue = double.Parse(args[0].ToString());
        sqrtValue = Math.Sqrt(argValue);
        Console.WriteLine(“–Demostrating Command Line Arguments–\n”);
        Console.WriteLine(“Square root of the Argument is: {0}\n”.sqrtValue);
    }
}
```

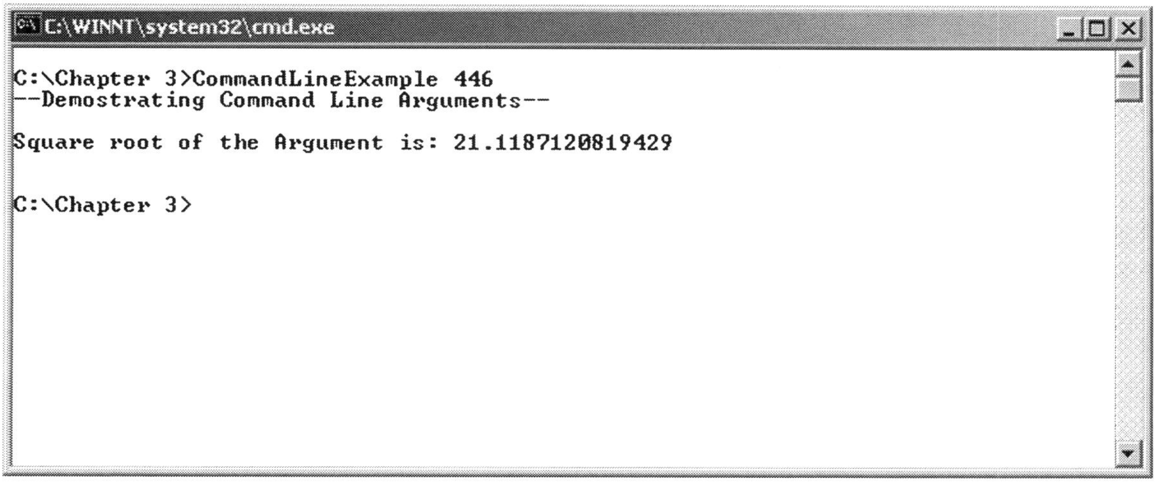

3.9 MAIN WITH A CLASS

Now that we have seen very simple programs, let us add a simple class to the class containing the **Main** method as shown in Program 3.8.

Program 3.8 A program with two classes

```
class TestClass      // class definition
{
    public void fun( )
    {
        System.Console.WriteLine("C# is modern");
    }
}

class SampleSeven
{
    public static void Main( )
    {
        TestClass test = new TestClass( );   // creating test object
        test.fun( );   // calling fun ( ) function
    }
}
```

This program has two class declarations, one for the **TestClass** and another for the **Main** method. **TestClass** contains only one method to print a string "C# is modern". The **Main** method in **SampleSeven** class creates an object of **TestClass** and uses it to invoke the method **fun ( )** contained in **TestClass** as follows:

```
TestClass test = new TestClass ( );
test.fun ( );
```

The object **test** is used to invoke the method **fun ( )** of **TestClass** with the help of the dot operator. Execution of Program 3.8 will produce the following output:

```
C# is modern
```

If we want, we could include the **TestClass** declaration within the class **SampleSeven**. That would mean creation of 'nested' classes. Classes, methods and creation of objects are discussed in detail in later chapters.

3.10 PROVIDING INTERACTIVE INPUT

So far we have seen two approaches for giving values to string objects:

- Using an assignment statement
- Through command line arguments

It is also possible to give values to string variables interactively through the keyboard at the time of execution. Program 3.9 shows how to get specific information from the user and print a message using that information.

Program 3.9 Interactive console input

```
using System;
class SampleEight
{
    public static void Main( )
    {
        Console.Write("Enter your name: ");
        string name = Console.ReadLine( );
        Console.WriteLine("Hello " + name);
    }
}
```

The method

```
Console.Write("Enter your name: ");
```

outputs the message "Enter your name:" and the method

```
Console.ReadLine( );
```

causes the execution to wait for the user to enter his name. The moment the user types his name and presses the 'Enter' key, the input is read into the string variable **name**. The second output line (WriteLine method) concatenates the name-string with 'Hello' and presents the resultant string to the user. The output will look like this:

```
Enter your name: John
Hello John
```

Change the contents of **Main** as shown below and try executing the program. Comment on the output.

```
public static void Main( )
{
    Console.Write ("What is your name? ");
    string s;
    s = Console.ReadLine( );
    Console.writeLine("Hello,"+ s)
    Console.Write("What is your age? ");
    s = Console.ReadLine( );
    Console.WriteLine("You look nice at "+ s);
}
```

3.11 USING MATHEMATICAL FUNCTIONS

Assume that we would like to compute and print the square root of a number. A C# program to accomplish this is shown in Program 3.10.

Program 3.10 Using mathematical functions

```
using System;
class SampleNine
{
    public static void Main( )
    {

        double x = 5.0; //Declaration and initialization
        double y; // Simple declaration
        y = Math.Sqrt(x);
        Console.WriteLine ( " y = " + y) ;
    }
}
```

This program contains more executable statements. The statement

```
double x = 5.0;
```

declares a variable **x** and initializes it to the value 5.0 and the statement

```
double y;
```

merely declares a variable **y**. Note that both of them have been declared as **double** type variables. **double** is a data type used to represent a floating-point number. Data types are discussed in the next chapter.

The statement

```
y = Math.Sqrt(x);
```

invokes the method **Sqrt()** of **Math** class which is a part of **System** namespace. The program produces the following output:

```
y = 2.23606
```

3.12 MULTIPLE MAIN METHODS

We have stated earlier that one of the classes must have the **Main** method as a member. This is what is normally expected. However, C# includes a feature that enables us to define more than one class with the **Main** method. Since **Main** is the entry point for program execution, there are now more than one entry points. In fact, there should be only one. This problem can be resolved by specifying which **Main** is to be used to the compiler at the time of compilation as shown below:

```
csc filename.cs/main:classname
```

Filename is the name of the file where the code is stored and *classname* is the name of the class containing the **Main** which we would like to be the entry point.

Program 3.11 shows a simple program that has two classes containing **Main** methods. We may save it in a file called **multimain.cs** and compile it using the switch

```
/main:ClassA or /main:ClassB
```

as required

Program 3.11 Application with multiple Main methods

```
using System;
class ClassA
{
    public static void Main( )
    {
        Console.WriteLine("Class A");
    }
}

class ClassB
{
    public static void Main( )
    {
        Console.WriteLine("Class B");
    }
}
```

3.13 COMPILE TIME ERRORS

What we have seen so far are very simple programs. But real-life applications will contain a large number of statements with complex logic. No matter how thoroughly the design is carried out, and no matter how much care is taken in coding, we can never say that a program is totally error-free. A program may contain two types of errors:

- Syntax errors
- Logic errors

While syntax errors will be caught by the compiler, logic errors should be eliminated by testing the program logic carefully.

When the compiler cannot interpret what we are attempting to convey through our code the result is syntax error. In such situations, the compiler will detect the error(s) and display the appropriate error message(s).

Let us consider a code as shown in Program 3.12. This program contains three errors. Save this in a file named **Errors.cs**

Program 3.12 Fixing syntax errors

```
// Program with syntax errors
using Systom;// Error here
class SampleTen
{

    public static void main( ) //Error here
    {
        Console.WriteLine("Hello, Errors")   //Error here
    }
}
```

Program 3.12, when compiled, will display the following output:

```
Errors.cs(2.7): error cs0234: The type or namespace name 'Systom'
does not exist in the class or namespace ' '
```

The compiler could not locate a namespace named 'System' and therefore produces an error message and then stops compiling. The error message contains the following information:

- Name of the file being compiled (Errors.cs)
- Line number and column position of the error (2.7)
- Error code as defined by the compiler (CS0234)
- Short description of the error

Here, the error is that **System** has been spelled wrongly. Once we correct this, the compiler will report the details of other errors in the same format.

Note: When a program contains syntax errors, we should not take compiler's messages at face value. Sometimes, they may be misleading. The actual problem may be something else or location may be somewhere else. We must look at the last few lines just before the point where the error was reported.

3.14 PROGRAM STRUCTURE

An executable C# program may contain a number coding blocks as shown in Fig. 3.3. The documentation section consists of a set of comments giving the name of the program, the author, date and other details, which the programmer (or other users) may like to use at a later stage. Comments must explain the

- Why and what of classes, and the
- How of algorithms

This would greatly help maintaining the program.

The **using** directive section will include all those namespaces that contain classes required by the application. **using** directives tell the compiler to look in the namespace specified for these unresolved classes.

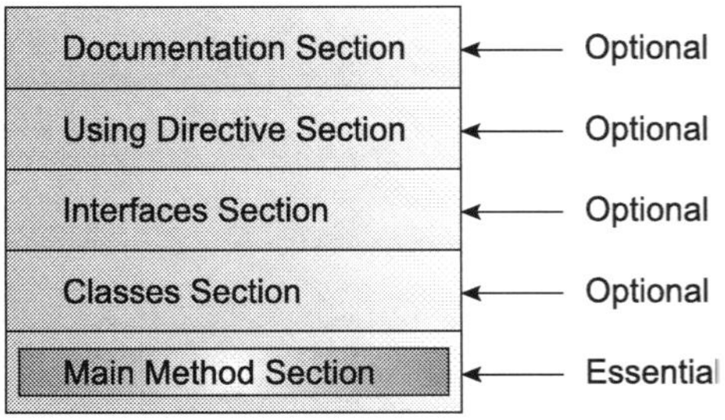

Figure 3.3 *C# program structure*

An interface is similar to a class but contains only abstract members. Interfaces are used when we want to implement the concept of multiple inheritance in a program. Interface is a new concept and is discussed in detail in Chapter 14.

A C# program may contain multiple class definitions. Classes are the primary and essential elements of a C# program. These classes are used to map the objects of real-world problems. The number of classes depends on the complexity of the problem.

Since every C# application program requires a **Main** method as its starting point, the class containing the **Main** is the essential part of the program. A simple C# program may contain only this part. The **Main** method creates objects of various classes and establishes communications between them. On reaching the end of **Main**, the program terminates and the control passes back to the operating system.

3.15 PROGRAM CODING STYLE

C# is a *freeform* language. We need not indent any lines to make the program work properly. C# does not care where on the line we begin coding. While this may be a license for bad programming, we should try to use this fact to our advantage for designing readable programs. Although several alternative styles are possible, we should select one and try to use it consistently.
For example, the statement

```
System.Console.WriteLine("Hello!");
```

can be written as

```
System.Console.WriteLine
("Hello!");
```

or, even as

```
System.Console.WriteLine
(
    "Hello!"
)
;
```

In this book, we normally follow the format used in the sample programs given in this chapter.

Case Study

Problem Statement

AUPart Enterprise is a manufacturing unit that is involved in the manufacturing of automobile parts. This company was established in 1980 and since then the staff of AUPart has been manually keeping the records of manufacturing parts. Now, it has been discovered by the management of AUPart Enterprise that manual record-keeping results in the occurrence of a large number of errors. In addition, it has also been discovered that manually keeping records of manufactured parts is time consuming. How can the management of AUPart Enterprise automate the task of maintaining records of manufactured parts?

Solution

After much research, the management of AUPart Enterprise comes to the conclusion that an application created in a high level programming language such as C# is needed to automate the task of maintaining the records of manufactured parts. Quick Soft Solutions Private Limited, which is a software development company, is contacted by AUPart Enterprise to develop the application for automating the task of maintaining the records of manufactured parts. The team of programmers at Quick Soft Solutions Private Limited first conducted an interview with the staff of AUPart Enterprise to find out their requirements. After requirement analysis, it was decided by the programmers at Quick Soft Solutions Private Limited that C# features such as constructors and variables must be used in the application for automating the task of keeping records of manufactured parts as illustrated below.

```
using System;
using System.Collections;

class Inventory
{
   string partName;
   double partPrice;
   int priceQty;

   public Inventory(string pName, double pPrice, int pQty)
   {
   partName = pName;
   partPrice = pPrice;
   priceQty = pQty;

   Console.WriteLine("\nPART NAME : {0,-10}",partName);
   Console.WriteLine("\tPrice : Rs.{0,5} \n\tQuantity: {1}",partPrice,
priceQty);
   }
}

public class StockData
{
   public static void Main()
   {
      Console.WriteLine("==ABC Spare Parts Ltd. Stock List==");
      new Inventory("Bolts", 2.25, 100);
      new Inventory("Shockers", 5.25, 30);
      new Inventory("Brake Pads", 3.50, 50);
      new Inventory("Steering", 120, 20);

   }
}
```

Remarks

In C#, constructors help initialize the objects of a class and variables are used to store data related to an object. The name of the constructor in C# must be the same as the name of the class.

Common Programming Errors

It is human to err. An error is simply a failure to adhere to the rules of the language. The greatest mistake is to assume that we never err. While designing and coding a program, we tend to commit certain errors, many a times by oversight. We list here some of the errors that are common to even the experienced programmers. After developing a program, we must check for such errors manually before attempting to execute it.

- *Omitting semicolons in statements.*
- *Mismatching of opening braces with the closing braces.*
- *Not using parentheses in the Main method.*
- *Misspelling words like Main, System, etc.,*
- *Mismatching of comment symbols /* and */.*
- *Not specifying return type for Main method.*
- *Not using proper using directives.*
- *Improper use of parameters in Math class methods.*
- *Misspelling of classes used from name spaces.*

Review Questions

3.1 State whether the following statements are true or false.
   (a) C# is the only language specially designed for the .NET platform.
   (b) C# code bear striking resemblances to C code.
   (c) In C#, everything must be placed inside a class.
   (d) Every class must include a Main method in its definition.
   (e) Some of the classes defined in the Base Class Library contain Main method.
   (f) Every application program must include the Main method in one of its classes.
   (g) The Main method never returns any value.
   (h) When saving a C# program in a file, the file name should be same as the name of the class containing the Main method.
   (i) We cannot insert a // style comment within a line of code.
   (j) It is possible to have more than one class containing the Main method.

3.2 Describe the structure of typical C# program.

3.3 Why do we use the using directive in a C# program?

3.4 What is the importance of the Main method in a C# program?

3.5 Why do we use comments in a program?

3.6 Describe the differences between the two styles of comments used in C#.

3.7 What are component libraries? How are they different from executable application programs?

3.8 What is an alias? Where and how it is used?

3.9 How does the Write( ) method differ from the WriteLine( ) method?

3.10 Describe in detail the steps involved in implementing an application program in C#.

3.11 What are command line arguments? How are they useful?

3.12 How does the ReadLine( ) method work?
3.13 How do we implement a program having more than one class with the Main method?
3.14 What is a syntax error? How do we detect them?
3.15 C# is a freeform language. Comment.
3.16 What are the approaches for assigning values to string objects in C#?
3.17 What is the termination status code? What is its purpose?

Debugging Exercises

3.1. Will the given program produce a compile-time or runtime error?

```
using System;
class MainExample
{
    public static void Main()
    {
        Console.WriteLine("Main Example");
        return 1;
    }

    }
```

3.2. Find error(s), if any, in the following program.

```
class Reading
{
    public static void Main ()
    {
        Console. Write ("Enter Name");
        string name - ReadLine ();
        Console. WriteLine (name);
    }
}
```

3.3. What is wrong with the following program?

```
class Maths
{
    public static void Main ()
    {
        double, y, x = 10.0;
        y = sqrt (x);
        System.Console.Write (y);
    }
}
```

Programming Exercises

3.1 Write a program to read two strings from the keyboard using the ReadLine statements and display them on one line using two Write statements.

3.2 Write a program that takes the line of text
John F Kennedy
as a command line input and displays the following output
Kennedy John F

3.3 Define two classes, one with a method to display the string "C Sharp" and the other to display the string "Programming". Write a program using these classes to display a single line output as follows:
C Sharp Programming

3.4 Write a program that assigns two double type values to two variables, computes their sum, assigns the result to a third variable, and displays all the three values in one line of output.

3.5 Write a program to display the following pattern on the screen.

```
  X
 XXX
XXXXX
 XXX
  X
```

Chapter 4

Literals, Variables and Data Types

4.1 INTRODUCTION

A programming language is designed to manipulate certain kinds of *data* consisting of numbers, characters and strings and provide useful output known as *information* to the user. The task of manipulating data is accomplished by executing a sequence of instructions constituting what is known as a *program*.

A C# program is basically a collection of classes. A class is defined by a set of declaration statements and methods containing instructions known as *executable statements*. These instructions are formed using certain symbols and words according to some rigid rules known as *syntax rules* (or grammar). The smallest, non-reducible, textual elements in a program are referred to as *tokens*. The compiler recognizes them for building up expressions and statements.

In simple terms, a C# program is a collection of tokens, comments and white spaces. C# includes the following five types of tokens:

- Keywords
- Identifiers
- Literals
- Operators
- Punctuators

White spaces and comments are not tokens, though they may act as separators for tokens.

Keywords are an essential part of a language definition. They implement specific features of the language. They are reserved, and cannot be used as identifiers except when they are prefaced by the @ character. Table 4.1 lists all the C# keywords.

Table 4.1 *C# keywords*

abstract	event	namespace	static
as	explicit	new	string
base	extern	null	struct
bool	false	object	switch
break	finally	operator	this
byte	fixed	out	throw
case	float	override	true
catch	for	params	try
char	foreach	private	typeof
checked	get	protected	unit
class	goto	public	ulong
const	if	readonly	unchecked
continue	implicit	ref	unsafe
decimal	in	return	ushort
default	int	sbyte	using
delegate	interface	sealed	value
do	internal	set	virtual
double	is	short	volatile
else	lock	sizeof	void
enum	long	stackalloc	while

Identifiers are programmer-designed tokens. They are used for naming classes, methods, variables, labels, namespaces, interfaces, etc. C# identifiers enforce the following rules:

- They can have alphabets, digits and underscore characters
- They must not begin with a digit
- Upper case and lower case letters are distinct
- Keywords in stand-alone mode cannot be used as identifiers

C# permits the use of keywords as identifiers when they are prefixed with the '@' character

Literals are the way in which the values that are stored in variables are represented. We shall discuss these in detail in the next section.

Operators are symbols used in expressions to describe operations involving one or more operands. Operators are considered in detail in Chapter 5.

Punctuators are symbols used for grouping and separating code. They define the shape and function of a program. Punctuators (also known as *separators*) in C# include:

- Parentheses ( )
- Braces { }
- Brackets [ ]
- Semicolon ;
- Colon :
- Comma ,
- Period .

Statements in C# are like sentences in natural languages. A statement is an executable combination of tokens ending with a semicolon. C# implements several types of statements. They include:

- Empty statements
- Labeled statements
- Declaration statements
- Expression statements
- Selection statements
- Interaction statements
- Jump statements
- The **try** statements
- The **checked** statements
- The **unchecked** statements
- The **lock** statements
- The **using** statements

These statements are discussed as and when they are encountered.

Program 4.1 demonstrates the application of @ as a prefix for using a keyword as an identifier. The program uses a variable @ if the print numbers from 0 to 9.

Program 4.1 Identifier example

```
using System;

class IdentifierExample
{
  public static void Main()
  {
    // Using if keyword as an identifier by prefixing @
    int @if;

    Console.WriteLine("-Demonstrating use of if keyword as an identifier by
prefixing @-\n");

    for(@if = 0; @if < 10; @if++)
      Console.WriteLine("The value of @if is: {0}",@if);
  }
}
```

```
C:\WINNT\system32\cmd.exe

C:\Chapter 4>IdentifierExample
--Demonstrating use of if keyword as an identifier by prefixing @--

The value of @if is: 0
The value of @if is: 1
The value of @if is: 2
The value of @if is: 3
The value of @if is: 4
The value of @if is: 5
The value of @if is: 6
The value of @if is: 7
The value of @if is: 8
The value of @if is: 9

C:\Chapter 4>_
```

4.2 LITERALS

Literals are value constants assigned to variables (or results of expressions) in a program. C# supports several types of literals as illustrated in Fig. 4.1.

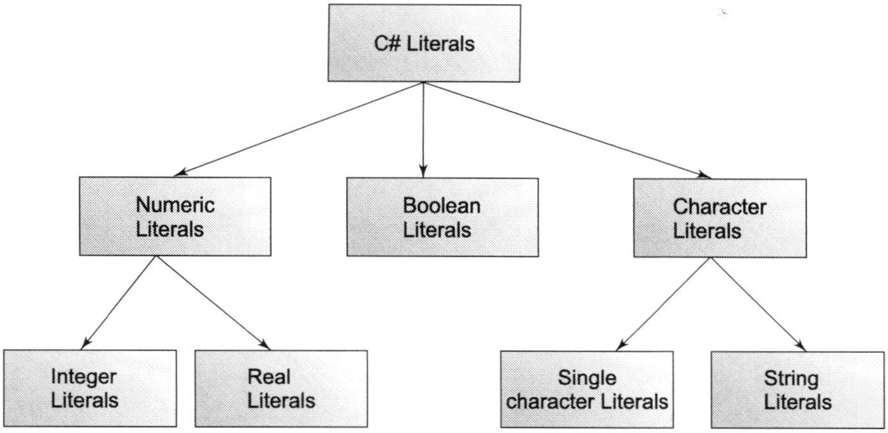

Figure 4.1 *C# literals*

4.2.1 Integer Literals

An *integer* literal refers to a sequence of digits. There are two types of integers, namely, *decimal* integers and *hexadecimal* integers.

Decimal integers consist of a set of digits, 0 through 9, preceded by an optional minus sign. Valid examples of decimal integer literals are:

```
123 -321 0 654321
```

Embedded spaces, commas and non-digit characters are not permitted between digits. For example,

```
15 750 20,000 $1000
```

are illegal numbers.

A sequence of digits preceded by 0x or 0X is considered as a *hexadecimal* integer (hex integer). It may also include alphabets A through F or 'a' through 'f'. A letter A through F represents the numbers 10 through 15. Following are the examples of valid hex integers.

```
0X2 0X9F 0Xbcd 0x
```

4.2.2 Real Literals

Integer literals are inadequate to represent quantities that vary continuously, such as distances, heights, temperatures, prices and so on. These quantities are represented by numbers containing fractional parts like 17.548. Such numbers are called *real* (or *floating point*) numbers. Further examples of real literals are:

```
0.0083 -0.75 435.36
```

These numbers are shown in *decimal notation*, having a whole number followed by a decimal point and the fractional part, which is an integer. It is possible that the number may not have digits before the decimal point, or digits after the decimal point. That is,

```
215.  .95  -.71
```

are all valid real literals.

A real literal may also be expressed in *exponential* (or *scientific*) *notation*. For example, the value 215.65 may be written as 2.1565e2 in exponential notation. e2 means multiply by 10^2. The general form is:

```
mantissa e exponent
```

The *mantissa* is either a real number expressed in *decimal notation* or an integer. The *exponent* is an integer with an optional *plus* or *minus* sign. The letter 'e' separating the mantissa and the exponent can be written in either lowercase or uppercase. Since the exponent causes the decimal point to 'float', this notation is said to represent a real number in *floating-point form*. Examples of legal floating-point literals are:

```
0.65e4  12e-2  1.5e+5  3.18E3  -1.2E-1
```

Embedded white (blank) space is not allowed in any numeric constant.

Exponential notation is useful for representing numbers that are either very large or very small in magnitude. For example, 7500000000 may be written as 7.5E9 or 75E8. Similarly, –0.000000368 is equivalent to –3.68E–7.

A floating-point literal may thus comprise four parts:

- a whole number
- a decimal point
- a fractional part
- an exponent

4.2.3 Boolean Literals

There are two Boolean literal values:

- true
- false

They are used as values of relational expressions.

4.2.4 Single Character Literals

A single-character literal (or simply character constant) contains a single character enclosed within a pair of single quote marks. Example of character in the examples above constants are:

```
'5'  'X'  ';'  ' '
```

Note that the character *constant* '5' is not the same as the *number* 5. The last constant in the example above is a blank space.

4.2.5 String Literals

A string literal is a sequence of characters enclosed between double quotes. The characters may be alphabets, digits, special characters and blank spaces. Examples are:

```
"Hello C#"  "2001"  "WELL DONE"  "?...!"  "5+3"  "X"
```

4.2.6 Backslash Character Literal

C# supports some special backslash character constants that are used in output methods. For example, the symbol '\n' stands for a new-line character. A list of such backslash character literals is given in Table 4.2. Note that each one represents one character, although they consist of two characters. These character combinations are known as *escape sequences*.

Table 4.2 *Backslash character literals*

Constant	Meaning
'\a'	alert
'\b'	back space
'\f'	form feed
'\n'	new-line
'\r'	carriage return
'\t'	horizontal tab
'\v'	vertical tab
'\''	single quote
'\"'	double quote
'\\'	backslash
'\o'	null

Program 4.2 prints the various numeric literals available in C#, such as **Integer, Double** and **Exponential**.

Program 4.2 Literal sample

```
using System;

class LiteralSample
{
    public static void Main()
    {
    System.Console.WriteLine("Integer Literal                         :
    {0}", 58);
    System.Console.WriteLine("Double Literal                          :
    {0}",1.6180339887498);
    System.Console.WriteLine("Decimal Literal                         :
    {0}",1.618033988734565m);
    System.Console.WriteLine("Exponential Notation for Decimal Literal:
    {0}",6.023E23f);
    System.Console.WriteLine("Hexadecimal Literal                     :
    {0}",0x003A);
    }
}
```

```
C:\WINNT\system32\cmd.exe

C:\Chapter 4>LiteralSample
Integer Literal                                : 58
Double Literal                                 : 1.6180339887498
Decimal Literal                                : 1.618033988734565
Exponential Notation for Decimal Literal: 6.023E+23
Hexadecimal Literal                            : 58

C:\Chapter 4>
```

4.3 VARIABLES

A *variable* is an identifier that denotes a storage location used to store a data value. Unlike constants that remain unchanged during the execution of a program, a variable may take different values at different times during the execution of the program. Every variable has a type that determines what values can be stored in the variable.

A variable name can be chosen by the programmer in a meaningful way so as to reflect what it represents in the program. Some examples of variable names are:

- average
- height
- total_height
- classStrength

As mentioned earlier, variable names may consist of alphabets, digits and the underscore (_), subject to the following conditions:

1. They must not begin with a digit.
2. Uppercase and lowercase are distinct. This means that the variable **Total** is not the same as **total** or **TOTAL**.
3. It should not be a keyword.
4. White space is not allowed.
5. Variable names can be of any length.

4.4 DATA TYPES

Every variable in C# is associated with a data type. Data types specify the size and type of values that can be stored. C# is a language rich in its *data types.* The variety available allows the programmer to select the type appropriate to the needs of the application.

The types in C# are primarily divided into two categories:

- Value types
- Reference types

Value types and reference types differ in two characteristics:

- Where they are stored in the memory
- How they behave in the context of assignment statements

Value types (which are of fixed length) are stored on the *stack*, and when a value of a variable is assigned to another variable, the value is actually copied. This means that two identical copies of the value are available in memory. Reference types (which are of variable length) are stored on the *heap*, and when an assignment between two reference variables occurs, only the reference is copied; the actual value remains in the same memory location. This means that there are two references to a single value.

A third category of types called *pointers* is available for use only in *unsafe code*. Value types and reference types are further classified as *predefined* and *user-defined* types as shown in Fig. 4.2.

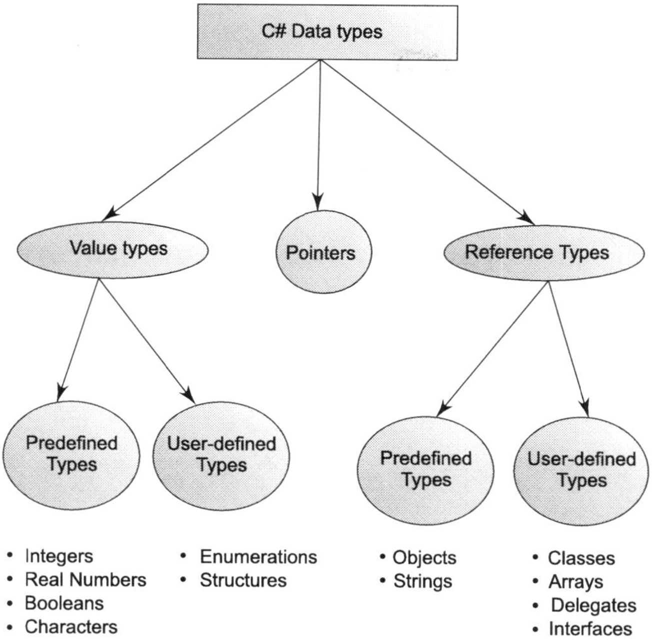

Figure 4.2 *Taxonomy C# data types*

4.5 VALUE TYPES

The value types of C# can be grouped into two categories (as shown in Fig. 4.2), namely,

- User-defined types (or complex types) and
- Predefined types (or simple types)

We can define our own complex types known as *user-defined* value types which include **struct** types and enumerations. They are discussed in Chapter 11.

Predefined value types which are also known as *simple types* (or primitive types) are further subdivided into:

- Numeric types,
- Boolean types, and
- Character types.

Numeric types include integral types, floating-point types and decimal types. These are shown diagrammatically in Fig. 4.3.

Note: C# 2.0 added a new type called *nullable* type. This type variable can hold an undefined value. Any value type variable can be defined as a nullable type.

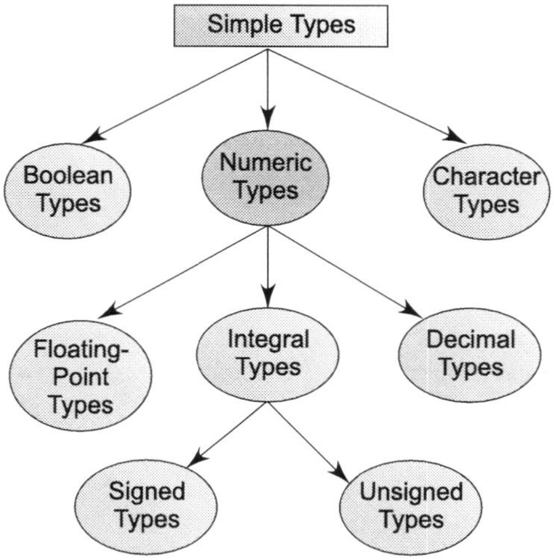

Figure 4.3 *Categories of simple type data*

4.5.1 Integral Types

Integral types can hold whole numbers such as 123, –96 and 5639. The size of the values that can be stored depends on the integral data type we choose. C# supports the concept of unsigned types and therefore it supports eight types of integers as shown in Figs 4.4 and 4.5.

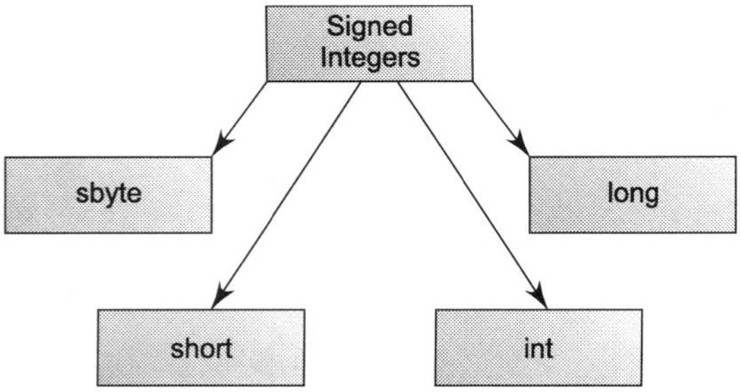

Figure 4.4 *Signed integer types*

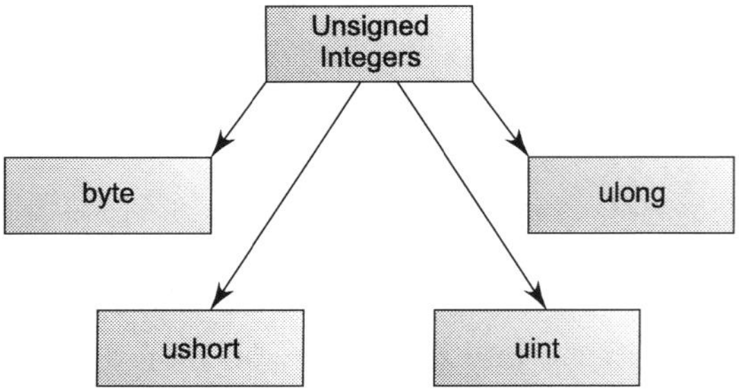

Figure 4.5 *Unsigned integer types*

4.5.2 Signed Integers

Signed integer types can hold both positive and negative numbers. Table 4.3 shows the memory size and range of all the four signed integer data types. (Note that one byte is equal to eight bits).

Table 4.3 *Size and range of signed integer types*

Type	Size	Minimum value	Maximum value
sbyte	One byte	–128	127
short	Two bytes	–32, 768	32,767
int	Four bytes	–2,147,483,648	2,147,483,647
long	Eight bytes	–9,223,372,036,854,775,808	9,223,372,036,854,775,807

It should be remembered that wider data types require more time for manipulation and therefore it is advisable to use smaller data types wherever possible. For example, instead of storing a number like 50 in an **int** type variable, we must use a **sbyte** variable to handle this number. This will improve the speed of execution of the program.

4.5.3 Unsigned Integers

We can increase the size of the positive value stored in an integer type by making it 'unsigned'. For example, a 16-bit **short** integer can store values in the range –32,768 to 32,767. However, by making it **ushort**, it can handle values in the range 0 to 65,535. Table 4.4 shows the size and range of all the four unsigned integer data types.

Table 4.4 *Size and range of unsigned integer types*

Type	Size	Minimum value	Maximum value
byte	One byte	0	255
ushort	Two bytes	0	65,535
uint	Four bytes	0	4,294,967,295
ulong	Eight bytes	0	18,446,744,073,709,551,615

All integers are by default **int** type. In order to specify which of the other integer types the values must take, we must append the characters U, L or UL as shown below:

123 U (for **uint** type)

123 L (for **long** type)

123 UL (for **ulong** type)

We may also use lower case u and l, although the letter l is confused with the number 1.

4.5.4 Floating-Point Types

Integer types can hold only whole numbers and therefore we use another type known as *floating-point* type to hold numbers containing fractional parts such as 27.59 and –1.375. There are two kinds of floating point storage in C# as shown in Fig. 4.6.

The **float** type values are *single-precision* numbers with a precision of seven digits. The **double** types represent *double-precision* numbers with a precision of 15/16 digits. Table 4.5 gives the size and range of these two types.

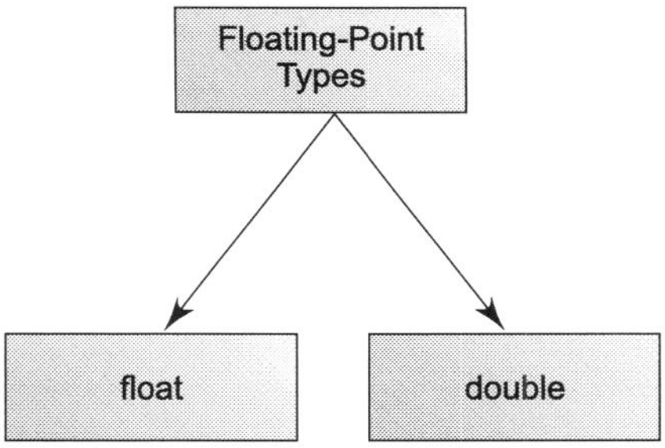

Figure 4.6 *Floating-point data types*

Table 4.5 *Size and range of floating-point types*

Type	Size	Minimum value	Maximum value
float	4 bytes	1.5×10^{-45}	3.4×10^{38}
double	8 bytes	5.0×10^{-324}	1.7×10^{308}

Floating-point numbers, by default, as double-precision quantities. To force them to be in single-precision mode, we must append f or F to the numbers.

Example:

1.23f

7.56923e5f

Double-precision types are used when we need greater precision in storage of floating-point numbers.

Floating-point data types support a special value known as Not-a-Number (NaN). NaN is used to represent the result of operations such as dividing zero by zero, where an actual number is not produced. Most operations that have NaN as an operand will produce NaN as a result.

4.5.5 Decimal Type

The decimal type is a high precision, 128-bit data type that is designed for use in financial and monetary calculations. It can store values in the range 1.0×10^{-28} to 7.9×10^{28} with 28 significant digits. Note that the precision is given in digits, not in decimal places.

To specify a number to be decimal type, we must append the character M (or m) to the value, like 123.45M. If we omit M, the value will be treated as **double**.

4.5.6 Character Type

In order to store single characters in memory, C# provides a character data type called **char**. The **char** type assumes a size of two bytes but, in fact it can hold only a single character. It has been designed to hold a 16-bit Unicode character, in which the 8-bit ASCII code is a subset.

4.5.7 Boolean Type

Boolean type is used when we want to test a particular condition during the execution of the program. There are only two values that a Boolean type can take: **true** or **false.** Remember, both these words have been declared as keywords. Boolean type is denoted by the keyword **bool** and uses only one bit of storage. In contrast to C and C++, in C#, we cannot use zero for **false** and non-zero for **true**. No conversion between **bool** type and other integer types is possible.

All comparison operators (see Chapter 5) return Boolean type values. Boolean values are often used in selection and iteration statements.

4.6 REFERENCE TYPES

As with value types, the reference types can also be divided into two groups:

- User-defined (or complex) types
- Predefined (or simple) types

User-defined reference types refer to those types which we define using predefined types. They include:

- Classes
- Interfaces
- Delegates
- Arrays

These complex types will be discussed in later chapters when we take up these topics individually.

Predefined reference types include two data types:

- Object type
- String type

The **object** type is the ultimate base type of all other intrinsic and user-defined types in C#. We can use an **object** reference to bind to an object of any particular type. We shall see later how we can use the **object** type to convert a value type on the stack to an **object** type to be placed on the heap.

C# provides its own string type for creating and manipulating strings. String literals discussed earlier can be stored in string objects as values. We can perform a number of operations such as copying, comparing and concatenation on these string objects. Strings and string objects are considered in more detail in Chapter 10.

4.7 DECLARATION OF VARIABLES

Variables are the names of storage locations. After designing suitable variable names, we must declare them to the compiler. Declaration does three things:

1. It tells the compiler what the variable name is.
2. It specifies what type of data the variable will hold.
3. The place of declaration (in the program) decides the scope of the variable.

A variable must be declared before it is used in the program.

A variable can be used to store a value of any data type. That is, the name has nothing to do with the type. C# allows any properly formed variable to have any declared data type. The declaration statement defines the type of variable. The general form of declaration of a variable is:

```
type    variable1, variable2, ........., variableN;
```

Variables are separated by commas. A declaration statement must end with a semicolon. Some valid declarations are:

```
int             count;
float           x, y;
double          pi;
byte            b;
char            c1, c2, c3;
decimal         d1, d2;
uint            m;
ulong           n;
```

4.8 INITIALIZATION OF VARIABLES

A variable must be given a value after it has been declared but before it is used in an expression. A simple method of giving value to a variable is through the assignment statement as follows:

```
variableName = value;
```

Examples:

```
initialValue  =  0;
finalValue    =  100;
yes           =  'x';
```

We can also string assignment expressions as shown below:

```
x = y = z = 0;
```

It is also possible to assign a value to a variable at the time of its declaration. This takes the form:

```
type variableName = value;
```

Examples:

```
int      finalValue    =  100;
char     yes           =  'X';
double   total         =  75.36;
bool     test          =  true;
```

The process of giving initial values to variables is known as *initialization.* The following are valid C# statements:

```
float x, y, z;     //declares three variables; compound declaration
int m = 5,n = 10;  //declares and initializes two int variables
int m, n = 10;     //declares m and n and initializes n
```

Consider the following initialization statements:

```
float      f    =  12.34F;
uint       ui   =  123U;
long       l    =  123L;
ulong      ul   =  123UL;
decimal    d    =  1.23M
```

All integer numbers by default represent **int** type values. If we want the value to represent any other integer type, we must append U to denote an unsigned type, L to denote long and UL to denote an unsigned long. Similarly, all floating-point numbers are **double** by default. Therefore we must append F or M to the numbers to denote **float** and **decimal** types respectively.

If any variables are not provided with initial values, the C# compiler will generate errors.

4.9 DEFAULT VALUES

A variable is either explicitly assigned a value or automatically assigned a default value. The following categories of variables are automatically initialized to their default values:

- Static variables
- Instance variables
- Array elements

The default value of a variable depends on the type of the variable. Table 4.6 shows the default values for the variables belonging to the above categories.

Table 4.6 *Default values of variables*

Type	Default Value
All integer types	0
char type	'\x000'
float type	0.0f
double type	0.0d
decimal type	0.0 m
bool type	false
enum type	0
All reference types	null

4.10 CONSTANT VARIABLES

The variables whose values do not change during the execution of a program are known as *constants*. For example, the variables representing the maximum number of rows and columns of a matrix or number of students in a class in a program may not change during execution of the program. Such variables can be made unmodifiable by using the **const** keyword while initializing them.

Example:

const int ROWS = 10;
const int COLS = 20;
const int NUM = 90;

Remember, constants must be declared and initialized simultaneously as shown above. For instance,

```
const int m;
m = 100;
```

is illegal. By definition, a constant cannot have its value changed later.

A constant variable can be initialized using an expression.

Example:

```
const int m = 10;
const int n = m * 5;
```

Remember, we cannot use non-const values in expressions. For instance,

```
int m = 10;              // non-constant value
const int n = m * 5;     // error
```

will result in an error.

In C and C++, such constants are known as *symbolic constants* and are defined using the **#define** statement. They are not declared for types.

Advantages of using constants are:

- They make our programs easier to read and understand.
- They make our programs easier to modify.
- They minimize accidental errors, like attempting to assign values to some variables which are expected to be constants.

Note:

1. The names of constant take the same form as variable names.
2. After declaration of constants, they should not be assigned any other value.
3. Constants are declared for types. This is not done in C and C++.
4. Constants cannot be declared inside a method. They should be declared only at class level.

4.11 SCOPE OF VARIABLES

The scope of a variable is the region of code within which the variable can be accessed. This depends on the type of the variable and place of its declaration. C# defines several categories of variables. They include:

- Static variables
- Instance variables
- Array elements
- Value parameters
- Reference parameters
- Output parameters
- Local variables

Consider the code shown below:

```
class ABC
{
        static int m;
        int n;
        void fun(int x, ref int y, out int z, int [ ] a)
        {
                int j = 10;
                . . . .
                . . . .
        }
}
```

This code contains the following variables:

- m as a static variable
- n as an instance variable
- x as a value parameter
- y as a reference parameter
- z as an output parameter
- a[0] as an array element
- j as a local variable

Static and instance variables are declared at the class level and are known as *fields* or field variables. The scope of these variables begins at the place of their declaration and ends when the **Main** method terminates.

The variables **x, y** and **z** are parameters of the method **fun( ).** The value parameter **x** will exist till the end of the method. The reference and output parameters (y and z) do not create new storage locations. Instead, they represent the same storage locations as the variables that are passed as arguments. The scope of these variables is always the same as the underlying-variables.

The elements of an array such as **a[0]** come into existence when an array instance is created, and cease to exist when there are no references to that array instance. Arrays declared at class level behave like fields.

Variables declared inside methods are called *local variables.* They are called so because they are not available for use outside the method definition. Local variables can also be declared inside program blocks that are defined between an opening brace { and a closing brace }. The scope of a local variable starts immediately after its identifier in the declaration and extends up to the end of the block containing the declaration. Within the scope of a local variable, it is an error to declare another local variable with the same name.

Consider a code segment containing nested blocks as shown below. Each block can contain its own set of local variable declarations. We cannot, however, declare a variable to have the same name as one in the outer block. The variable **x** declared in Block1 is available in all the three blocks. However, the variable **n** declared in Block2 is available only in Block2 because it goes out of scope at the end of Block2. Similarly, the variable **m** is accessible only in Block3. Note that we cannot declare the variable name **x** again in either Block2 or Block3.

```
            .....
            .....
          ┌─ {
          │        int x=0;
          │        .....
          │     ┌─{
          │     │       .....
          │Block2│       int n=5;     //Ok
          │     │       int x=10;  //wrong, x is already defined
          │     │       .....
          │     └─}
Block1    │
          │        n=n+1;  // wrong, n not available here
          │     ┌─{
          │     │       .....
          │Block3│       .....
          │     │       int   m=20; //ok
          │     │          x=m;     //ok
          │     │       .....
          │     └─}
          │        .....
          └─ }
```

C# makes a fundamental distinction between the field variables declared at the class level and the local variables declared within methods. Consider the code shown below:

```
using System;
class Scope
{
        int x=10;
        static int y=20;

                public static void Main( )
                {
                    int x=100;
                    int y=200;
                    Console.WriteLine(x); //local x
                    Console.WriteLine(y); //local y
                    Console.WriteLine(this.x); //instance x
                    Console.WriteLine(Scope.y); //static y
        }
}
```

This program when executed will produce the following output:

```
100
200
10
20
```

Although we have declared two variables named **x** and two variables named **y**, the code will compile and produce the output. In this case, the variables declared inside the **Main** method hide the class-level variables with the same name. Note that we have used the dot operator with the class name to access the static variable **y** and with the keyword **this** to access the instance variable **x**. More about static and instance variables will be considered later while discussing classes.

Let us consider another example code as shown below:

```
using System;
class Scope
{
        public static void Main( )
        {
            int m=100;
            for (int i=0; i<=4; i++)
            {

                    int m=200; // Error
                    Console.WriteLine (m*i); //ok m = 100
            }
        }
}
```

This code will not compile because the compiler cannot understand the statement.

```
int m=200;
```

placed inside the **for** loop. The variable **m** defined before the **for** loop statement is still in scope inside the **for** loop and goes out of scope only when the **Main** has finished executing. If we delete the statement

```
int m = 100;
```

then the program will compile and produce the following output:

```
0
100
200
300
400
```

4.12 BOXING AND UNBOXING

In object-oriented programming, methods are invoked using objects. Since value types such as **int** and **long** are not objects, we cannot use them to call methods. C# enables us to achieve this through a technique known as *boxing* . Boxing means the conversion of a value type on the stack to a **object** type on the heap. Conversely, the conversion from an **object** type back to a value type is known as *unboxing*.

4.12.1 Boxing

Any type, value or reference can be assigned to an object without an explicit conversion. When the compiler finds a value type where it needs a reference type, it creates an object 'box' into which it places the value of the value type. The following code illustrates this:

```
int m = 100;
object om = m; // creates a box to hold m
```

When executed, this code creates a temporary reference_type 'box' for the object on heap. We can also use a C-style cast for boxing.

```
int m = 100;
object om = (object)m; //C-style casting
```

Note that the boxing operation creates a copy of the value of the **m** integer to the object **om**. Now both the variables **m** and **om** exist but the value of **om** resides on the heap. This means that the values are independent of each other. Consider the following code:

```
int m = 10:
object om = m;
m = 20;
Console.WriteLine(m);  // m = 20
Console.WriteLine(om); //om = 10
```

When a code changes the value of **m**, the value of **om** is not affected.

4.12.2 Unboxing

Unboxing is the process of converting the object type back to the value type. Remember that we can only unbox a variable that has previously been boxed. In contrast to boxing, unboxing is an explicit operation using C-style casting.

```
int m = 10;
object om = m;    //box m
int n = (int)om;  //unbox om back to an int
```

When performing unboxing, C# checks that the value type we request is actually stored in the object under conversion. Only if it is, the value is unboxed.

When unboxing a value, we have to ensure that the value type is large enough to hold the value of the object. Otherwise, the operation may result in a runtime error. For example, the code

```
int m = 500;
object om = m;
byte n = (byte)om;
```

will produce a runtime error.

Notice that when unboxing, we need to use explicit cast. This is because in the case of unboxing, an object could be cast to any type. Therefore, the cast is necessary for the compiler to verify that it is valid as per the specified value type.

Case Study

Problem Statement

ForEx Enterprises is a medium sized organisation that is involved in the money exchange business. This means that the main function of ForEx is to convert the money of one currency to the money of another currency for its customers. ForEx Enterprises was started in 2000 and since then it is has been manually calculating the value of foreign currency into domestic currency and vice versa. Manual calculation of money value in a specific currency is recently leading to the occurrence of a large number of errors. In addition, the staff of ForEx Enterprises is spending a lot of its time in the process of converting the currency value of a certain amount of money into another currency value. How can the management of ForEx Enterprises automate the task of converting the money value of a specific currency into the money value of another currency?

Solution

To automate the task of converting currency value of specific amount of money into a value of another currency, ForEx Enterprises needs an application created in a high level programming language such as Java or C#. For this, ForEx Enterprises hires the services of a software development company, InfoTech Solutions. After analysing the requirements of ForEx Enterprises, the programmers working in InfoTech decided to use operators such as * used for multiplication in a C# application that will help in converting the currency value of money of one currency into the currency value of another currency. The operators in C# help perform calculations such as arithmetic calculations and binary operations. The team of programmers at InfoTech Solutions created the following application in C#.

```
using System;

public class CurrencyConvertor
{
    static void Main(string [] args)
    {
        int num=Int32.Parse(args[0]);
        double poundValue=(double)num*83.2;
        double dollarValue=(double)num*40;

        Console.WriteLine ("==Currency Convertor==\n");
        Console.WriteLine ("Currency Data Formatted Using the #
character\n");
        Console.WriteLine ("\tRupee Amount  : {0,0:Rs ###.##}", num);
        Console.WriteLine ("\tUSD Value     : {0,0:$ ###.##}", dollarValue);
        Console.WriteLine("\tGBP Value:{0,0:£###.##}\n",poundValue );

        Console.WriteLine ("Currency Data Formatted Using the $
character\n");
```

```
        Console.WriteLine ("\tRupee Amount : {0,0:Rs 000.00}", num);
        Console.WriteLine ("\tUSD Value    : {0,0:$ 000.00}", dollarValue);
        Console.WriteLine ("\tGBP Value    : {0,0:£ 000.00}\n", poundValue);

        Console.WriteLine ("Currency Data Formatted Using the comma
only\n");
        Console.WriteLine ("\tRupee Amount : {0,0:Rs 000,000}",num);
        Console.WriteLine ("\tUSD Value    : {0,0:$ 0,000.000}",
dollarValue);
        Console.WriteLine ("\tGBP Value    : {0,0:£ 0,000.000}",
poundValue);
    }
}
```

Remarks

C# supports various types of operators such as logical, arithmetic and bitwise that help perform different operations. For example, the arithmetic operators help perform operations such as addition and multiplication on specific data. The logical operations help perform logical operations such as AND and OR on data.

Note: Formatting outputs is discussed in detail in Chapter 17.

Common Programming Errors

- *Forgetting to append F or f for floating point literals.*
- *Forgetting to append U, L or UL to integer literals when they represent unsigned integers or long integers or unsigned integers.*
- *Not appending the letter M or m to floating point values when they represent decimal type.*
- *Forgetting to declare a variable.*
- *Wrong declaration of types.*
- *Forgetting to initialize a variable.*
- *Not initializing constant variables at the time of declaration.*
- *Not declaring type for a constant.*
- *Trying to unbox an object that has not been boxed earlier.*
- *Assuming wrong scope of a variable in the program.*
- *Attempting to modify the variable declared as constants.*

Review Questions

4.1 State whether the following statements are true or false.
- (a) A C# program is basically a collection of classes.
- (b) In C#, comments are referred to as tokens.
- (c) Keywords cannot be used as variable names.
- (d) Boolean type variables can be assigned any integer type values.
- (e) Pointers are predefined value types.

(f) Value type data are stored on the stack.
(g) Like classes, structures are reference types.
(h) Floating point numbers, by default, are treated as double type quantities.
(i) In C#, zero means false and nonzero means true.
(j) Constants can be declared anywhere in the program.
(k) Wider data types require more time for manipulation.
(l) Decimal type data takes double the space of double type data.

4.2 List the five types of tokens available in C#.
4.3 What is a literal?
4.4 What is a constant?
4.5 What is a variable?
4.6 How are literals and variables are important in developing a program?
4.7 Why do we need to use constants in a program?
4.8 There are thirteen simple value types. List them.
4.9 List the two predefined reference types.
4.10 What is a local variable?
4.11 What do you mean by scope of a variable?
4.12 State the scope of the following variables.
(a) Local variables (b) Field variables (c) Method parameters
4.13 What is initialization? Why is it necessary?
4.14 When dealing with a very small or very large numbers, what steps would you take to improve the accuracy of computations?
4.15 What are the applications of backslash character literals?
4.16 How do the value types differ from reference types in terms of their storage?
4.17 State the rules that govern the naming of variables.
4.18 What does a declaration of a variable accomplish?
4.19 Which of the following are invalid literals and why?

0.0001	5 * 1.5	RS 75.50
+100	75.45E-2	"15.75"
–45.6	–1.45e(+4)	0,000001234

4.20 Which of the following are invalid variable names and why?

Minimum	first.Name	n1+n2
doubles	3rd-row	N$
float	Sum Total	Total-Marks

4.21 Find errors, if any, in the following statements.

```
Int x,
float length, HEIGHT
double = p, q
character C1;
short int TOTAL;
float pi = 3.142;
long int m;
bool m = 1;
byte fixed;
```

4.22 What is boxing? Why do we do it?
4.23 What is unboxing? How is it achieved?
4.24 What do you mean by default values?

4.25 State default values of the following objects:

(a) char type
(b) decimal type
(c) bool type
(d) class type

Debugging Exercises

4.1 Find errors, in any, in the following program.

```
using System;
class constants
{
    public static void Main ()
    {
        Const int m;
        int n = 10;
        const int k = n*5;
        m = 100;
        Console.WriteLine (m+k);
        }
    }
```

4.2 What is wrong with the following program?

```
using System;
class scope
}
    int n = 100;
    public static void Main ()
    {
        for (int i = 0; i < 5, i++)
        {
            int m = 10;
            Console.WriteLine (m*i);
        }
        Console.WriteLine (m,n,i);
    }
}
```

Programming Exercises

4.1 Write a program that would illustrate the concepts of declaration and initialization of value type variables. Use the following literals for initializing the variables:

(a) 'A'
(b) 50
(c) 123456789
(d) 1234567654321
(e) true
(f) 0.000000345
(g) 1.23e.5

4.2 Write a program to perform the following tasks.
   (a) Declare two variables x and y as float type variables
   (b) Declare m as an integer variable
   (c) Assign the values 75.86 to x and 43.48 to y
   (d) Assign the sum of x and y to m
   (e) Display the value of m

   Comment on the output.

4.3 Execute the programs illustrated in Section 4.11 and analyse the output.

4.4 Develop a program to demonstrate the concepts of boxing and unboxing.

4.5 Declare three variables b1, b2 and b3 of type byte. Assign a value of 100 to b1 and 200 to b2, and the sum of b1 and b2 to b3. Develop a program to achieve these and printout the values of all the three variables.

   If the compiler gives any error message, modify the program, till you obtain the correct output. Comment on any changes you made to the original program.

4.6 Write a program to implement the following output statements

   Console.WriteLine(“Hello,\”RAM\”!”);

   Conole.WriteLine(“*\n**\n***\n****\n”);

   Comment on the output.

4.7 Write a program that prints an upward arrow as shown below:

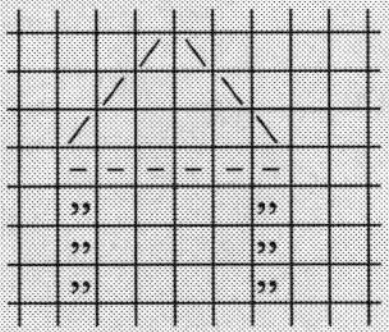

   Use backslash characters to print \ and “ characters.

4.8 Write a program to implement the following statements:

```
int m = 100;
long n = 200;
long l = m + n;
Console.WriteLine("l = " + l);
```

   If the output is an error message, correct the error and execute the program again.

4.9 What is the output of the following program?

```
using System;

class CheckNaN
{
    public static void Main()
        {
                Double zero = 0;
                Console.WriteLine("The result of division by zero
in C# is: {0}", (0 / zero));
        }
}
```

4.10 What is the output of the given program?

```
     using System;
     class DataTypeSample
     {
          public static void Main()
          {
          System.Console.WriteLine("Integer Literal                    :
          {0}", 58);
          System.Console.WriteLine("Double Literal                     :
          {0}",1.6180339887498);
          System.Console.WriteLine("Decimal Literal                    :
          {0}",1.618033988734565m);
          System.Console.WriteLine("Exponential Notation for Decimal
Literal:
          {0}",6.023E23f);
          System.Console.WriteLine("Hexadecimal Literal                :
          {0}",0x003A);
           }
     }
```

Chapter

5

Operators and Expressions

5.1 INTRODUCTION

An operator is a symbol that tells the computer to perform certain mathematical or logical manipulations. Operators are used in programs to manipulate data and variables. They usually form a part of mathematical or logical expressions.

C# supports a rich set of operators. C# operators can be classified into a number of related categories as below:

1. Arithmetic operators
2. Relational operators
3. Logical operators
4. Assignment operators
5. Increment and decrement operators
6. Conditional operators.
7. Bitwise operators
8. Special operators

5.2 ARITHMETIC OPERATORS

C# provides all the basic arithmetic operators. They are listed in Table 5.1. The operators + , – , * and / all work the same way as they do in other languages. These can operate on any built-in numeric data type. We cannot use these operators on Boolean type. The unary minus operator, in effect, multiplies its single operand by –1. Therefore, a number preceded by a minus sign changes its sign.

Table 5.1 *Arithmetic operators*

Operator	Meaning
+	Addition or unary plus
–	Subtraction or unary minus
*	Multiplication
/	Division
%	Modulo division

Arithmetic operators are used as shown below:

a – b	a + b
a * b	a / b
a % b	–a * b

Here **a** and **b** may be variables or constants and are known as *operands.*

5.2.1 Integer Arithmetic

When both the operands in a single arithmetic expression such as a+b are integers, the expression is called an *integer expression*, and the operation is called *integer arithmetic*. Integer arithmetic always yields an integer value. In the above examples, if a and b are integers, then for a = 14 and b = 4 we have the following results:

a – b	=	10	
a + b	=	18	
a * b	=	56	
a / b	=	3	(decimal part truncated)
a % b	=	2	(remainder of integer division)

a/b, when **a** and **b** are integer types, gives the result of division of **a** by **b** after truncating the divisor. This operation is called *integer division.*

For *modulo division*, the sign of the result is always the sign of the first operand (the dividend). That is

–14 % 3	= –2
–14 % –3	= –2
14 % –3	= 2

*(Note that modulo division is defined as: a%b = a–(a/b)*b, where a/b is the integer division).*

5.2.2 Real Arithmetic

An arithmetic operation involving only real operands is called *real arithmetic*. A real operand may assume values either in decimal or exponential notation. Since floating-point values are rounded to the number of significant digits permissible, the final value is an approximation of the correct result.

Unlike C and C++, the modulus operator % can be applied to the floating-point data as well. The floating-point modulus operator returns the floating-point equivalent of an integer division. What this means is that the division is carried out with both floating-point operands, but the resulting divisor is treated as an integer, resulting in a floating-point remainder. The result of a % b is computed as a–n*b, where n is the largest possible integer that is less than or equal to a/b. Program 5.1 shows how arithmetic operators work on floating-point values.

Program 5.1 Floating-point arithmetic

```
class FloatPoint
{
    public static void Main()
    {
        float a = 20.5F, b = 6.4F;
```

```
        System.Console.WriteLine("  a = " + a);
        System.Console.WriteLine("  b = " + b);
        System.Console.WriteLine("  a+b = " + (a+b));
        System.Console.WriteLine("  a-b = " + (a-b));
        System.Console.WriteLine("  a*b = " + (a*b));
        System.Console.WriteLine("  a/b = " + (a/b));
        System.Console.WriteLine("  a%b = " + (a%b));
    }
}
```

The output of Program 5.1 is follows:

```
a = 20.5
b = 6.4
a+b = 26.9
a-b = 14.1
a*b = 131.2
a/b=3.203125
a%b = 1.3
```

5.2.3 Mixed-mode Arithmetic

When one of the operands is real and the other is integer, the expression is called a *mixed-mode arithmetic* expression. If either operand is of the real type, then the other operand is converted to real and real arithmetic is performed. The result will be a real. Thus

15/10.0 produces the result 1.5

whereas

15/10 produces the result 1

5.3 RELATIONAL OPERATORS

We often compare two quantities and depending on their relation, take certain decisions. For example, we may compare the age of two persons or the price of two items and so on. These comparisons can be done with the help of *relational operators.* An expression such as

```
a < b or x < 20
```

containing a relational operator is termed as a *relational expression.* The value of a relational expression is either true or false. For example, if x = 10, then

```
x < 20 is true
```

while

```
20 < x is false.
```

C# supports six relational operators in all. These operators and their meanings are shown in Table 5.2.

Table 5.2 *Relational operators*

Operator	Meaning
<	is less than
<=	is less than or equal to
>	is greater than
>=	is greater than or equal to
==	is equal to
!=	is not equal to

A simple relational expression contains only one relational operator and is of the following form:

`ae-1 relational operator ae-2`

ae-1 and *ae-2* are arithmetic expressions, which may be simple constants, variables or combination of them. Table 5.3 shows some examples of simple relational expressions and their values.

Table 5.3 *Relational expressions*

Expression	Value
4.5 <= 10	true
4.5 < –10	false
–35 >= 0	false
10 < 7+5	true
5.0 != 5	false
a + b == c+d	true*

*only if the sum of the values of a and b is equal to the sum of the values of c and d.

When arithmetic expressions are used on either side of a relational operator, the arithmetic expressions will be evaluated first and then the results compared. That is, arithmetic operators have a higher priority over relational operators. Program 5.2 shows the implementation of relational operators.

Program 5.2 Implementation of relational operators

```
using System;

class RelationalOperators
{
    public static void Main( )
    {
        float a = 15.0F, b = 20.75F, c = 15.0F;
        Console.WriteLine(" a = " + a);
        Console.WriteLine(" b = " + b);
        Console.WriteLine(" c = " + c);
        Console.WriteLine(" a < b is " + (a<b));
```

```
        Console.WriteLine(" a > b is " + (a>b));
        Console.WriteLine(" a == c is " + (a==c));
        Console.WriteLine(" a <= c is " + (a<=c));
        Console.WriteLine(" a >= b is " + (a>=b));
        Console.WriteLine(" b != c is " + (b!=c));
        Console.WriteLine(" b == a+c is " + (b==a+c));
    }
}
```

The output of Program 5.2 would be:

```
a = 15
b = 20.75
c = 15
a < b is true
a > b is false
a == c is true
a <= c is true
a >= b is false
b != c is true
b == a+c is false
```

Relational expressions are used in *decision statements* such as **if** and **while** to decide the course of action of a running program. Decision statements are discussed in detail in Chapters 6 and 7.

5.4 LOGICAL OPERATORS

In addition to the relational operators, C# has six logical operators, which are given in Table 5.4.

Table 5.4 *Logical operators*

Operator	Meaning
&&	logical AND
||	logical OR
!	logical NOT
&	bitwise logical AND
|	bitwise logical OR
^	bitwise logical exclusive OR

The logical operators && and || are used when we want to form compound conditions by combining two or more relations. An example is:

```
a > b && x == 10
```

An expression of this kind which combines two or more relational expressions is termed as a *logical expression* or a *compound relational expression*. Like the simple relational expressions, a logical expression also yields a value of true or false, according to the *truth table* shown in Table 5.5. The logical expression

given above is true only if both **a>b** and **x == 10** are true. If either (or both) of them are false, the expression is false.

Table 5.5 *Truth table*

op-1	op-2	Value of the Expression	
		op-1 && op-2	op-1 || op-2
true	true	true	true
true	false	false	true
false	true	false	true
false	false	false	false

Note:

- *op-1 && op-2 is true if both op-1 and op-2 are true and false otherwise.*
- *op-1 || op-2 is false if both op-1 and op-2 are false and true otherwise.*

Some examples of the usage of logical expressions are:

1. **if** (age > 55 && salary < 1000), give increment.
2. **if** (number < 0 || number > 100), stop.

5.5 ASSIGNMENT OPERATORS

Assignment operators are used to assign the value of an expression to a variable. We have seen the usual assignment operator, '='. In addition, C# has a set of 'shorthand' assignment operators which are used in the form

```
v op= exp
```

where v is a variable, *exp* is an expression and *op* is a C# binary operator. The operator **op =** is known as the *shorthand assignment operator.*

The assignment statement

```
v op= exp
```

is equivalent to

```
v = v op(exp);
```

with v accessed only once. Consider an example

```
x += y+1;
```

This is same as the statement

```
x = x+(y+1);
```

The shorthand operator **+=** means 'add y+1 to x' or 'increment x by y+1'. For y = 2, the above statement becomes

```
x += 3;
```

and when this statement is executed, 3 is added to x. If the old value of x is, say 5, then the new value of x is 8. Some of the commonly used shorthand assignment operators are illustrated in Table 5.6.

Table 5.6 *Shorthand assignment operators*

Statement with simple assignment operator	Statement with shorthand operator
a = a+1	a + = 1
a = a–1	a – = 1
a = a*(n+1)	a * = n+1
a = a/(n+1)	a / = n+1
a = a%b	a % = b

The use of shorthand assignment operators has three advantages:

1. What appears on the left-hand side need not be repeated and therefore it becomes easier to write.
2. The statement is more concise and easier to read.
3. The use of shorthand operators results in a more efficient code.

5.6 INCREMENT AND DECREMENT OPERATORS

C# has two very useful operators not generally found in many other languages. These are the increment and decrement operators:

```
++ and --
```

The operator ++ adds 1 to the operand while –– subtracts 1. Both are unary operators and are used in the following form:

```
++m; or m++;
--m; or m--;
++m; is equivalent to m = m + 1; (or m + = 1;)
--m; is equivalent to m = m - 1; (or m - = 1;)
```

We use the increment and decrement operators extensively in **for** and **while** loops.

While ++m and m++ mean the same thing when they form statements independently, they behave differently when they are used in expressions on the right-hand side of an assignment statement. Consider the following:

```
m = 5;
y = ++m;
```

In this case, the value of y and m would be 6. Suppose, if we rewrite the above statement as

```
m = 5;
y = m++;
```

then, the value of y would be 5 and m would be 6. A prefix operator first adds 1 to the operand and then the result is assigned to the variable on left. On the other hand, a postfix operator first assigns the value to the variable on left and then increments the operand. Program 5.3 illustrates this.

Program 5.3 Increment operator illustrated

```
class IncrementOperator
{
    public static void Main()
    {
        int m = 10, n = 20;
        System.Console.WriteLine(" m = " + m);
        System.Console.WriteLine(" n = " + n);
        System.Console.WriteLine(" ++m = " + ++m);
        System.Console.WriteLine(" n++ = " + n++);
        System.Console.WriteLine(" m = " + m);
        System.Console.WriteLine(" n = " + n);
    }
}
```

The output of Program 5.3 is as follows:

```
m = 10
n = 20
++m = 11
n++ = 20
m = 11
n = 21
```

Similar is the case when we use ++ (or – –) in subscripted variables. That is, the statement

```
a[i++] = 10
```

is equivalent to

```
a[i]   = 10
i = i+1
```

5.7 CONDITIONAL OPERATOR

The character pair **? :** is a *ternary operator* available in C#. This operator is used to construct conditional expressions of the form

exp1 ? exp2 : exp3

where *exp1*, *exp2* and *exp3* are expressions.

The operator **? :** works as follows: *exp1* is evaluated first. If it is true, then the expression *exp2* is evaluated and becomes the value of the conditional expression. If *exp1* is false, *exp3* is evaluated and its value becomes the value of the conditional expression. Note that only one of the expressions (either *exp2* or *exp3*) is evaluated. For example, consider the following statements:

```
a = 10;
b = 15;
x = (a>b) ? a : b;
```

In this example, x will be assigned the value of b. This can be achieved using the **if...else** statement as follows:

```
if(a > b)
    x = a;
else
    x = b;
```

5.8 BITWISE OPERATORS

C# supports operators that may be used for manipulation of data at bit level. These operators may be used for testing the bits or shifting them to the right or left. Bitwise operators may not be applied to floating-point data. Table 5.7 lists the bitwise logical and shift operators.

Table 5.7 *Bitwise operators*

Operator	Meaning
&	bitwise logical AND
|	bitwise logical OR
^	bitwise logical XOR
~	one's complement
<<	shift left
>>	shift right

Program 5.4 performs BitWise Operations available in C#, such as AND, OR, XOR, SHIFT LEFT, SHIFT RIGHT and ONE's COMPLEMENT.

Program 5.4 BitWise operators illustrated

```
using System;

class BitWiseOperator
{
 public static void Main()
 {
  byte bwAnd, bwOr, bwXor;
  int bwLeft,bwRight,bwComp;

  bwAnd = 15 & 3;
  bwOr = 15 | 3;
  bwXor = 15 ^ 3;
  bwLeft=15<<5;
  bwRight=15>>3;
```

```
  bwComp=~15;

  System.Console.WriteLine("–To demonstrate the use of Bitwise Operators
available in C#–\n");
  System.Console.WriteLine("Bitwise AND Operator Result: = {0} \nBitwise OR
Operator Result : = {1}",bwAnd, bwOr);
  System.Console.WriteLine("Bitwise XOR Operator Result: = {0} \nBitwise
SHIFT LEFT Operator Result: = {1}",bwXor,bwLeft);
  System.Console.WriteLine("Bitwise SHIFT LEFT Operator Result: =
{0}\nBitwise ONE'S COMPLEMENT Operator Result: = {1}",bwRight,bwComp);
 }
}
```

```
C:\WINNT\system32\cmd.exe

C:\Chapter 5>BitWiseOperator
--To demonstrate the use of Bitwise Operators available in C#--

Bitwise AND Operator Result: = 3
Bitwise OR Operator Result : = 15
Bitwise XOR Operator Result: = 12
Bitwise SHIFT LEFT Operator Result: = 480
Bitwise SHIFT LEFT Operator Result: = 1
Bitwise ONE'S COMPLEMENT Operator Result: = -16

C:\Chapter 5>_
```

5.9 SPECIAL OPERATORS

C# supports the following special operators.

is	(relational operator)
as	(relational operator)
typeof	(type operator)
sizeof	(size operator)
new	(object creator)
.(dot)	(member-access operator)
checked	(overflow checking)
unchecked	(prevention of overflow checking)

These operators will be discussed as and when they are encountered.

5.10 ARITHMETIC EXPRESSIONS

An arithmetic expression is a combination of variables, constants and operators arranged as per the syntax of the language. We have used a number of simple expressions in the examples discussed so far. C# can handle any complex mathematical expressions. Some of the examples of C# expressions are shown in Table 5.8. Remember that C# does not have an operator for exponentiation.

Table 5.8 *Expressions*

Algebraic expression	C# expression
axb–c	a*b–c
(m+n)(x+y)	(m+n)*(x+y)
$\frac{ab}{c}$	a*b/c
$3x^2+2x+1$	3*x*x+2*x+1
$\frac{x}{y}+c$	x/y+c

Program 5.5 coducts basic division and modulo division of various numeric types available in C#, such an **integer, float, double** and **decimal**.

Program 5.5 Divisions illustrated

```
using System;
class DivisionExample
{
   static void Main(  )
   {
      int firstInt, secondInt;
      float firstFloat, secondFloat;
      double firstDouble, secondDouble;
      decimal firstDecimal, secondDecimal;

      firstInt = 12;
      secondInt = 5;
      firstFloat = 12;
      secondFloat = 5;
      firstDouble = 12.55;
      secondDouble = 5;
      firstDecimal = 12;
      secondDecimal = 5;

    Console.WriteLine("–Division Operator usage in Different Types of Numeric Data–");
```

```
        Console.WriteLine( "\nInteger Division: {0}",firstInt / secondInt );
        Console.WriteLine( "Float Division   : {0}",firstFloat / secondFloat);
        Console.WriteLine( "Double Division : {0}",firstDouble / secondDouble);
        Console.WriteLine("Decimal Division: {0}",firstDecimal / secondDecimal);

          Console.WriteLine("\n—Modulo Division Operator usage in Different Types of Numeric
  Data—");
        Console.WriteLine( "\nInteger Modulo Division: {0}",firstInt % secondInt );
        Console.WriteLine( "Float Modulo Division   : {0}", firstFloat % secondFloat );
        Console.WriteLine( "Double Modulo Division : {0}",firstDouble % secondDouble );
        Console.WriteLine( "Decimal Modulo Division: {0}", firstDecimal % secondDecimal );


    }
  }
```

```
C:\WINNT\system32\cmd.exe

C:\Chapter 5>DivisionExample
--Division Operator usage in Different Types of Numeric Data--

Integer Division: 2
Float Division  : 2.4
Double Division : 2.51
Decimal Division: 2.4

--Modulo Division Operator usage in Different Types of Numeric Data--

Integer Modulo Division: 2
Float Modulo Division  : 2
Double Modulo Division : 2.55
Decimal Modulo Division: 2

C:\Chapter 5>_
```

5.11 EVALUATION OF EXPRESSIONS

Expressions are evaluated using an assignment statement of the form

```
variable = expression;
```

variable is any valid C# variable name. When the statement is encountered, the *expression* is evaluated first and the result then replaces the previous value of the variable on the left-hand side. All variables used in the expression must be assigned values before evaluation is attempted. Examples of evaluation statements are

```
x = a*b-c;
y = b/c*a;
z = a-b/c+d;
```

The blank space around an operator is optional and is added only to improve readability. When these statements are used in a program, the variables **a, b, c** and **d** must be declared and assigned values before they are used in the expressions. Similarly, the variables **x, y** and **z** must have been declared earlier for their types.

5.12 PRECEDENCE OF ARITHMETIC OPERATORS

An arithmetic expression without any parentheses will be evaluated from left to right using the rules of precedence of operators. There are two distinct priority levels of arithmetic operators in C#:

High priority * / %
Low priority + –

The basic evaluation procedure includes two left-to-right passes through the expression. During the first pass, the high priority operators (if any) are applied as they are encountered. During the second pass, the low priority operators (if any) are applied as they are encountered. Consider the following evaluation statement:

x = a–b/3+c*2–1

When a = 9, b = 12, and c = 3, the statement becomes

x = 9–12/3+3*2–1

and is evaluated as follows:

First pass

Step1: x = 9–4+3*2–1 (12/3 evaluated)
Step2: x = 9–4+6–1 (3*2 evaluated)

Second pass

Step3: x = 5+6–1 (9–4 evaluated)
Step4: x = 11–1 (5+6 evaluated)
Step5: x = 10 (11–1 evaluated)

However, the order of evaluation can be changed by introducing parentheses into an expression. Consider the same expression with parentheses as shown below:

9–12/(3+3)*(2–1)

Whenever the parentheses are used, the expressions within parentheses assume highest priority. If two or more sets of parentheses appear one after another as shown above, the expression contained in the left-most set is evaluated first and the right-most in the last. Given below are the new steps.

First pass

Step1: 9–12/6*(2–1)
Step2: 9–12/6*1

Second pass

Step3: 9–2*1
Step4: 9–2

Third pass

Step5: 7

This time, the procedure consists of three left-to-right passes. However, the number of evaluation steps remain the same as 5 (i.e., equal to the number of arithmetic operators).

Parentheses may be nested, and in such cases, evaluation of the expression will proceed outward from the innermost set of parentheses. Just make sure that every opening parentheses has a matching closing one. For example,

9–(12/(3+3)*2)–1 = 4

whereas

9–((12/3)+3*2)–1 = –2

While parentheses allow us to change the order of priority, we may also use them to improve understandability of the program. When in doubt, we can always add an extra pair just to make sure that the priority assumed is the one we require.

5.13 TYPE CONVERSIONS

We often encounter situations where there is a need to convert a data of one type to another before it is used in arithmetic operations or to store a value of one type into a variable of another type. For example, consider the code below:

```
byte b1 = 50;
byte b2 = 60;
byte b3 = b1 + b2;
```

This code attempts to add two byte values and to store the result into a third byte variable. But this will not work. The compiler will give an error message:

`"cannot implicitly convert type int to type byte."`

Strange message! We are not using any **int** type data in the code. The reason is as follows: When we add two byte values, the compiler automatically converts them into **int** types and the result of addition is an **int** value, not another **byte**.

Why should the compiler do it? This is because the sum of two byte values may very easily result in a value that is much larger than the range of a **byte**.

Since the result is of type **int**, we cannot assign it to a **byte** variable. We need to convert the result back to a **byte** type, if we want to store it as a byte value. Otherwise, we may assign the result to an **int** type variable:

```
int b3 = b1 + b2; // no error
```

Here, the compiler does the conversion for us, without out explicit request to do so.

In C#, type conversions take place in two ways (as shown in Fig. 5.1)

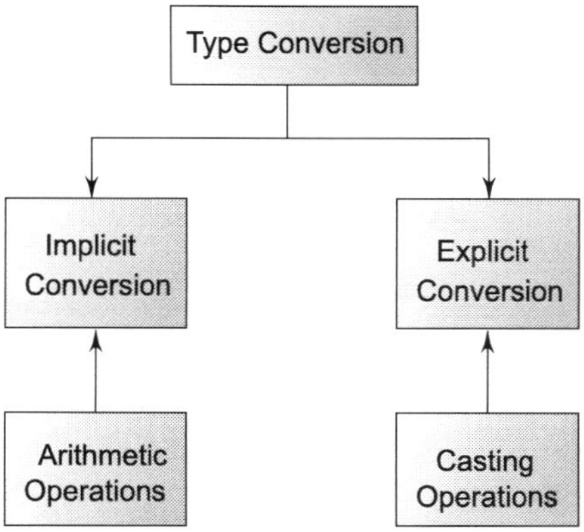

Figure 5.1 *Type conversions in C#*

- Implicit conversions
- Explicit conversions

5.13.1 Implicit Conversions

Implicit conversions are those that will always succeed. That is, the conversion can always be performed without any loss of data. For numeric types, this implies that the *destination* type can fully represent the range of the *source* type. For example, a **short** can be converted implicitly to an **int**, because the **short** range is a subset of the **int** range. Therefore,

```
short b = 75;
int a = b, // implicit conversion.
```

are valid statements. C# does the conversion automatically. An implicit conversion is also known as *automatic* type conversion.

The process of assigning a smaller type to a larger one is known as *widening* or *promotion* and that of assigning a larger type to a smaller one is known as *narrowing*. Note that narrowing may result in loss of information.

For numeric types, widening implicit conversions exist for all the signed and unsigned types. Figure 5.2 shows the conversion hierarchy chart.

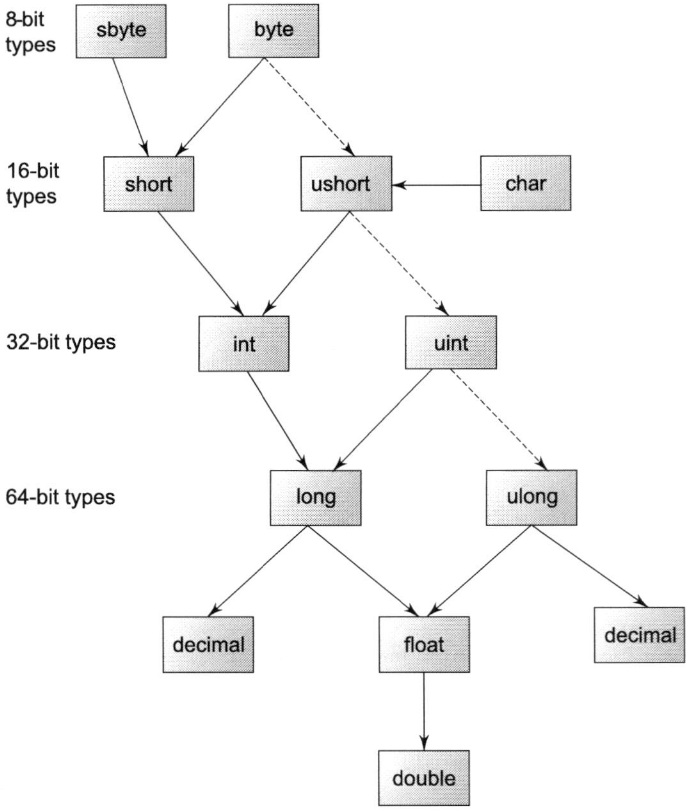

Figure 5.2 *C# conversion hierarchy chart*

In Fig. 5.2, arrows indicate the direction of implicit conversion from source to destination. If a path of arrows can be followed from a source type to a destination type, then an implicit conversion is possible between them. For example, implicit conversions are possible in the following cases:

- From **byte** to **decimal**
- From **uint** to **double**
- From **ushort** to **long**

Note that the conversions in these cases take place in a single operation, not by converting them in stages. Some examples of implicit conversion are:

```
byte x1 = 75;
short x2 = x1;
int x3 = x2;
long x4 = x3;
float x5 = x4;
decimal x6 = x4;
```

5.13.2 Explicit Conversions

There are many conversions that cannot be implicitly made between types. If we attempt such conversions, the compiler will give an error message. For example, the following conversions cannot be made implicitly:

- **int** to **short**
- **int** to **uint**
- **uint** to **int**
- **float** to **int**
- **decimal** to any numeric type
- any numeric type to **char**

However, we can explicitly carry out such conversions using the 'cast' operator. The process is known as *casting* and is done as follows:

```
type variable1 = (type) variable2;
```

The destination type is placed in parentheses before the source variable.

Examples:

```
int m = 50;
byte n = (byte) m;

long x = 1234L;
int y = (int) x;

float f = 50.0F;
long y = (long) f;
```

Casting is often necessary when a method returns a type other than the one we require. (Note that in C#, type casting is always done using the old C_style syntax. The new C++ style cannot be used).

Casting into a smaller type may result in loss of data. For example, casting a floating-point value to an integer will result in the loss of the fractional part. *We should bear in mind that all casts are potentially unsafe.*

We can use casting to round-off a floating-point number to the nearest integer.

Examples:

```
double x = 65.40;
double y = 79.60;
int x = (int) (x+0.5); // x=65
int y = (int) (y+0.5); // y=80
```

C# provides methods to convert between numeric values and strings.

Example:

```
int m = 100;
sring s = m.Tostring();
```

This returns the string representation of 100. That is, **s** contains the string '100'. Similarly, we can do the reverse:

```
string s = "100";
int n = int.Parse(s);
int m = n+50;
```

Now, the value of m would be 150.

5.13.3 Casting in Expressions

There could be instances where we may have to force a type conversion in one of the operands of an expression. Consider, for example, the calculation of ratio of females to males in a town.

```
ratio = female_number/male_number
```

Since **female_number and male_number** are declared as integers in the program, the decimal part of the result of the division would be lost and the **ratio** would not represent a correct figure. This problem can be solved by converting locally one of the variables to the floating point as shown below:

```
ratio = (float) female_number/male_number
```

The operator (**float**) converts the **female_number** to floating point for the purpose of evaluation of the expression. Then, using the rule of automatic conversion, the division is performed in floating-point mode, thus retaining the fractional part of the result.

Note that in no way does the operator (**float**) affect the value of the variable **female_number**. And also, the type of **female_number** remains as **int** in the other parts of the program.

Some examples of casts and their actions are shown in Table 5.9.

Table 5.9 *Use of casts*

Example	Action
x = (**int**) 7.5	7.5 is converted to integer by truncation.
a =(**int**)21.3/(**int**) 4.5	Evaluated as 21/4 and the result would be 5.
b = (**double**) sum/n	Division is done in floating-point mode.
y = (**int**) (a+b)	The result of a + b is converted to integer.
z = (**int**) a+b	**a** is converted to integer and then added to b.
p = cost ((**double**) x)	Converts x to double before using it as parameter.

Program 5.6 illustrates the use of casting in evaluating the equation.

$$\text{Sum} = \sum_{i=1}^{n} \frac{1}{i}$$

Program 5.6 The use of casting operation

```
using System;
class Casting
{
    public static void Main ( )
    {
        float sum;
        int i ;
        sum = 0.0F;
        for (i = 1; i <= 10 ; i++)
        {
            sum = sum + 1/ (float)i;
            Console.Write (" i = " + i );
            Console.WriteLine (" Sum = " + sum) ;
        }
    }
}
```

The output of Program 5.6:

```
i = 1    Sum = 1
i = 2    Sum = 1.5
i = 3    Sum = 1.833333        i = 8     Sum = 2.717857
i = 4    Sum = 2.083333        i = 9     Sum = 2.828969
i = 5    Sum = 2.283334        i = 10    Sum = 2.928968
i = 6    Sum = 2.45
i = 7    Sum = 2.592857
```

5.14 OPERATOR PRECEDENCE AND ASSOCIATIVITY

Each operator in C# has a *precedence* associated with it. This precedence is used to determine how an expression involving more than one operator is evaluated. There are distinct levels of precedence and an operator may belong to one of the levels. The operators at the higher level of precedence are evaluated first. The operators of the same precedence are evaluated either from left to right or from right to left, depending on the level. This is known as the *associativity* property of an operator. Table 5.10 provides a complete list of operators, their precedence levels, and their rules of association. The groups are listed in the order of decreasing precedence (rank 1 indicates the highest precedence level and 14 the lowest). The list also includes those operators which we have not yet discussed.

Table 5.10 *Summary of C# operators*

Operator	Description	Associativity	Rank
. ( ) [ ] x++ x− − new typeof sizeof checked unchecked	Member selection Function call Array element reference Increment Decrement Object creator Type operator Size operator Overflow checking Presentation of overflow checking	Left to right	1
+ − + + x − − x ! ~ (type)	Unary plus Unary minus Increment Decrement Logical negation Ones complement Casting	Right to left	2
* / %	Multiplication Division Modulus	Left to right	3
+ −	Addition Subtraction	Left to right	4
<< >>	Left shift Right shift	Left to right	5
< <= > >= is as	Less than Less than or equal to Greater than Greater than or equal to Relational Operator Relational Operator	Left to right	6
== !=	Equality Inequality	Left to right	7
& ^ | && || ? : = op=	Logical Bitwise AND Logical Bitwise OR Logical Bitwise OR Logical AND Logical OR Conditional operator Assignment operators Shorthand assignment	Left to right Left to right Left to right Left to right Left to right Right to left Right to left	8 9 10 11 12 13 14

It is very important to note carefully, the order of precedence and associativity of operators. Consider the following conditional statement:

```
if(x == 10+15 && y<10)
```

The precedence rules say that the addition operator has a higher priority than the logical operator (&&) and the relational operators (== and <). Therefore, the addition of 10 and 15 is executed first. This is equivalent to:

```
if(x == 25 && y<10)
```

The next step is to determine whether **x** is equal to 25 and **y** is less than 10. If we assume a value of 20 for x and 5 for y, then

```
x == 25 is false
y < 10 is true
```

Note that since the operator < enjoys a higher priority compared to ==, y < 10 is tested first and then x ==25 is tested.

Finally we get:

```
if(false && true)
```

Because one of the conditions is **false**, the compound condition is **false.**

When an operand occurs between two operators with the same precedence, the *associativity* of the operators controls the order in which the operations are performed:

- Except for the assignment operators, all binary operators are *left-associative*. For example, x + y + z is evaluated as (x + y) + z
- The assignment operators and the conditional operator are *right-associative*. For example, x = y = z is evaluated as x = (y = z)

Precedence and associativity can be controlled using parentheses. For example, x + y * z and (x + y) * z will give different results. In the first case, y * z is done first and in the second case, x + y is done first.

Program 5.7 illustrates the use of different types of expressions.

Program 5.7 Demonstration of C# expressions

```
using System;
class Expressions
{
    public static void Main( )
    {
    // Declaration and Initialization
    int a = 10, b = 5, c = 8, d = 2;
    float x = 6.4F, y = 3.0F;

    // Order of Evaluation
    int answer1 = a * b + c / d;
    int answer2 = a * (b + c) /d;
```

```
        // Modulo Operations
        int answer3 = a % c;
        float answer4 = x % y;

        // Logical Operations
        bool bool1 = a > b && c > d;
        bool bool2 = a < b && c > d;
        bool bool3 = a < b || c > d;
        bool bool4 = !(a-b == c);

        Console.WriteLine("Order of Evaluation");
        Console.WriteLine("  a * b + c / d = " + answer1);
        Console.WriteLine("  a * (b + c) / d = " + answer2);

        Console.WriteLine("Modulo Operations");
        Console.WriteLine("  a % c = " + answer3);
        Console.WriteLine("  x % y = " + answer4);

        Console.WriteLine("Logical Operations");
        Console.WriteLine("  a > b && c > d  = " + bool1);
        Console.WriteLine("  a < b && c > d  = " + bool2);
        Console.WriteLine("  a < b || c > d  = " + bool3);
        Console.WriteLine("  !(a-b) == c  = " + bool4);
    }
}
```

Program 5.7 outputs the following:

```
Order of Evaluation
   a * b + c / d = 54
   a * (b + c) / d = 65
Modulo Operations
   a % c = 2
   x % y = 0.4
Logical Operations
   a > b && c > d = true
   a < b && c > d = false
   a < b || c > d = true
   !(a-b) == c = true
```

5.15 MATHEMATICAL FUNCTIONS

Often, we may need to use trigonometric functions and logarithms. The **System** namespace defines a class known as **Math** class with a rich set of static methods that makes math-oriented programming easy and efficient. It also contains two static members, **E** and **PI**, representing the values of mathematical constants e and p. Table 5.11 lists some of the mathematical methods contained in the **Math** class.

Table 5.11 *Main mathematical methods*

Method	Description
Sin ( )	sine of an angle in radians
Cos ( )	cosine of an angle in radians
Tan ( )	tangent of an angle in radians
Asin ( )	inverse of sine
Acos ( )	inverse of cosine
Atan ( )	inverse of tangent
Atan2 ( )	inverse tangent, with x and y co-ordinates specified
Sinh ( )	hyperbolic sine
Cosh ( )	hyperbolic cosine
Tanh ( )	hyperbolic tangent
Sqrt ( )	square root
Pow ( )	number raised to a given power
Exp ( )	exponential
Log ( )	natural logarithm (base e)
Log10 ( )	base 10 logarithm
Abs ( )	absolute value of a number
Min ( )	lower of two numbers
Max ( )	higher of two numbers
Sign ( )	sign of the number

Note that all the methods take **double** type parameters and return a **double** type value. However, the parameters of **Abs ( ), Min ( ), Max ( )** , and **Sign ( )** may of any numeric type.

Program 5.8 demonstrates the use of Math class with a simple application.

Program 5.8 Use of Math class methods

```
using System;
class MathTest
{
      public static void Main ( )
      {
         Console.WriteLine("sines of angles from 0 to 90 degrees);
         for ( double theta = 0.0; theta < = 90.00; theta + = 15)
         {
            double x = Math.Sin (theta * Math.PI/180);
            Console.Write ("Sin" + theta);
```

```
            Console.WriteLine(" ={0:F4}", x);
        }
    }
}
```

Program 5.8 will produce the following output.

```
Sines of angles from 0 to 90 degrees
Sin05 = 0.0000
Sin15 = 0.2588
Sin30 = 0.5000
Sin45 = 0.7071
Sin60 = 0.8660
Sin75 = 0.9659
Sin90 = 1.0000
```

{0:F4} is an output format that prints output with four decimal places. Output formats are discussed in Chapter 17.

Since the data members and methods of **Math** class are **static** type, they are accessed using the class name itself instead of its objects. Note that we need to convert degrees to radians by multiplying by (π/180) since all trigonometric functions take radians as parameters.

At present, there is no support for complex numbers and vectors. However, we can build these types using the concept of operator overloading. See Chapter 15 for details on operator overloading.

Case Study

Problem Statement

Martin is working as a junior programmer in HighFi Software Solutions which is a software development company involved in developing applications in C, Java and C# programming languages. The Project Manager of Martin has asked him to create an application for converting a number into its equivalent binary format. Martin has to develop the application for an overseas client, TechSoft Solutions. How should Martin create the application?

Solution

To satisfy the overseas client, Martin should use an object oriented language such as C# to create the application for converting a number into a binary format. C# provides bitwise operators that can be easily used in a C# application to perform manipulation on binary data. After much research, Martin develops the following application in C# for automating the task of converting a number into binary format.

```
using System;

class Convert
{
  public int nobits;

 public Convert(int n)
 {
    nobits = n;
  }

 public void ConvertToBinary(int num)
 {
    int mask = 1;
    mask <<= nobits-1;

    int spaceVal = 0;
    for(; mask != 0; mask >>= 1)
     {
      if((num & mask) != 0)
           Console.Write("1");
       else
           Console.Write("0");
      spaceVal++;
      if((spaceVal % 8) == 0)
      {
       Console.Write(" ");
        spaceVal = 0;
      }
    }
    Console.WriteLine();
  }
}

public class BitWiseDemo
{
 public static void Main()
 {

    Console.WriteLine("Enter a Number for Conversion to Binary Form: ");
    int num=int.Parse(Console.ReadLine());
```

```
        Convert con = new Convert(8);
        Console.WriteLine("The Number in Binary Form is: ");
        con.ConvertToBinary(num);

    }
}
```

The bitwise operators can be easily used in a C# application to shift the bits from left to right and vice versa. Manually performing the task of bit manipulation is time-consuming and error-prone.

Common Programming Errors

- *Forgetting to declare a variable used in an expression.*
- *Mixing data types in expressions without clear understanding of their effect on the result.*
- *Using a variable in an expression before a value has been assigned.*
- *Attempting to store one data type in a variable declared for different type.*
- *Using expressions that may result in round-off errors.*
- *Using expressions that may result in overflow errors.*
- *Applying integer division accidently.*
- *Applying implicit conversions incorrectly.*
- *Unbalanced parentheses.*
- *Improper use of parentheses.*
- *Not using parentheses, where necessary.*
- *Improper understanding of operators precedence.*
- *Lack of understanding of associativity of operators.*
- *Improper construction of logical or a compound relational expression.*
- *Improper use of increment or decrement operators in expressions.*

Review Questions

5.1 State whether the following statements are true or false.
  (a) The sign of the result of the module division is always positive.
  (b) The modulus operator % can be applied to both the integer type and floating point type data.
  (c) When one of the operand is real and the other is integer is an expression, the integer type is converted to real, before the expression is evaluated.
  (d) All the bitwise operators have the same level of precedence in C#.
  (e) If a = 10 and b = 15, then the statement x = (a>b)?a:b assigns the value 15 to x.
  (f) In evaluating a logical expression of type (boolean_expression1 & & boolean_expression2) both the boolean expressions are not always evaluated.
  (g) In evaluating the expression (x ==y && a<b), the boolean expression x ==y is evaluated first and then a<b is evaluated.
  (h) The behaviour of expressions m++ and ++m are identical always.
  (i) All arithmetic operators enjoy the same level of predence.
  (j) Casting operations never result in loss of data.

5.2 Arrange the given operators in the order of precedence as defined in C#:

- !=
- &
- ?:
- ++
- &&

5.3 What are the special operators available in C#?

5.4 Which of the following arithmetic expressions are valid? If valid, give the value of the expression; otherwise give reason.

(a) 25/3 % 2
(b) +9/4 + 5
(c) 7.5%3
(d) 14% 3 + 7% 2
(e) –14%3
(f) 15.25+–5.0
(g) (5/3)*3+5%3
(h) 21%(int)4.5

5.5 Write C# assignment statements to evaluate the following equations:

(a) Area = $pr^2 + 2prh$

(b) Torque = $\frac{2m_1m_2}{m_1 + m_2} * g$

(c) Side = $\sqrt{a^2 + b^2 - 2ab\cos(x)}$

(d) Energy = $m\left(a * h + \frac{v^2}{2}\right)$

5.6 Identify unnecessary parentheses in the following arithmetic expressions.

(a) (x–(y/5)+z) % 8) + 25
(b) ((x–y) * p) + q
(c) (m*n) + (–x/y)
(d) x/(3*y)

5.7 Find errors, if any, in the following assignment statements and rectify them.

(a) x = y = z = 0.5, 2.0 – 5.75;
(b) m = ++a * 5;
(c) y = sqrt(100);
(d) p * = x/y;
(e) s = /5;
(f) a = b++ – c * 2

5.8 Determine the value of each of the following logical expressions if a = 5, b = 10 and c = –6

(a) a > b && a < c
(b) a < b && a > c
(c) a == c || b > a
(d) b > 15 && c < 0 || a > 0
(e) (a/2.0 == 0.0 && b/2.0 ! =0.0) || c < 0.0

5.9 Explain what each of the following code segments computes:

(a) x = 4;
y = x+x;

(b) x = "4";
y = x+x;

5.10 How do you convert a string to a number?

5.11 How do you convert a number to string?

5.12 Where do we use the modulo division?
5.13 What is a mixed-mode arithmetic expression? How is it evaluated?
5.14 What is a shorthand assignment operator? What are the advantages of using a shorthand operator?
5.15 Why do we require type conversion operations?
5.16 What do you mean by implicit conversion? Give an example of application.
5.17 What do you mean by explicit conversion? Give an example of application.
5.18 Give an example of integer overflow. Will the same example work correctly, if we use the floating point values?
5.19 Give an example of floating point round-off error. How would you overcome this problem?
5.20 Given the value of a variable, write a C# statement (without using if construct) that will produce the absolute value of the variable.
5.21 What will be the output of the following code?

```
string s;
System.Console.WriteLine("s =" + s);
```

5.22 What will be output of the following code snippet?

```
int x = 10;
int y = 20;
if ((x<y) || (x=5)>10)
   Console.WriteLine(x);
else
   Console.WriteLine(y);
```

Debugging Exercises

5.1 Find error, if any, in the following segment.

```
for (int m = 1; m < 100; m)
{
   Console.WriteLine (m, m * m);
   if (m = 10) break;
}
```

5.2 Which of the conversions in the given program are invalid?

```
using System;

class ConversionExample
{
  public static void Main()
  {

  int iInt= 22;
  long lLongInt = 44;
  double dDouble = 1.406;
```

```
    lLongInt = iInt;
    dDouble= iInt;
    iInt = lLongInt;
    lLongInt = dDouble;

    }

}
```

Programming Exercises

5.1 *Write a program to read interactively two integer values using the methods* ***Console.ReadLine( )*** *and* ***int.Parse( )*** *and then display their*

- *Sum*
- *Difference*
- *Product*
- *Integer division*
- *Modulus division*

5.2 *Write a program to read interactively two floating-point values using* ***Console.ReadLine( )*** *and* ***double.Parse( )*** *methods and then display their*

- *Sum*
- *Product*
- *Difference*
- *Modulus division*

5.3 *Write a program to compute and display the average of the numbers 25, 75 and 100.*

5.4 *Given the radius of the circle as 12.5 centimeters, write a program to compute its circumference and area and display their values.*

5.5 *State why the expression*

$$x - y = 100$$

is invalid but the expression

$$x - (y=100)$$

is valid . Execute a program to demonstrate your answer.

5.6 *Write a program to read the price of an item in decimal form (like 75.95) and print the output in paise (like 7595 paise).*

5.7 *Write a program to convert the given temperature in fahrenheit to celsius using the following conversion formula.*

$$C = \frac{F - 32}{1.8}$$

and display the values in a tabular form.

5.8 *The straight-line method of computing the yearly depreciation of the value of an item is given by*

$$\text{Depreciation} = \frac{\text{Purchase price} - \text{Salvage value}}{\text{Years of service}}$$

Write a program to determine the salvage value of an item when the purchase price, years of service and the annual depreciation are given.

5.9 Write a program that will read a real number from the keyboard and print the following output in one line:

Smallest integer not less than than the number	The given number	Largest integer not greater than the number

5.10 The total distance travelled by a vehicle in t seconds is given by

$$distance = ut + (at^2)/2$$

where u is the initial velocity (metres per second), a is the acceleration (metres per second^2). Write a program to evaluate the distance travelled at regular intervals of time, given the values of u and a. The program should provide the flexibility to the user to select his own time intervals and repeat the calculations for different values of u and a.

5.11 In inventory management, the Economic Order Quantity for a single item is given by

$$EOQ = \sqrt{\frac{2 * demand\ rate * setup\ costs}{holding\ cost\ per\ item\ per\ unit\ time}}$$

and the optimal Time Between Orders

$$TBO = \sqrt{\frac{2 * setup\ costs}{demand\ rate\ *\ holding\ cost\ per\ item\ per\ unit\ time}}$$

Write a program to computer EOQ and TBO, given demand rate (items per unit time), setup costs (per order), and the holding cost (per item per unit time).

5.12 For a certain electrical circuit with an inductance L and resistance R, the damped natural frequency is given by

$$Frequency = \sqrt{\frac{1}{LC} - \frac{R^2}{4C^2}}$$

It is desired to study the variation of this frequency with C (capacitance). Write a program to calculate the frequency for different values of C starting from 0.01 to 0.1 in steps of 0.01.

5.13 What is the output of the following program?

```
using System;

class DecrementOperator
{
    static void Main(  )
    {
        int var = 15;
        int resultVar;

Console.WriteLine("–Program to demonstrate the use of Decrement
Operatior in C#–\n");
Console.WriteLine("–var: First Decrement the value, then assigns it...");
        resultVar = –var;
```

```
        Console.WriteLine( "After -var: {0}, {1}", var,
        resultVar );

Console.WriteLine("\nvar-: First, assigns the value and then decrements
it...");
        resultVar = var-;
        Console.WriteLine( "After var-: {0}, {1}",
        var, resultVar );
    }
}
```

Chapter 6

Decision Making and Branching

6.1 INTRODUCTION

A C# program is a set of statements that are normally executed sequentially in the order in which they appear. This happens when options or repetitions of certain calculations are not necessary. However, in practice, we have a number of situations, where we may have to change the order of execution of statements based on certain conditions, or repeat a group of statements until certain specified conditions are met. This involves a kind of decision-making to see whether a particular condition has occurred or not and then to direct the computer to execute certain statements accordingly.

When a program breaks the sequential flow and jumps to another part of the code, it is called *branching*. When the branching is based on a particular condition, it is known as *conditional branching*. If branching takes place without any decision, it is known as *unconditional branching*.

C# language possesses such decision-making capabilities and supports the following statements known as *control* or *decision-making* statements.

1. **if** statement
2. **switch** statement
3. Conditional operator statement

In this chapter, we shall discuss the features, capabilities and applications of these statements which are classified as *selection* statements.

6.2 DECISION MAKING WITH IF STATEMENT

The **if** statement is a powerful decision-making statement and is used to control the flow of execution of statements. It is a *two-way* decision statement and is used in conjunction with an expression. It takes the following form:

```
If(expression)
```

It allows the computer to evaluate the *expression* first and then, depending on whether the value of the *expression* (relation or condition) is 'true' or 'false', it transfers the control to a particular statement. The program at this point has two paths to follow, one for the ***true*** condition and the other for the ***false*** condition as shown in Fig. 6.1.

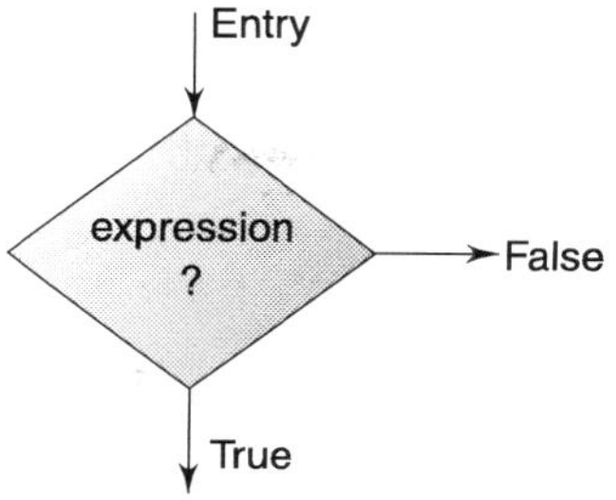

Figure 6.1 *Two-way branching*

Some examples of decision making, using the **if** statement are:

1. **if** (bank balance is zero)
   borrow money
2. **if** (room is dark)
   put on lights
3. **if** (code is 1)
   person is male
4. **if** (age is more than 55)
   person is retired

The **if** statement may be implemented in different forms depending on the complexity of the conditions to be tested.

1. Simple **if** statement
2. **if..else** statement
3. Nested **if..else** statement
4. **else if** ladder

6.3 SIMPLE IF STATEMENT

The general form of a simple **if** statement is

```
if(boolean-expression)
{
   statement-block;
}
statement-x;
```

The 'statement-block' may be a single statement or a group of statements. If the *boolean-expression* is true, the *statement-block* will be executed; otherwise the statement-block will be skipped and the execution will jump to the *statement-x*.

It should be remembered that when the condition is true both the statement-block and the statement-x are executed in sequence. This is illustrated in Fig. 6.2.

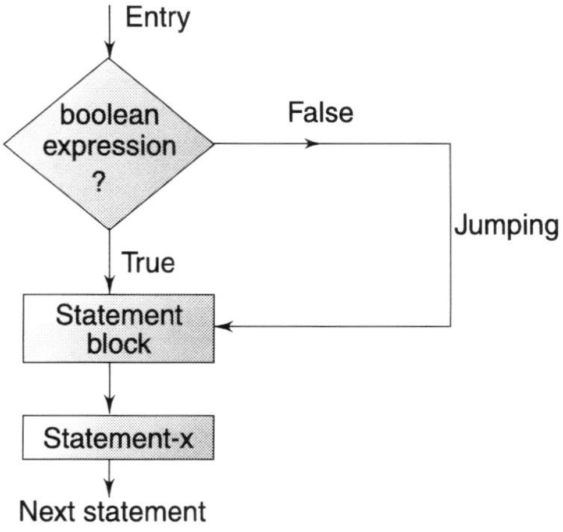

Figure 6.2 *Flowchart of simple if control*

Consider the following segment of a program that is written for processing of marks obtained in an entrance examination.

```
..........
..........
if(category == SPORTS)
{
```

```
        marks = marks + bonus_marks;
    }
    System.Console.WriteLine(marks);
    ..........
    ..........
```

The program tests the type of category of the student. If the student belongs to the SPORTS category, then additional bonus_marks are added to his marks before they are printed. For others, bonus_marks are not added.

Consider a case having two test conditions, one for weight and another for height. This is done using the compound relation

```
if(weight < 50 && height > 170) count = count +1;
```

This could have been done equivalently using two **if** statements as follows:

```
if(weight<50)
if(height>170)
count = count+1;
```

If the value of **weight** is less than 50, then the following statement is executed, which in turn is another **if** statement. This if statement tests **height** and if the **height** is greater than 170, then the **count** is incremented by 1. Program 6.1 illustrates the implementation of the above statement.

Program 6.1 Counting with if statement

```
using System;
class IfTest
{
  public static void Main( )
   {
    int i, count, count1, count2;
    float[] weight = { 45.0F,55.0F,47.0F,51.0F,54.0F };
    float[] height = { 176.5F,174.2F,168.0F,170.7F,169.0F };

    count  = 0;
    count1 = 0;
    count2 = 0;

    for (i = 0; i <= 4; i++)
     {
      if(weight[i] < 50.0 && height[i] > 170.0)
       {
       count1 = count1 + 1;  // Executed when condition is true
       }
      count = count + 1;      // Always executed

     }
```

```
    count2 = count - count1;
    Console.WriteLine("Number of persons with ... ");
  Console.WriteLine("Weight<50 and height>170 = "+count1);
  Console.WriteLine("Others = " + count2);

 }
}
```

The output of Program 6.1 will be:

```
Number of persons with ...
Weight<50 and height>170 = 1
Others = 4
```

C and C++ programmers must note that in C# the expression in an **if** statement must resolve to a bool value. We cannot use integer values, like if (x).... Arrays are discussed in Chapter 9.

6.4 THE IF... ELSE STATEMENT

The **if....else** statement is an extension of the simple **if** statement. The general form is

```
if(boolean_expression)
{
   True-block statement(s)
}
else
{
   False-block statement(s)
}
statement-x
```

If the *boolean_expression* is true, then *the true-block statement(s),* immediately following the **if** statement, are executed; otherwise, the *false-block statement(s)* are executed. In either case, either *true-block* or *false-block* will be executed, not both. This is illustrated in Fig. 6.3. In both the cases, the control is transferred subsequently to the statement-x.

Let us consider an example of counting the number of boys and girls in a class. We use code 1 for a boy and 2 for a girl. The program statements to do this may be written as follows:

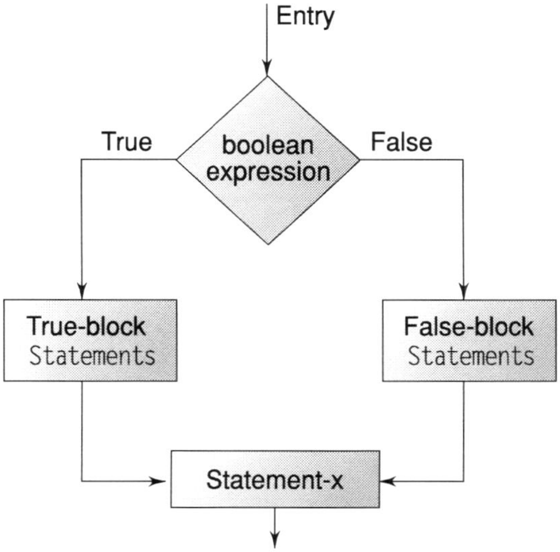

Figure 6.3 *Flowchart of if....else control*

```
..........
..........
if(code == 1)
    boy = boy + 1;
if(code == 2)
    girl = girl + 1;
..........
..........
```

The first test determines whether or not the student is a boy. If yes, the number of boys is increased by 1 and the program continues to the second test. The second test again determines whether the student is a girl. This is unnecessary. Once a student is identified as a boy, there is no need to test again for a girl. A student can be either a boy or a girl, not both. The above program segment can be modified using the **else** clause as follows:

```
..........
..........
if(code == 1)
    boy = boy +1;
else
    girl = girl + 1;
xxx;
..........
```

Here, if the code is equal to 1, the statement **boy = boy + 1;** is executed and the control is transferred to the statement **xxx**, after skipping the **else** part. If the code is not equal to 1, the statement **boy = boy + 1;** is skipped and the statement in the **else** part **girl = girl + 1;** is executed before the control reaches the statement **xxx.**

Program 6.2 counts the even and odd numbers in a list of numbers using the **if....else** statement.

Program 6.2 Experimenting with if....else statement

```
using System;
class IfElseTest
{
  public static void Main( )
  {
    int[] number = { 50, 65, 56, 71, 81 };
    int even = 0, odd = 0;

    for (int i = 0; i < number.Length; i++)
    {
     if ((number[i] % 2) == 0)  // use of modulus operator
      {
      even += 1;  // counting EVEN numbers
      }
     else
      {
      odd += 1;  // counting ODD numbers

      }
```

```
    }

   Console.WriteLine("Even Numbers : " + even);
   Console.WriteLine("Odd Numbers : " + odd);
  }
}
```

Output of Program 6.2 would be:

```
Even Numbers  : 2
Odd Numbers   : 3
```

Note that number.Length gives the length of number array.

Program 6.3 presents different ways of using if statement in C#, including a single **if** statement, **if-else** construct and multiple **else if** statements. The program takes a number as an input from the user and provides information about the number using the various forms of **if** statement.

Program 6.3 If else illustrated further

```
using System;

class IfElseExample
{
   public static void Main()
   {
      string numInput;
      int numValue;

      Console.Write("Please enter a number: ");
      numInput = Console.ReadLine();
      numValue = Int32.Parse(numInput);

      // A Single Decision with its Action
      if (numValue > 0)
      {
         Console.WriteLine("The number {0} is greater than zero.", numValue);
      }

      // A Single Decision with its Action without brackets
      if (numValue < 0)
         Console.WriteLine("The number {0} is less than zero.", numValue);

      // A Decision using Logical Operator
      if (numValue != 0)
      {
         Console.WriteLine("The number {0} is not equal to zero.", numValue);
      }
      else
```

```
          {
            Console.WriteLine(“The number {0} is equal to zero.”, numValue);
          }

        // Multiple Else If Conditions
        if (numValue < 0 || numValue == 0)
          {
            Console.WriteLine(“The number {0} is less than or equal to zero.”,
numValue);
          }
        else if (numValue > 0 && numValue <= 10)
          {
            Console.WriteLine(“The number {0} falls in the range from 1 to 10.”,
numValue);
          }
        else if (numValue > 10 && numValue <= 20)
          {
            Console.WriteLine(“The number {0} falls in the range from 11 to 20.”,
numValue);
          }
        else if (numValue > 20 && numValue <= 30)
          {
            Console.WriteLine(“The number {0} falls in the range from 21 to 30.”,
numValue);
          }
          else
          {
            Console.WriteLine(“The number {0} is greater than 30.”, numValue);
          }
    }
}
```

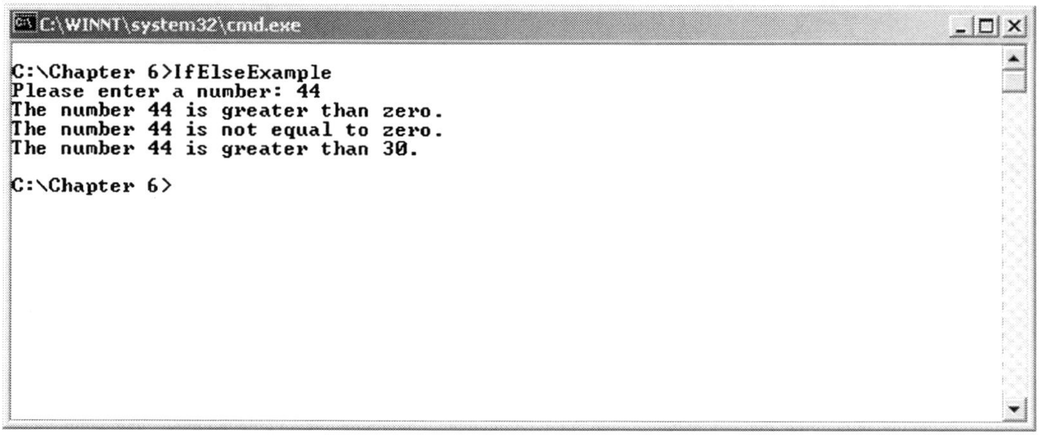

6.5 NESTING OF IF....ELSE STATEMENTS

When a series of decisions are involved, we may have to use more than one **if....else** statement in *nested* form as follows:

```
if (test condition1)
{
    if (test condition2)
    {
        statement-1;
    }
    else
    {
        statement-2;
    }

}
else
{
        statement-3;
}
statement-x;
```

The logic of execution is illustrated in Fig. 6.4. If the *condition-1* is false, the statement-3 will be executed; otherwise it continues to perform the second test. If the *condition-2* is true, then statement-1 will be evaluated; otherwise statement-2 will be evaluated and the control transferred to statement-x.

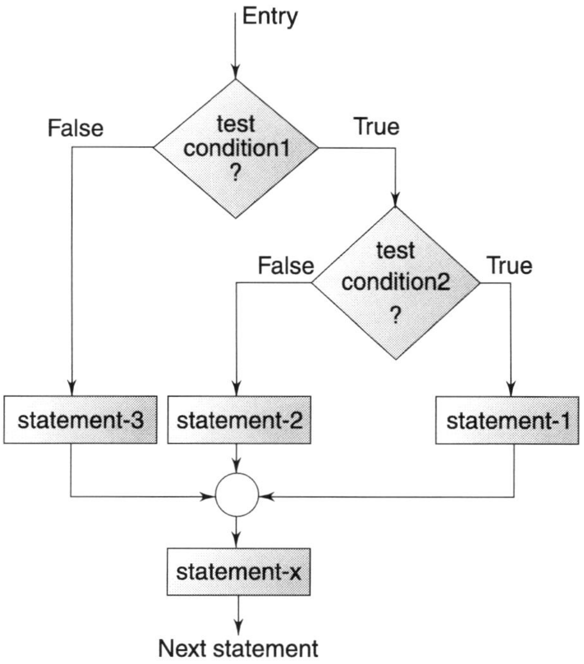

Figure 6.4 *Flowchart of nested if....else statements*

A commercial bank has introduced an incentive policy of giving a bonus to all its deposit holders. The policy is as follows: A bonus of 2 per cent of the *balance* held on 31st December is given to every one, irrespective of their balances, and 5 per cent is given to female account holders if their balance is more than Rs 5000. This logic can be coded as follows:

```
..........
if(sex is female)
{
    if (balance > 5000)
        bonus = 0.05 * balance;
    else
        bonus = 0.02 * balance;
}
else
{
    bonus = 0.02 * balance;
}
balance = balance + bonus;
..........
..........
```

When nesting, care should be exercised to match every **if** with an **else.** Consider the following alternative to the above program (which looks right at first sight):

```
if(sex is female)
    if(balance > 5000)
        bonus = 0.05 * balance;
else
    bonus = 0.02 * balance;
    balance = balance + bonus;
```

There is an ambiguity as to over which **if** the **else** belongs to. In C# an **else** is linked to the closest non-terminated **if.** Therefore, the **else** is associated with the inner **if** and there is no **else** option for the outer **if**. This means that the computer is trying to execute the statement

```
    balance = balance + bonus;
```

without really calculating the bonus for the male account holders.

Consider another alternative which also looks correct:

```
if(sex is female)
{
    if (balance > 5000)
        bonus = 0.05 * balance;
}
else
    bonus = 0.02 * balance;
    balance = balance + bonus;
```

In this case, **else** is associated with the outer **if** and therefore bonus is calculated for the male account holders. However, bonus for the female account holders, whose balance is equal to or less than 5000 is not calculated because of the missing **else** option for the inner **if.**

Program 6.4 employs nested **if....else** statements to determine the largest of three given numbers.

Program 6.4 Nesting if....else statements

```
using System;
class IfElseNesting
{
  public static void Main( )
  {
    int a = 325, b = 712, c = 478;

    Console.Write("Largest value is : ");

    if (a > b)
    {
      if (a > c)
      {
       Console.WriteLine(a);
      }
      else
      {
       Console.WriteLine(c);
      }
    }

    else
    {
      if (c > b)
      {
       Console.WriteLine(c);
      }
      else
      {
       Console.WriteLine(b);
      }
    }
  }
}
```

Output of Program 6.4:

```
Largest value is : 712
```

6.6 THE ELSE IF LADDER

There is another way of putting **ifs** together when multipath decisions are involved. A multipath decision is a chain of **ifs** in which the statement associated with each **else** is an **if**. It takes the following general form:

```
if (condition 1)
    statement-1;
else if (condition 2)
    statement-2;
    else if (condition 3)
        statement-3;
    else if (condition n)
        statement-n;
    else
       default-statement;
 statement-x;
```

This construct is known as the **else if** ladder. The conditions are evaluated from the top (of the ladder), downwards. As soon as the true condition is found, the statement associated with it is executed and the control is transferred to the *statement-x* (skipping the rest of the ladder). When all the n conditions become false, then the final **else** containing the *default-statement* will be executed. Figure 6.5 shows pictorially the logic of execution of **else if** ladder statements.

Let us consider an example of grading the students in an academic institution. The grading is done according to the following rules:

Average marks	Grade
80 to 100	Honours
60 to 79	First Division
50 to 59	Second Division
40 to 49	Third Division
0 to 39	Fail

This grading can be done using the **else if** ladder as follows:

```
if(marks > 79)
    grade = "Honours";
else if(marks > 59)
    grade = "First Division";
    else if(marks > 49)
        grade = "Second Division";
```

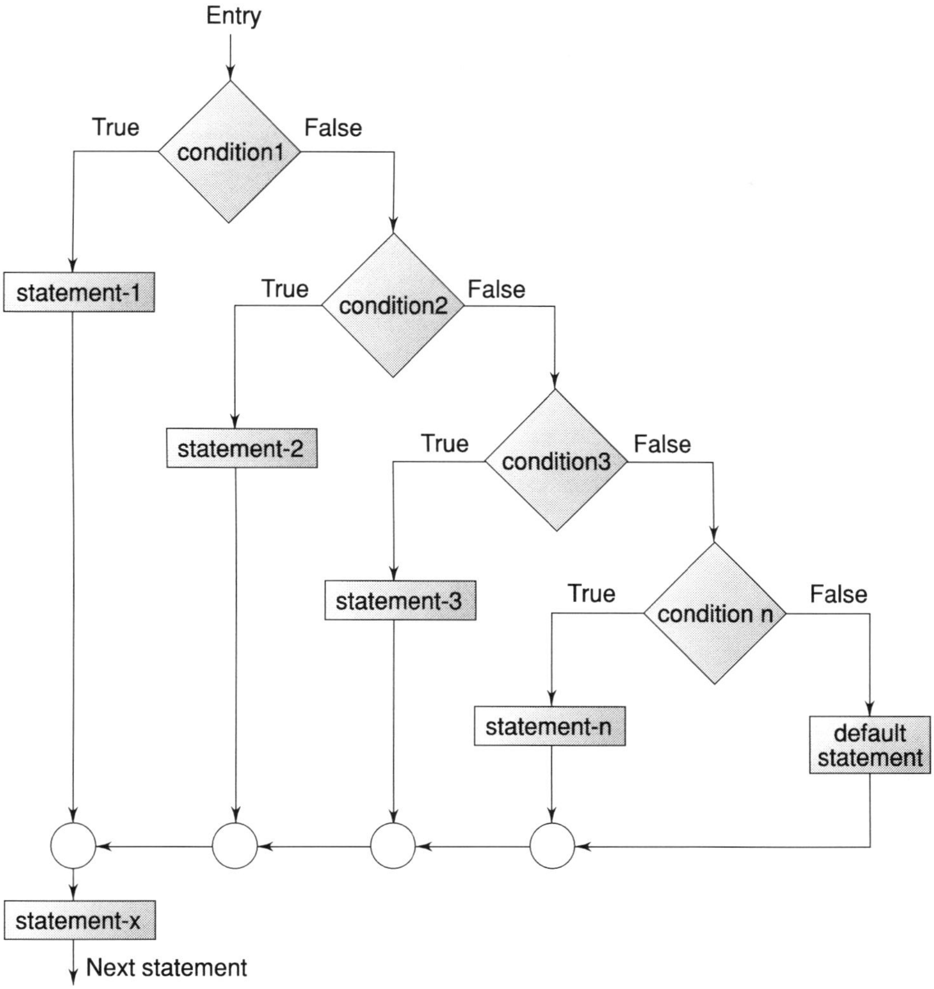

Figure 6.5 *Flowchart of else...if ladder*

```
        else if(marks > 39)
           grade = "Third Division";

           else
           grade = "Fail";
Console.WriteLine("Grade: " + grade);
```

Consider another example given below:

```
..........
..........
if (code == 1)
   colour = "RED";
```

```
else if (code == 2 )
    colour = "GREEN";

    else if (code == 3)
        code = "WHITE";

        else
        colour = "YELLOW";
..........
..........
```

Code numbers other than 1, 2 or 3 are considered to represent YELLOW. The same results can be obtained by using nested **if....else** statements.

```
if(code != 1)
    if (code != 2)
        if (code != 3)
            colour = "YELLOW";
        else
        colour = "WHITE";
    else
    colour = "GREEN";
else
    colour = "RED";
```

In such situations, the choice of the method is left to the programmer. However, in order to choose an **if** structure that is both effective and efficient, it is important that the programmer is fully aware of the various forms of an **if** statement and the rules governing their nesting. Program 6.5 demonstrates the use of **if....else** ladder in analysing a marks list.

Program 6.5 Demonstration of else if ladder

```
using System;
class ElseIfLadder
{
  public static void Main( )
  {
    int[] rollNumber = { 111, 222, 333, 444 };
    int[] marks = { 81, 75, 43, 58 };

    for (int i = 0; i < rollNumber.Length; i++)
    {
      if (marks[i] > 79)
        Console.WriteLine(rollNumber[i] + " Honours");

      else if (marks[i] > 59)
```

```
            Console.WriteLine(rollNumber[i] + " I Division");

        else if (marks[i] > 49)
            Console.WriteLine(rollNumber[i] + " II Division");

            else
            Console.WriteLine(rollNumber[i] + " FAIL");
    }
  }
}
```

Program 6.5 produces the following output:

```
111 Honours
222 I Division
333 FAIL
444 II Division
```

6.7 THE SWITCH STATEMENT

We have seen that when one of many alternatives has to be selected, we can design a program using **if** statements to control the selection. However, the complexity of such a program increases dramatically when the alternatives increase. The program becomes difficult to read and follow. At times, it may confuse even the designer of the program. Fortunately, C# has a built-in multiway decision statement known as a **switch**. The **switch** statement tests the value of a given variable (or expression) against a list of **case** values and when a match is found, a block of statements associated with that **case** is executed. The general form of the **switch** statement is as shown below:

```
switch(expression)
{
    case value-1:
        block-1
        break;
    case value-2:
        block-2
        break;
    ......
    ......
    default:
        default-block
        break;
}
statement-x;
```

The *expression* must be an integer type or **char** or **string** type. *value-1*, *value-2* .... are constants or constant expressions and are known as *case labels*. Each of these values should be unique within a **switch** statement. *block-1*, *block-2* .... are statement lists and may contain zero or more statements. There is no need to put braces around these blocks but it is important to note that case labels end with a colon (:).

The **switch** statement is executed in the following order:

1. The **expression** is evaluated first.
2. The value of the expression is successively compared against the values, *value-1, value-2, ...* If a **case** is found whose value matches the value of the *expression*, then the block of statements that follows the case are executed.
3. The **break** statement at the end of each block signals the end of a particular case and causes an exit from the **switch** statement, transferring the control to the *statement-x* following the **switch.**
4. The **default** is an optional case. When present, it will be executed if the value of the expression does not match any of the case values. If not present, no action takes place when all matches fail and the control goes to the *statement-x*.

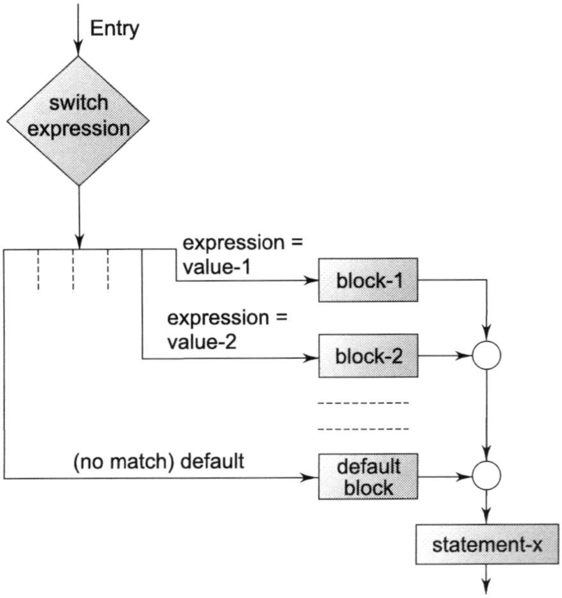

Figure 6.6 *Selection process of the switch statement*

The selection process of **switch** statement is illustrated in the flowchart shown in Fig. 6.6.

The **switch** statement can be used to grade the students as discussed in the last section. This is illustrated below:

```
..........
..........
index = marks/10;
switch(index)
{
    case 10:
    case 9:
    case 8:
        grade = "Honours";
        break;
    case 7:
    case 6:
        grade = "First Division";
        break;
    case 5:
        grade = "Second Division";
```

```
            break;
        case 4:
            grade = "Third Division";
            break;
        default:
            grade = "Fail";
            break;
    }
    Console.WriteLine(grade);
    ..........
    ..........
```

Note that we have used a conversion statement

```
index = marks/10;
```

where, index is defined as an integer. The variable index takes the following integer values.

Marks	Index
100	10
90 – 99	9
80 – 89	8
70 – 79	7
60 – 69	6
50 – 59	5
40 – 49	4
30 – 39	3
20 – 29	2
10 – 19	1
0 – 9	0

This segment of the program illustrates two important features. First, it uses empty cases. The first three cases will execute the same statements

```
grade = "Honours";
break;
```

Same is the case with case 7 and case 6. Second, the default condition is used for all other cases where marks are less than 40.

Program 6.6 illustrates the use of **switch** for a menu-driven interactive program.

Program 6.6 Testing the switch statement

```
using System;
class CityGuide
{
    public static void Main( )
    {
        Console.WriteLine("Select your choice");
```

```
        Console.WriteLine("London");
        Console.WriteLine("Bombay");
        Console.WriteLine("Paris");
        Console.WriteLine("Type your choice");
        String name = Console.ReadLine ( );

        switch (name)
        {
            case "Bombay":
                Console.WriteLine("Bombay:Guide 5");
                break;
            case "London":
                Console.WriteLine("London:Guide 10");
                break;
            case "Paris":
                Console.WriteLine("Paris:Guide 15");
                break;
            default:
                Console.WriteLine ("Invalid choice");
                break;
        }
    }
}
```

Output of Program 6.6:

```
Select your choice
London
Bombay
Paris
Type your choice
London
London : Guide 10
```

6.7.1 Fallthrough in Switch Statement

In the absence of the break statement in a case block, if the control moves to the next case block without any problem, it is known as 'fallthrough'. Fallthrough is permitted in C, C++ and Java. But C# does not permit automatic fallthrough, if the case block contains executable code. However, it is allowed if the case block is empty as discussed earlier. For instance,

```
Switch (m)
{
    case 1:
        x = y;
    case 2:
        x = y + m;
```

```
    default:
        x = y - m;
}
```

is an error in C#. If we want two consecutive case blocks to be executed continuously, we have to force the process by using the **goto** statement. For example:

```
Switch (m)
{
    case 1:
        x = y;
        goto case 2;
    case 2:
        x = y + m
        goto default;
    default:
        x = y - m;
        break;
}
```

The **goto** mechanism enables us to jump backward and forward between cases and therefore arrange labels arbitrarily. *Example*:

```
Switch (m)
{
    default:
        x = y-m;
        break;
    case 2:
        x = y + m;
        goto default;
    case 1:
        x = y;
        goto case 2;
}
```

is perfectly valid.

Program 6.7 uses switch case to allow users to select the size of a cola can. Based on the user selection, the program prints the amount the user must pay for the particular type of can selected.

Program 6.7 Another switch example

```
using System;

class SwitchExample
{
  public static void Main()
  {
    Console.WriteLine("Select your Cola Can:");
    Console.WriteLine("S=SMALL M=MEDIUM L=LARGE");
```

```
    Console.Write("Please enter your option: ");
    String s = Console.ReadLine();
    int cost = 0;

    switch(s)
    {
      case "S":
        cost += 10;
        break;

      case "M":
        cost += 15;
        goto case "S";

      case "L":
        cost += 30;
        goto case "S";

      default:
        Console.WriteLine("Invalid option. Please select either S, M, or L.");
      break;
    }

    if( cost != 0 )
      Console.WriteLine("Please pay {0} Rupees.", cost);

  }
}
```

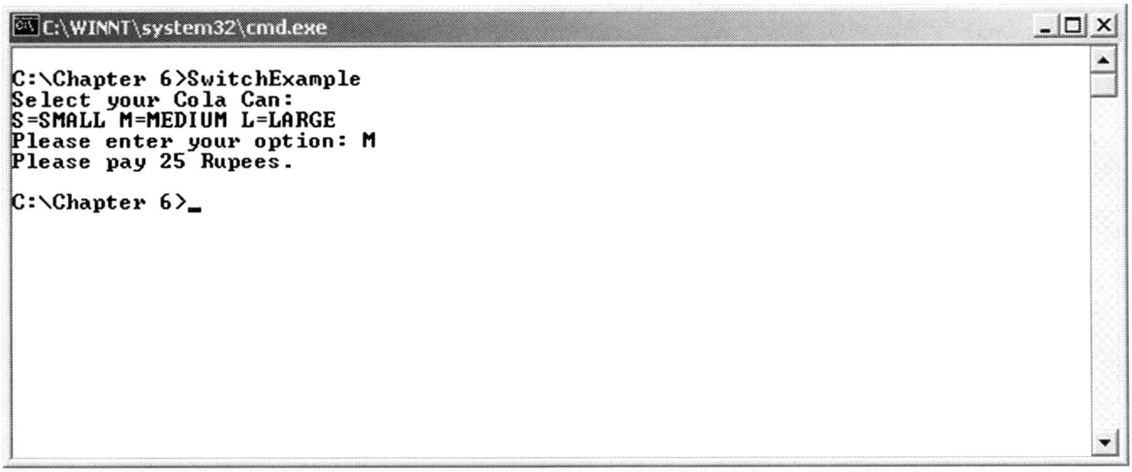

6.8 THE ? : OPERATOR

C# has an unusual operator, useful for making two-way decisions; it is a combination of ? and :, and takes three operands. This operator is popularly known *as the conditional operator*. The general form of use of the *conditional operator* is as follows:

```
conditional expression ? expression1 : expression2
```

The *conditional expression* is evaluated first. If the result is true, *expression 1* is evaluated and is returned as the value of the conditional expression. Otherwise, *expression2* is evaluated and its value is returned. For example, the segment

```
if (x < 0)
    flag = 0;
else
    flag = 1;
```

can be written as

```
flag = (x<0) ? 0 : 1;
```

Consider the evaluation of the following function:

```
y = 1.5x + 3   for x <= 2
y = 2x + 5     for x > 2
```

This can be evaluated using the conditional operator as follows:

```
y = (x>2) ? (2*x+5) : (1.5*x+3);
```

The conditional operator may be nested for evaluating more complex assignment decisions. For example, consider the weekly salary of a salesgirl who is selling some domestic products. If x is the number of products sold in a week, her weekly salary is given by

```
          {  4x + 100      for x < 40
salary =  {  300           for x = 40
          {  4.5x + 150    for x > 40
```

This complex equation can be written as

```
salary = (x!=40) ? ((x<40)?(4*x+100):(4.5*x+150)) : 300;  // nesting
```

The same can be evaluated using **if....else** statements as follows:

```
if (x<=40)
    if (x<40)
        salary = 4*x+100;
    else
        salary = 300;
else
    salary = 4.5*x+150;
```

When the conditional operator is used, the code becomes more concise and perhaps, more efficient. However, the readability is poor. It is better to use **if** statements when more than a single nesting of the conditional operator is required.

Case Study

Problem Statement

SafeMoney Bank provides loans to individuals at simple and compound interest for a period ranging from 1 to 15 years. Till recently, the staff of the SafeMoney Bank has been manually calculating simple and compound interest for loans provided to individuals by the bank. It has been found by the manager of SafeMoney Bank that for the past two months, a lot of errors have been occurring during the calculation of simple and compound interest by the staff of the Bank. In addition, there has been a lot of delay in approving of loans. As a result, the bank is losing valuable customers. How can the manager of SafeMoney Bank solve this problem?

Solution

The task of calculating simple and compound interest must be automated in order to decrease the errors in the calculations. The manager of SafeMoney Bank asks Binoy who is a senior programmer in the IT department of SafeMoney Bank to create an application for automating the task of calculating simple and compound interest so that it can be performed efficiently by the bank staff. After a requirement analysis, Binoy decides that decision-making statements such as **If Else** supported by C# must be used in the application for calculating simple and compound interest. Binoy creates the following C# application.

```
using System;

class InterestCalculator
{
      public static void Main()
      {
        Console.WriteLine("==Interest Calculator==");
        Console.WriteLine("Calculate Simple (S) / Compound (C) Interest: ");
        string choice=Console.ReadLine();

        Console.WriteLine("Enter Principal Amount: ");
        double p=double.Parse(Console.ReadLine());
        Console.WriteLine("Enter Rate of Interest: ");
        double r=double.Parse(Console.ReadLine());
        Console.WriteLine("Enter Duration: ");
        int t=Int32.Parse(Console.ReadLine());

        if(choice=="S")
```

```
        {
           double simpleInterest=(double)((p*r*t)/100);
           Console.WriteLine("The Simple Interest is: {0.0:Rs
###.##}",simpleInterest);
        }
        else
        {
           double compoundInterest=(double)p*(Math.Pow(((r/100)+1),t));
           Console.WriteLine("The Compound Interest is: {0.0:Rs
###.##}",compoundInterest);
        }
      }
}
```

Remarks

The **If Else** statement in C# can be used to evaluate a condition and on the basis of the result of the evaluation, the control flow of the program must be decided. If the condition given with the **If** statement evaluates to **true** then a specific set of program statements must be executed. However if the condition evaluates to **false** then another set of program statements must be executed.

Common Programming Errors

- *Using = in place of = = sign.*
- *Dangling **else** in nested **if** statements.*
- *Forgetting to initialize variables in some paths of a multiple branch construct.*
- *Improper and confused use of && and \\ conditional operators.*
- *Not applying the operators precedence properly.*
- *Using multiple relational operators like (a <= x <= b).*
- *Comparing floating point expressions in if statements like if (x = = y).*
- *Forgetting to use colon for case labels in switch statement.*
- *Not using **goto** for fallthrough cases in **switch** statement.*
- *Using operators <=, >=, and != as =<, =>, and +!*
- *Providing spaces between the symbols of operators ==, <=, >=, and !=.*
- *Not using braces when the **if** statement blocks contain multiple statements.*
- *Using integer values in the test expression of an if statement, like **if** (x)……*
- *Placing semicolon immediately after the right parentheses after the condition in a **if** statement.*
- *Improper nesting of **if…else** statements.*

Review Questions

6.1 Determine whether the following are true or false:
   (a) When **if** statements are nested, the last **else** gets associated with the nearest **if** without an **else**.
   (b) One **if** can have more than one **else** clause.
   (c) A **switch** statement can always be replaced by a series of **if....else** statements.
   (d) A **switch** expression can be of any type.

(e) A program stops its execution when a **break** statement is encountered.
(f) The expression in an **if** statement must always resolve to an integer value.
(g) A set of nested **if....else** statements can be replaced by an **else if** ladder.
(h) Every case label in a **switch ( )** should have an associated break statement.
(i) The **break** statement is optional in the **default** case, when it is the last one.
(j) The conditional operator ?: can be replaced by an **if else** construct.
(k) The **default** in **switch** must be the last statement in the list of cases.
(l) The **default** is an optional case in **switch** statement.

6.2 What is the sequence of execution of **switch** statement?

6.3 What are the various forms of **if** statement and their specific uses?

6.4 In what ways does a **switch** statement differ from an **if** statement?

6.5 Find errors, if any, in each of the following **if** segments:
(a) if(x+y = z && y>0)
(b) if(p<0) || (q<0)

6.6 The following is a segment of a program:

```
    x = 1;
    y = 1;
if(n>0)
    x = x+1;
    y = y-1;
```

What will be the values of x and y if n assumes a value of (a)1 and (b) 0.

6.7 Rewrite each of the following without using compound relations:

(a)
```
if(grade<=59 && grade>=50)
second = second + 1;
```
(b)
```
if(number>100 || number<0)
    System.Console.Write("Out of range");
else
    sum = sum + number;
```
(c)
```
if  ((M1>60 && M2>60) || T>200)
    y=1;
    else
    y=0;
```

6.8 What is "fallthrough" in switch statement" How is it achieved in C#?

6.9 When do you consider it essential to use a **goto** statement?

6.10 Write a pseudocode for each of the following:
(a) Read two numbers from the keyboard and display the larger of the two numbers
(b) Read three numbers from the keyboard and display the lowest of the three numbers.

6.11 When will you use the conditional operator?: compared to **if...else** statements?

6.12 Identify the errors in the following statements.

(a)
```
if (code = = 1)
    Console.WriteLine("Pass");
else;
    Console.WriteLine("Fail");
```

(b)
```
if (code > 1);
    a = b + c
else
    a = 0
```

Debugging Exercises

6.1 Find the error in the following program.

```
using System;

class TerneryExample
{
 static void Main()
 {
       int num = 3;
       string result = (num < 2) ? "True" : "False";
       Console.WriteLine(result);
 }
}
```

6.2 Debug the program given below.

```
using System;

class StudentRecord
{
    static void Main(string[] args)
    {
        Console.WriteLine("Enter the name of the student: ");
        string name = Console.ReadLine();

        Console.WriteLine("Enter the grade for the student: ");
        string grade = Console.ReadLine();

        if (grade == "a")
            Console.WriteLine("{0} is an Outstanding student.",name);
            Console.WriteLine("You have received a scholarship.");
        else if (grade == "b")
            Console.WriteLine("{0} is a Good student.",name);
        else if (grade == "c")
            Console.WriteLine("{0} is an Average student.",name);
        else if (grade == "d")
            Console.WriteLine("{0} Needs more practice.",name);
        else
            Console.WriteLine("{0} Need a lot of practice.",name);
    }
}
```

Programming Exercises

6.1 *Write a program to add all the odd numbers from 0 to 20. Use a simple* **if** *and* **goto** *statements to form a loop of operations*

6.2 *Modify the above program so that it also adds all the even numbers between 0 and 20*

6.3 *Write a program to find the number of and sum of all integers greater than 100 and less than 200 that are divisible by 7.*

6.4 *A set of two linear equations with two unknowns x_1 and x_2 is given below:*

$$ax_1 + bx_2 = m$$

$$cx_1 + dx_2 = n$$

The set has a unique solution

$$x_1 = \frac{md + bn}{ad - cb}$$

$$x_2 = \frac{na + mc}{ad - cb}$$

provided the denominator ad – cb is not equal to zero.

Write a program that will read the values of constants a, b, c, d, m and n and compute the values of x_1 and x_2. An appropriate message should be printed if ad – cb = 0.

6.5 *Given a list of marks ranging from 0 to 100, write a program to compute and print the number of students who have obtained marks*

(a) in the range 81 to 100,
(b) in the range 61 to 80,
(c) in the range 41 to 60, and
(d) in the range 0 to 40.

The program should use a minimum number of **if** *statements.*

6.6 *Admission to a professional course is subject to the following conditions:*

(a) Marks in mathematics	*>= 60*
(b) Marks in physics	*>= 50*
(c) Marks in chemistry	*>= 40*
(d) Total in all three subjects	*>= 200*
(or)	
Total in mathematics and physics	*>= 150*

Given the marks in the three subjects, write a program to process the applications to list the eligible candidates.

6.7 *Write a program to print a two-dimensional Square Root Table as shown below, to provide the square root of any number from 0 to 9.9. For example, the value x will give the square root of 3.2 and y the square root of 3.9.*

Square Root Table

Number	0.0	0.1	0.2	....	0.9
0.0 1.0 2.0 3.0 . . 9.0			x		y

6.8 Shown below is a Floyd's triangle

1
2 3
4 5 6
7 8 9 10
11 15
.
.
79.. 91

(a) Write a program to print this triangle.
(b) Modify the program to produce the following form of Floyd's triangle:

1
0 1
1 0 1
0 1 0 1
1 0 1 0 1

6.9 A cloth showroom has announced the following seasonal discounts on purchase of items:

Purchase Amount	Discount	
	Mill cloth	Handloom items
0 – 100	—	5.0%
101 – 200	5.0%	7.5%
201 – 300	7.5%	10.0%
Above 300	10.0%	15.0%

Write a program using **switch** and **if** statements to compute the net amount to be paid by a customer.

6.10 Write a program that will read the value of x and evaluate the following function

```
        {   1 for x > 0
  y =   {   0 for x = 0
        {  -1 for x < 0
```

using
(a) nested if statements
(b) else if statements, and
(c) conditional operator?

Chapter 7

Decision Making and Looping

7.1 INTRODUCTION

A computer is well suited to perform repetitive operations. It can do so tirelessly ten, hundred or even ten thousand times. Every computer language must have features that instruct a computer to perform such repetitive tasks. The process of repeatedly executing a block of statements is known as *looping*. The statements in the block may be executed any number of times, from zero to an *infinite* number. If a loop continues forever, it is called an *infinite loop*. C# supports such looping features that enable us to develop concise programs containing repetitive processes without using unconditional branching statements like the **goto** statement.

In looping, a sequence of statements are executed until some conditions for the termination of the loop are satisfied. A *program loop* therefore consists of two segments, one known as the *body of the loop* and the other known as the *control statement*. The control statement tests certain conditions and then directs the repeated execution of the statements contained in the body of the loop.

Depending on the position of the control statement in the loop, a control structure may be classified either as the *entry-controlled loop* or as the *exit-controlled loop*. The flowcharts in Fig. 7.1 illustrate these structures. In the *entry-controlled loop*, the control conditions are tested before the start of the loop execution. If the conditions are not satisfied, then the body of the loop will not be executed. In the case of an exit-controlled loop, the test is performed at the end of the body of the loop and therefore the body is executed unconditionally for the first time.

The test conditions should be carefully stated in order to perform the desired number of loop executions. It is assumed that the test condition will eventually transfer the control out of the loop. If for some reason it does not do so, the control sets up an *infinite loop* and the body is executed over and over again.

A looping process, in general, would include the following four steps:

1. Setting and initialization of a counter.
2. Execution of the statements in the loop.
3. Test for a specified condition for execution of the loop.
4. Incrementing the counter.

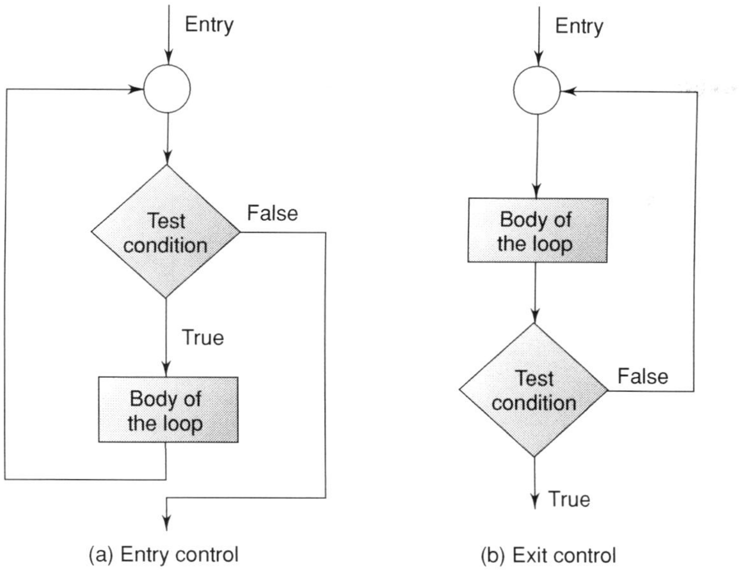

Figure 7.1 *Loop control structures*

The test may be either to determine whether the loop has been repeated the specified number of times or to determine whether a particular condition has been met with.

The C# language provides for four constructs for performing loop operations. They are:

1. The **while** statement
2. The **do** statement
3. The **for** statement
4. The **foreach** statement

These statements are known as *iteration* or *looping* statements. We shall discuss the features and applications of each of these statements in this chapter.

7.2 THE WHILE STATEMENT

The simplest of all the looping structures in C# is the **while** statement. The basic format of the **while** statement is

```
initialization;
while(test condition)
{
    Body of the loop
}
```

while is an *entry-controlled loop* statement. The *test condition* is evaluated and if the condition is true, then the body of the loop is executed. After execution of the body, the test condition is once again evaluated and if it is true, the body is executed once again. This process of repeated execution of the body continues until the test condition finally becomes false and the control is transferred out of the loop. On exit, the program continues with the statement immediately after the body of the loop.

The body of the loop may have one or more statements. The braces are needed only if the body contains two or more statements. However, it is a good practice to use braces even if the body has only one statement.

Consider the following code segment:

```
..........
..........
sum = 0;
n = 1;  // counter
while(n <= 10)
{
    sum = sum + n * n;
    n = n+1;  // incrementing the number
}
System.Console.WriteLine("Sum = "+ sum);
..........
..........
```

The body of the loop is executed 10 times for n = 1, 2, ..., 10 each time adding the square of the value of n, which is incremented inside the loop. The test condition may also be written as n < 11; the result would be the same. Program 7.1 illustrates the use of the **while** loop. The program prints all the odd numbers between 1 and 10.

Program 7.1 Application of while loop

```
using System;
class WhileTest
{
    public static void Main ( )
    {
        int n = 1;
        while (n<=10)
        {
            if (n%2==0)
            {
                n++;
            }
            else
            {
                Console.Write("  "+n);
                n++;
            }
        }
    }
}
```

Given below is the output of Program 7.1:

1 3 5 7 9

7.3 THE DO STATEMENT

The **while** loop construct that we have discussed in the previous section makes a test condition *before* the loop is executed. Therefore, the body of the loop may not be executed at all if the condition is not satisfied at the very first attempt. On some occasions it might be necessary to execute the body of the loop before the test is performed. Such situations can be handled with the help of the **do** statement. This takes the form:

```
initialization;
do
{
    Body of the loop
}
while (test condition);
```

On reaching the **do** statement, the program proceeds to evaluate the body of the loop first. At the end of the loop, the *test condition* in the **while** statement is evaluated. If the condition is true, the program continues to evaluate the body of the *loop* once again. This process continues as long as the condition is true. When the condition becomes false, the loop will be terminated and the control goes to the statement that appears immediately after the **while** statement.

Since the *test condition* is evaluated at the bottom of the loop, the **do....while** construct provides an *exit-controlled loop* and therefore the body of the loop is always executed at least once.

Consider an example:

```
..........
..........
i = 1;
sum = 0;
do
{
    sum = sum + i;
    i = i+2;
}
while(sum < 40 || i < 10);  // semicolon here
..........
..........
```

The loop will be executed as long as one of the two relations is true. Program 7.2 illustrates the use of **do....while** loops for printing a multiplication table.

Program 7.2 Printing the multiplication table using do....while loop

```
class DoWhileTest
{
    public static void Main ( )
    {
        int row, column, y;
        row = 1;
        System.Console.WriteLine("Multiplication Table \n");

        do
        {
            column = 1;

            do
            {
                y = row * column;
                System.Console.Write("  " + y);
                column = column + 1;
            }

            while (column <= 3);

            System.Console.WriteLine("\n");
            row = row + 1;
        }
        while (row <= 3);
    }
}
```

Program 7.2 uses two **do-while** loops in nested form and produces the following output:

```
Multiplication Table
    1    2    3
    2    4    6
    3    6    9
```

Note that the program uses two **do** loops in nested form.

Program 7.3 generates a menu using switch case, which is repeated for user selection, till the user presses 'C' key to signal the running program to close. The repetition of menu is achieved using **do while** loop available in C#.

Program 7.3 Another do example

```
using System;

class DoExample
{
    public static void Main()
    {
        string option;

        do
       {
            Console.WriteLine("—Address Book—\n");

            Console.WriteLine("N - Add New Entry");
            Console.WriteLine("U - Update Entry");
            Console.WriteLine("R - Remove Entry");
            Console.WriteLine("S - Show All Entries");
            Console.WriteLine("C - Close\n");

            Console.WriteLine("Please Enter Your Choice: ");
           option = Console.ReadLine();

            switch(option)
            {
                case "N":
                case "n":
                    Console.WriteLine("You have selected the Add New Entry
option.");
                   break;
                case "U":
                case "u":
                    Console.WriteLine("You have selected the Update Entry
option.");
                   break;
                case "R":
                case "r":
                    Console.WriteLine("You have selected the Remove Entry
option.");
                   break;
                case "S":
                case "s":
                    Console.WriteLine("You have selected the Show All Entries
option.");
```

```
                        break;
                    case "C":
                    case "c":
                        Console.WriteLine("The program will close now..");
                        break;
                    default:
                         Console.WriteLine("You have not entered a valid choice.",
option);
                         break;
                }

                Console.Write("");
                Console.ReadLine();
            } while (option != "C" && option != "c");
        }
}
```

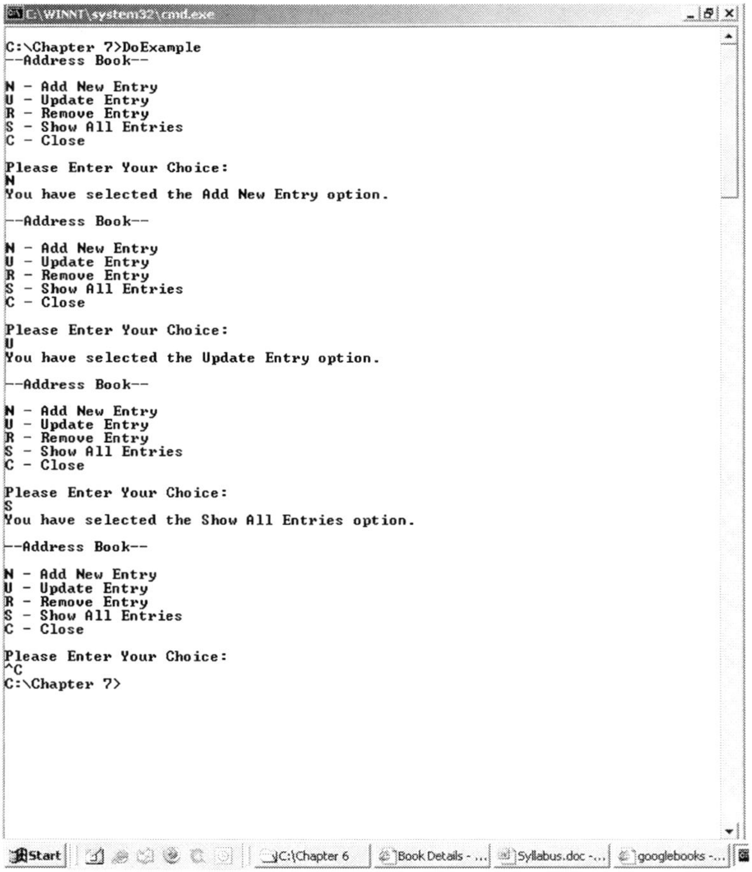

7.4 THE FOR STATEMENT

for is another *entry-controlled loop* that provides a more concise loop-control structure. The general form of the **for** loop is

```
for (initialization ; test condition ; increment)
{
    Body of the loop
}
```

The execution of the **for** statement is as follows:

1. *Initialization* of the *control variables* is done first, using assignment statements such as i = 1 and count = 0. The variables **i** and **count** are known as loop-control variables.
2. The value of the control variable is tested using the *test condition*. The test condition is a relational expression, such as i < 10, which determines when the loop will exit. If the condition is *true*, the body of the loop is executed; otherwise the loop is terminated and the execution continues with the statement that immediately follows the loop.
3. When the body of the loop is executed, the control is transferred back to the **for** statement after evaluating the last statement in the loop. Now, the control variable is *incremented* using an assignment statement such as i = i+1 and the new value of the control variable is again tested to see whether it satisfies the loop condition. If the condition is satisfied, the body of the loop is again executed. This process continues till the value of the control variable fails to satisfy the test condition.

Consider the following segment of a program

```
for (x = 0 ; x <= 9 ; x = x+1)
{
    System.Console.WriteLine(x);
}
```

This **for** loop is executed ten times and prints the digits 0 to 9. The three sections enclosed within parentheses must be separated by semicolons. Note that there is no semicolon at the end of the *increment* section, **x = x+1**.

The **for** statement allows for *negative increments*. For example, the loop discussed above can be written as follows:

```
for ( x = 9 ; x >= 0 ; x = x-1)
    System.Console.WriteLine(x);
```

This loop is also executed ten times, but the output would be from 9 to 0 instead of 0 to 9. Note that braces are optional when the body of the loop contains only one statement.

Since the conditional test is always performed at the beginning of the loop, the body of the loop may not be executed at all, if the condition fails at the start. For example,

```
for (x = 9; x < 9;x = x-1)
{
..........
..........
}
```

will never be executed because the test condition fails at the very beginning itself.

Let us consider the problem of the sum of squares of integers discussed in Section 7.2. This problem can be coded using the **for** statement as follows:

```
..........
..........
sum = 0;
int n;
for (n = 1; n <= 10; n = n+1)
{
    sum = sum + n*n;
}
..........
..........
```

The body of the loop

```
sum = sum + n*n;
```

is executed ten times for n = 1, 2, ..., 10 each time incrementing the **sum** by the square of the value of **n**.

One of the important points about the **for** loop is that all the three actions, namely *initialization, testing* and *incrementing*, are placed in the **for** statement itself, thus making them visible to the programmers and users in one place. The **for** statement and its equivalent of **while** and **do** statements are shown in Table 7.1.

Table 7.1 *Comparison of the three loops*

for	while	do

for(n=1;n<=10;++n)
{
....
....
}
	n = 1;
	while (n<=10)
	{
	....
	....
	n = n+1;
	}
		n = 1;
		do
		{
		....
		....
		n = n+1;
		}
		while (n<=10);

Program 7.4 illustrates the use of the **for** loop for computing and printing the "power of 2" table.

Program 7.4 Computing the 'power of 2' using for loop

```
using System;
class ForTest
{
    public static void Main ( )
    {
```

```
        long p;
        int n;
        double q;
        Console.WriteLine("2 to power -n      n      2 to power n");

        p = 1L;
        for (n = 0; n < 10; ++n)
        {
            if (n == 0)
                p = 1L;
            else
                p = p * 2;
            q = 1.0 / (double)p;
            Console.WriteLine ("{0:F6} {1:D} {2:D}" , q,n,p);
        {
    }
}
```

Output of Program 7.4 would be:

```
2 to power -n      n      2 to power n
     1.000000      0            1
     0.500000      1            2
     0.250000      2            4
     0.125000      3            8
     0.062500      4           16
     0.031250      5           32
     0.015625      6           64
     0.003906      7          128
     0.001953      8          256
     0.001953      9          512
```

Numeric formatted outputs are discussed in Chapter 17.

7.4.1 Additional Features of the for Loop

The **for** loop has several capabilities that are not found in other loop constructs. For example, more than one variable can be initialized at a time in the **for** statement. The statements

```
p = 1;
for (n=0; n<17; ++n)
```

can be rewritten as

```
for (p=1, n=0; n<17; ++n)
```

Notice that the initialization section has two parts p = 1 and n = 0 separated by a *comma.*

Like the initialization section, the increment section may also have more than one part. For example, the loop

```
for (n=1, m=50; n<=m; n=n+1, m=m-1)
{
    ..........
    ..........
}
```

is perfectly valid. The multiple arguments in the increment section are separated by *commas.*

The third feature is that the test condition may have any compound relation and the testing need not be limited only to the loop control variable. Consider the example below:

```
sum = 0 ;
for (i = 1, i < 20 && sum < 100; ++i)
{
    ..........
    ..........
}
```

The loop uses a compound test condition with the control variable **i** and external variable **sum**. The loop is executed as long as both the conditions i < 20 and sum < 100 are true. The sum is evaluated inside the loop.

It is also permissible to use expressions in the assignment statements of the initialization and increment sections. For example, a statement of the type

```
for (x = (m+n)/2; x > 0; x = x/2)
```

is perfectly valid.

Another unique aspect of the **for** loop is that one or more sections can be omitted, if necessary. Consider the following statements:

```
..........
..........
m = 5;
for ( ;m != 100 ; )
{
    System.Console.WriteLine(m);
    m = m+5;
}
..........
..........
```

Both the initialization and increment sections are omitted in the **for** statement. The initialization has been done before the **for** statement and the control variable is incremented inside the loop. In such cases, the sections are left blank. However, the semicolons separating the sections must remain. If the test condition is not present, the **for** statement sets up an infinite loop.

We can set up time-delay loops using the null statement as follows:

```
for (j = 1000; j > 0; j = j-1)
; //simple semicolon
```

This loop is executed 1000 times without producing any output; it simply causes a time delay. Notice that the body of the loop contains only a semicolon, known as a *empty* statement. This can also be written as

```
for (j=1000; j > 0; j = j-1);
```

This implies that the compiler will not give an error message if we place a semicolon by mistake at the end of a **for** statement. The semicolon will be considered as an *empty* statement and the program may produce some nonsense.

Program 7.5 uses **for** looping construct to loop through two numbers, one in increasing order and the other in decreasing order. The program also compares the values of the two numbers at each loop iteration and uses ternary operator to inform the whether the first number is greater or smaller than the second number.

Program 7.5 Additional features of for loop

```
using System;

class forExample
{
  public static void Main()
  {
    int num1=0, num2=10;
    for(; ((num1<=10) && (num2>=0)); num2–, num1++)
   {
    System.Console.WriteLine("First Number {0} is {1} than Second Number {2}.\t",
           num1, (num1>num2? '>' : '<'), num2);
   }
  }
}
```

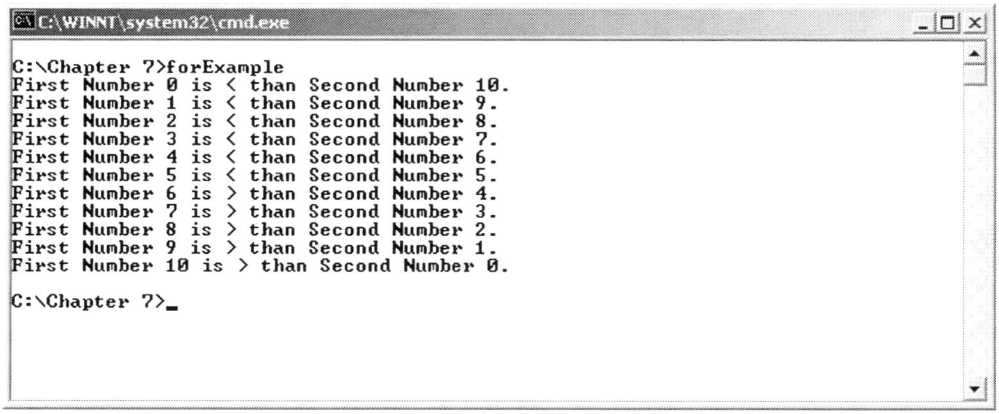

7.4.2 Nesting of for Loops

Nesting of loops, that is, one **for** statement within another **for** statement, is allowed in C#. We have used this concept in Program 7.2. Similarly, **for** loops can be nested as follows:

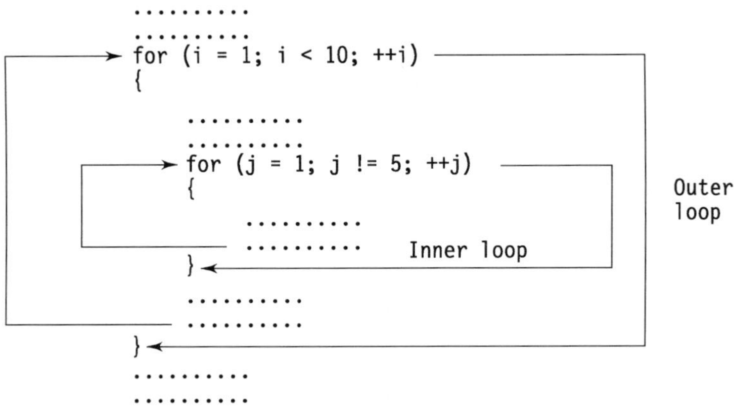

The loops should be properly indented so as to enable the reader to easily determine which statements are contained within each **for** statement.

A program segment to print a multiplication table using **for** loops is shown below:

```
    ..........
    ..........
    for (row = 1; row <= ROWMAX; ++row)
    {
    for (column = 1; column <= COLMAX; ++ column)
    {
        Y = row * column
        System.Console.Write(" " + Y);
    }
    System.Console.WriteLine(" ");
}
..........
..........
```

The outer loop controls the rows while the inner one controls the columns.

7.5 THE FOREACH STATEMENT

The **foreach** statement is similar to the **for** statement but implemented differently. It enables us to iterate the elements in arrays and collection classes such as **List** and **HashTable**. The general form of the **foreach** statement is:

```
foreach (type variable in expression)
{
```

```
        Body of the loop
    }
```

The *type* and *variable* declare the *iteration* variable. During execution, the iteration variable represents the array element (or collection element in case of collections) for which an iteration is currently being performed. **in** is a keyword.

The *expression* must be an *array* or *collection* type and an explicit conversion must exist from the element type of the collection to the type of the iteration variable. *Example:*

```
public static void Main (string [ ] args)
{
    foreach ( string s in args)
    {
        Console.WriteLine(s);
    }
}
```

This program segment displays the command line arguments. The same may be achieved using the **for** statement as follows:

```
public static void Main (string[ ] args)
{
    for ( int i = 0; i < args.Length ; i++ )
    {
        Console.WriteLine(args [i]);
    }
}
```

Program 7.6 illustrates the use of **foreach** statement for printing the contents of a numerical array.

Program 7.6 Printing array values using foreach statement

```
using System;
class ForeachTest
{
    public static void Main ( )
    {
        int[ ] arrayInt = { 11, 22, 33, 44 };
        foreach ( int m in arrayInt)
        {
            Console.Write(" " + m);
        }
        Console.WriteLine( );
    }
}
```

Program 7.6 will display the following output:

```
11  22  33  44
```

An important point to note is that we cannot change the value of the iteration variable during execution. For instance, the code

```
int[ ] x = { 1, 2, 3 };
    foreach ( int i in x )
    {
        i++;
        Console.Write(i);
    }
```

will not work. If we need to change the values of an item during the iteration process, we may use the **for** loop construct.

The advantage of **foreach** over the **for** statement is that it automatically detects the boundaries of the collection being iterated over. Further, the syntax includes a built-in iterator for accessing the current element in the collection.

7.6 JUMPS IN LOOPS

Loops perform a set of operations repeatedly until the control variable fails to satisfy the test condition. The number of times a loop is repeated is decided in advance and the test condition is written to achieve this. Sometimes, when executing a loop it becomes desirable to skip a part of the loop or to leave the loop as soon as a certain condition occurs. For example, consider the case of searching for a particular name in a list containing, say, a hundred names. A program loop written for reading and testing the names a hundred times must be terminated as soon as the desired name is found. C# permits a jump from one statement to the end or beginning of a loop as well as a jump out of a loop.

7.6.1 Jumping Out of a Loop

An early exit from a loop can be accomplished by using the **break** and **goto** statements. We have already seen the use of the **break** in the **switch** statement. These statements can also be used within **while**, **do** or **for** loops for an early exit.

When the **break** statement is encountered inside a loop, the loop is immediately exited and the program continues with the statement immediately following the loop. When the loops are nested, the break would only exit from the loop containing it. That is, the break will exit only a single loop. Figure 7.2 illustrates the use of **break** for exiting a loop.

7.6.2 Skipping a Part of a Loop

During the loop operations, it may be necessary to skip a part of the body of the loop under certain conditions. For example, in processing of applications for some job, we might like to exclude the processing of data of applicants belonging to a certain category. On reading the category code of an applicant, a test is made to see whether his application should be considered or not. If it is not to be considered, the part of the program loop that processes the application details is skipped and the execution continues with the next loop operation.

Like the **break** statement, C# supports another similar statement called the **continue** statement. However, unlike **break** which causes the loop to be terminated, the **continue** statement, as the name implies, causes the loop to continue with the next iteration after skipping any statements in between. The continue statement tells the compiler. "SKIP THE FOLLOWING STATEMENTS AND CONTINUE WITH THE NEXT ITERATION". The format of the **continue** statement is simply

```
continue;
```

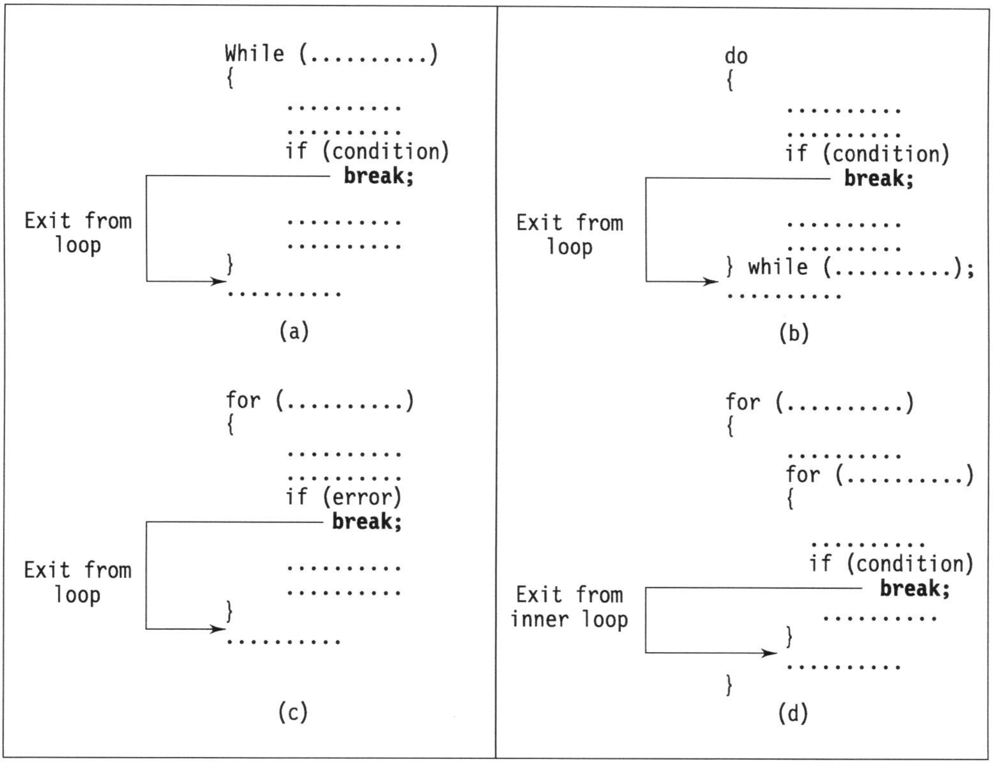

Figure 7.2 *Exiting a loop with break statement*

The use of the **continue** statement in loops is illustrated in Fig. 7.3. In **while** and **do** loops, **continue** causes the control to go directly to the *test condition* and then to continue the iteration process. In the case of **for** loop, the *increment* section of the loop is executed before the *test condition* is evaluated. Program 7.7 shows how **continue** and **break** statements are implemented.

Program 7.7 Use of continue and break statements

```
using System;
class ContinueBreak
{
    public static void Main ( )
    {
```

```
      int n = 10;
      while ( n < 200 )
      {
         if ( n < 50 )
         {
            Console.Write(" " + n);
            n = n + 10;
            continue;

         }

         If (n == 50)
         {
         Console.WriteLine( );
            n = n + 10;
            continue;
         }
         if (n > 90)break;
         Console.Write(" " + n);
            n = n + 10;
   }
   Console.WriteLine( );
      }
   }
```

Program 7.7 will generate the following output.

```
10    20    30    40
60    70    80    90
```

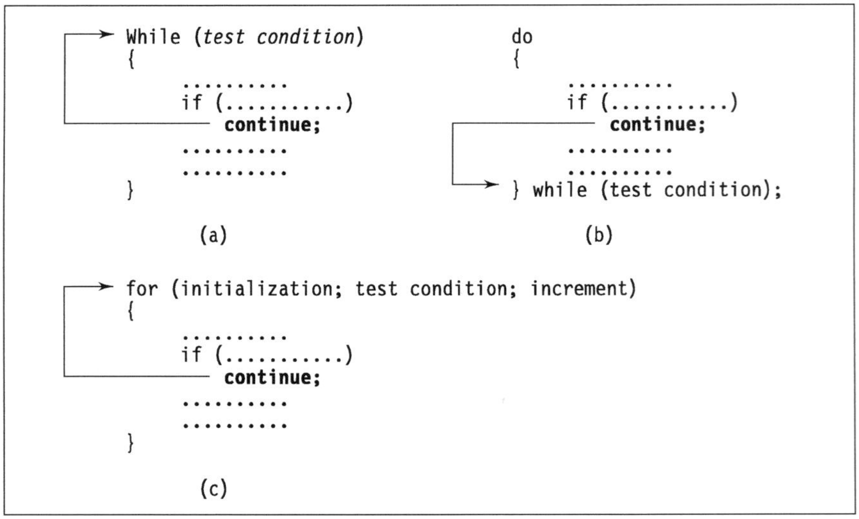

Figure 7.3 *Bypassing and continuing in loops*

7.6.3 Labelled Jumps

We have seen that the **break** will enable us to come out of only the nearest loop and the **continue** will enable us to restart the current loop. If we want to jump a set of nested loops or to continue a loop that is outside the current one, we may have to use the concept **labelling** and the **goto** statement.

C# has retained the infamous **goto** and also permits us to use labels. A label is any valid C# variable name ending with a colon. We can use labels anywhere in a program and use the **goto** statement to transfer the control to the statement marked by the label. *Example:*

```
    public static void Main(String [ ] args)
{
    if (args.Length == 0)
        goto end;
    Console.WriteLine(args.Length);
    end: // Label name
    Console.WriteLine ("end");
}
```

We can use a **goto** statement and a label to transfer control out of a nested loop as shown in the code below:

```
for ( int i = 0; i < 10; i++)
{
    while ( x < 100 )
    {
        y = i * x;
        if (y > 500)
            goto out;
        . . . .
        . . . .
    }
}
out:
. . . .
. . . .
```

Here, the label **out** is outside the **for** loop and therefore the statement.

```
goto out;
```

causes the execution to break out of both the loops. Note that a simple **break** cannot achieve this. Program 7.8 illustrates the use of **continue**, **break** and **goto** statements.

Program 7.8 Use of goto and label statements

```
using System;
class GotoLabel
{
```

```
    public static void Main ( )
    {
        for (int i = 1; i < 100; i++)
        {
            Console.WriteLine(" ");
            if (i >= 10)
                break;
            for (int j = 1; j < 100; j++)
            {
            Console.Write(" * ");
              if (j == i)
                goto loop1;
            }
            loop1:continue;
        }
        Console.WriteLine("Termination by BREAK");
    }
}
```

Program 7.8 produces the following output:

```
*
* *
* * *
* * * *
* * * * *
* * * * * *
* * * * * * *
* * * * * * * *
* * * * * * * * *
Termination by BREAK
```

Case Study

Problem Statement

St. Peter's School is a convent school that provides education from KG the 12th class. Quarterly, half yearly and final exams are conducted every year for students of all classes in the school. The class teacher of each class has to prepare a student report for every student after a particular exam has been conducted. Recently, it has been found by the principal of St. Peter's School that a lot of time is being taken by class teachers in preparing the student reports. The principal has also found out that the efficiency of the teachers have decreased because they have been manually performing the task of preparing the student reports for each student. St. Peter's has 600 students studying in its different classes. Preparing student reports for at least 30 students per class has become a tedious task for the class teacher of a specific class. How can the principal of St. Peter's School solve the problem of the teachers?

Solution

After holding meetings with senior teachers of the school, the principal of St. Peter's School has come to the conclusion that the task of calculating total marks for preparing a student report must be automated through a software application running on a computer system. After coming to this conclusion, the principal of St. Peter's School asks Nandita, who is teaching computer science to students of higher classes such as 9th , 10th, 11th and 12th to create the application for automating the calculaton of total marks, the task that is performed by class teachers while preparing the student report. For this, Nandita first conducts an interview with the class teachers of different classes to understand their requirements. After the analysis of the requirements, Nandita decides to use the decision making and looping concepts of C# in an application to automate the task of calculating the total marks for each student of a class. Decision making and looping allow a programmer to decide the flow of control for a program or application. Nandita creates the following application for automating the task of calculating the total marks.

```
using System;
class StudentMarks
{
      static void Main(string[] args)
      {
       int[] marks = new int[5];
       int totalMarks = 0;

       Console.WriteLine("==Student Information==\n");
       for (int i = 0; i < 5; i++)
       {
        Console.WriteLine("Enter Your Marks In Subject {0}: ",i+1);
        marks[i] = int.Parse(Console.ReadLine());
       }
       Console.WriteLine();
       // Displaying Student Marks
       Console.Write("The marks you have entered are: ");
       for (int i = 0; i < 5; i++)
       {
        if(i<4)
           Console.Write(marks[i] + ", ");
        else
           Console.Write(marks[i]);
       }
```

```
        // Calculating and Displaying Total Marks of the Student
        for (int i = 0; i < 5; i++)
        {
         totalMarks += marks[i];
        }
        Console.WriteLine();
        Console.WriteLine("Your total marks are: {0}", totalMarks);
        // Calculating and Displaying Maximum Marks of the Student
        int maxMarks = marks[0];
        int index=1;
        for (int i = 1; i < marks.Length; i++)
        {
         if (marks[i] > maxMarks)
         {
          maxMarks = marks[i];
          index=i+1;
         }
        }
        Console.WriteLine("You have scored the maximum marks {0} in Subject {1}.",
maxMarks,index);
        // Calculating and Displaying Minimum Marks of the Student
        int minMarks = marks[0];
        index=1;
        for (int i = 1; i < marks.Length; i++)
        {
         if (marks[i] < minMarks)
         {
          minMarks = marks[i];
          index=i+1;
         }
        }
        Console.WriteLine("You have scored the minimum marks {0} in Subject {1}.",
minMarks,index);
        // Calculating and Displaying Percentage of the Student
        double percentage;
        double val=0.0;
        val=(double)totalMarks / 500;
        percentage = (double) val*100;
        Console.WriteLine("Your percentage is: {0}%", percentage);
        Console.ReadLine();
      }
}
```

Remarks

In looping, a set of program staements are executed based on a condition. The program statements are executed until a specific condition is true. When the condition becomes false, the exeuction of the program statements is terminated.

Common Programming Errors

- *Creating an infinite loop by accident.*
- *Using semicolon after the header of for loop.*
- *Using semicolon at the end of while loop.*
- *Forgetting semicolon at the end of do loop.*
- *Spelling the word while like While.*
- *Improper increment or decrement steps.*
- *Improper operands in relational expressions.*
- *Improper relational and conditional operators.*
- *Setting wrong upper or lower limits for control variables.*
- *Using floating point values in relational expressions.*
- *Using commas instead of semicolons in a for header.*
- *Not initializing the control variable properly.*
- *Using fractions as step values for increment or decrement thus causing imprecise counter values and inaccurate tests for termination.*
- *Incorrect use of break and continue statements.*

Review Questions

7.1 State whether the following statements are true or false:
   (a) The while loop is an entry controlled loop.
   (b) The body of a do loop is always executed at least once.
   (c) The task that can be achieved by a do loop cannot be achieved using a while loop.
   (d) The body of a for loop is always executed at least once.
   (e) The body of a loop statement must always be contained within the opening and closing braces.
   (f) It is legal to have negative increments in for statement.
   (g) We can have compound test conditions in for statement.
   (h) We cannot use expressions in the initialization section of a for statement.
   (i) It is legally valid to omit one or more sections of a for statement.
   (j) A loop construct may be implemented using if and goto statements.

7.2 Compare in terms of their functions, the following pairs of statements:
   (a) while and do…while
   (b) while and for
   (c) break and continue
   (d) for and foreach

7.3 What are the basic steps in the looping process?

7.4 What are the ways to create an infinite loop that does not end until we specify a break statement?

7.5 Analyze each of the program segments that follow and determine how many times the body of each loop will be executed.

(a)
```
x = 5;
y = 50;
while (x <= y)
{
        x = y/x;
        . . . . . . . ... .
        . . . . . . . . . .
}
```

(b)
```
m = 1;
do
{
        . . . . . . . . ...
        . . . . . . . . . .
        m = m + 2;
}
while (m < 10);
```

(c)
```
int i;
for (i = 0; i <= 5; i = i+2/3)
{
        . . . . . . . . . .
        . . . . . . . . . .
}
```

(d)
```
int m = 10;
int n = 7;
while (m % n >=0)
{
        . . . . . . . . . . .
        m = m + 1;
        n = n + 2;
        . . . . . . . ... .
}
```

7.6 Find errors, if any, in each of the following looping segments. Assume that all the variables have been declared and assigned values.

(a)
```
while (count ! = 10);
{
        count = 1;
        sum = sum + x;
        count = count+1;
}
```

(b)
```
name = 0,
do ( name = name + 1;
Console.WriteLine("My name is John\n";}
While (name=1)
```

(c)
```
for (x = 1, x > 10; x=x+1)
{
    . . . . . . . .
    . . . . . . . ..
}
```
(d)
```
m = 1;
n = 0;
for (; m + n < 19; ++n)
Console.WriteLine("Hello\n");
m = m +10;
```

7.7 What is an empty statement? How is it useful?

7.8 What is wrong with the following while loop constructs?

(a)
```
while (i >0)
{
    i - - ;
    other statements
}
```
(b)
```
int i = 10;
while (i)
{
    body statements
}
```

7.9 How many times the body of the following for loop is executed?
```
for (int i=10, int j=0; i+j>5; i=i-2, j++)
{
 body statements
}
```

7.10 What is wrong with the following code snippet?
```
int j = 0;
while (j < 10)
{
        j++;
        if (j = = 5)
                continue loop;
        Console.WriteLine("j is " + j);
}
```

7.11 What is the output of the following code?
```
int m = 100;
int n = 300;
while (++m < –n)
        Console.WriteLine(m);
```

7.12 What is the output of the following code?
```
int m = 100;
while(true)
{
```

```
        if(m<10)
        break;
        m = m-10;
}
Console.WriteLine("m is " + m);
```

7.13 What are nested loops? Give an example where a nested loop is typically used.

7.14 How does the foreach statement differ from the for statement? Give a typical example where a foreach is used.

7.15 What is a break statement? How and when a break is typically used?

7.16 What is a continue statement? How and when a continue is typically used?

7.17 What is a label? How and when a label is used in a program?

Debugging Exercises

7.1 Given below is a program to print values from 0 to 9 using a while loop. Will the program generate the desired output? If not, why?

```
using System;
class WhileExample
{
    static void Main()
    {
        int count = 0;
        while (count <= 10)
        {
            Console.WriteLine(count);
        }
    }
}
```

7.2. In the following program for printing numbers from 0 to 9, the loop must break when the value of the number becomes 5. Place the break statement at the appropriate position to achieve the result.

```
using System;
public class StopCounter
{
    public static void Main(   )
    {
      for ( int index = 0; index < 10; index++ )
      {
         Console.WriteLine("Counter Index: {0} ", index );
         if ( index == 5 )
```

```
        {
            {
                Console.WriteLine( "Stopping and Exiting the loop." );
            }
        }
        Console.WriteLine( "For loop Iteration." );
    }
  }
}
```

Programming Exercises

7.1 *Given a number, write a program using while loop to reverse the digits of the number. For example, the number*
12345
should be written as
54321
(Hint: Use modulus operator to extract the last digit and the integer division by 10 to get the n–1 digit number from the n digit number)

7.2 *The factorial of an integer m is the product of consecutive integers from 1 to m. That is*
*Factorial m = m! = m * (m–1)*.....*1.*
Write a program that computes and prints a table of factorials for any given m.

7.3 *Write a program to compute the sum of the digits of a given integer number.*

7.4 *The numbers in the sequence*
11 2 3 5 8 13 21 ......
are called Fibonacci numbers. Write a program using a do ... while loop to calculate and print the first m Fibonacci numbers.
(Hint: After the first two numbers in the series, each number is the sum of the two preceding numbers)

7.5 *Write a program to evaluate the following investment equation*

$$V = P(1+r)^n$$

and print the tables which would give the value of V for various combination of the following values of P, r and n.

P : 1000, 2000, 3000, ..., 10,000
r : 0.10, 0.11, 0.12, ...0.20
n : 1,2,3,.....10

(Hint: P is the principal amount and V is the value of money at the end of n years. This equation can be recursively written as

$$V = P\,(1+r)$$
$$P = V$$

That is, the value of money at the end of first year becomes the principal amount for the next year and so on.

7.6 *Write a program to print the following outputs using for loops.*

```
(a) 1              (b) $ $ $ $ $        (c)         1
    2 2                $ $ $ $                    2   2
    3 3 3              $ $ $                    3   3   3
    4 4 4 4            $ $                    4   4   4   4
    5 5 5 5 5          $                    5   5   5   5   5
```

7.7 *Write a program using a for that accepts five values in US dollars, one at a time, and converts each value entered to its Indian rupees equivalent before the next value is requested.*

7.8 *Repeat the Exercise 7.7 using a do statement.*

7.9 *Illustrate the application of the foreach statement through a simple program.*

7.10 *Demonstrate the typical use of the following jump statements:*

- *beak*
- *continue*
- *goto*

in a single program.

Chapter

8

Methods in C#

8.1 INTRODUCTION

In object-oriented programming, objects are used as building blocks in developing a program. They are the runtime entities. They may represent a person, a place, a bank account, a table of data or any item that the program handles.

Objects encapsulate data, and code to manipulate that data. The code designed to work on the data is known as *methods* in C#. Methods give objects their behavioral characteristics. They are used not only to access and process data contained in the object but also to provide responses to any messages received from other objects.

C# is a pure object-oriented language and therefore, every method must be contained within a class. Remember, in other languages like VB, C and C++, we can define global functions that are not associated with a particular class. In this chapter, we shall see how methods are defined and used in different situations. More about their applications in the development of object-oriented system will be discussed in Chapter 12.

8.2 DECLARING METHODS

Methods are declared inside the body of a class, normally after the declaration of data fields. The general form of a method declaration is

```
modifiers type methodname (formal-parameter-list)
{
   method _ body
}
```

Method declaration has five parts:

- Name of the method (*methodname*)

- Type of value the method returns (*type*)
- List of parameters (*formal-parameter-list*)
- Body of the method
- Method modifiers (*modifier*)

The *methodname* is a valid C# identifier. The *type* specifies the type of value the method will return. This can be a simple data type such as **int** as well as any class type. If the method does not return anything, we specify a return type of **void.** Note that we cannot omit the return type altogether.

The *formal-parameter-list* is always enclosed in parentheses. This list contains variable names and types of all the values we want to give to the method as input. The parameters are separated by commas. In the case where no input data are required, the declaration must still include an empty set of parantheses( ) after the method name. Examples are:

```
int Fun1 (int m, float x, float y) //three parameters
void Display ( ) //no parameters
```

The *body* enclosed in curly braces actually describes the operations to be performed on the data. For example, the code segment below computes the product of two integer values and returns the result.

```
int Product ( int x, int y )
{
      int m = x * y; //operation, m is a local variable
      return(m);     // returns the result (int type)
}
```

The **return** statement in the body may be omitted if the method does not return any value.

Example:

```
void Display ( int x )
{
      Console.WriteLine(x);
}
```

Remember that the formal parameters should be declared for their types individually. For example, the method header

```
int Product ( int m, float x, y )
```

is invalid.

The *modifiers* specify keywords that decide the nature of accessibility and the mode of application of the method. A method can take one or more of the modifiers listed in Table 8.1. Use of these modifiers will be discussed in later chapters as and when they are required.

Table 8.1 *List of method modifiers*

Modifier	Description
new	The method hides an inherited method with the same signature
public	The method can be accessed from anywhere, including outside the class
protected	The method can be accessed from within the class to which it belongs, or a type derived from that class

(*Contd*)

(*Contd*)

Modifier	Description
internal	The method can be accessed from within the same program
private	The method can only be accessed from inside the class to which it belongs
static	The method does not operate on a specific instance of the class
virtual	The method can be overridden by a derived class
abstract	A virtual method which defines the signature of the method, but doesn't provide an implementation
override	The method overrides an inherited virtual or abstract method
sealed	The method overrides an inherited virtual method, but cannot be overridden by any classes which inherit from this class. Must be used in conjunction with override
extern	The method is implemented externally, in a different language

8.3 THE MAIN METHOD

As mentioned earlier, C# programs start execution at a method named **Main( ).** This method must be the static method of a class and must have either **int** or **void** as return type

```
public static int Main( )
Or
public static void Main( )
```

The modifier **public** is used as this method must be called from outside the program. The **Main** can also have parameters which may receive values from the command line at the time of execution.

```
public static int Main (string [ ] args)
public static void Main (string [ ] args)
```

More about method modifiers will be discussed later in Chapter 12.

8.4 INVOKING METHODS

Once methods have been defined, they must be activated for operations. The process of activating a method is known as *invoking* or *calling*. The invoking is done using the dot operator as shown below:

```
objectname.methodname( actual-parameter-list );
```

Here, *objectname* is the name of the object on which we are calling the method *methodname*. The *actual-parameter-list* is a comma separated list of 'actual values' (or expressions) that must match in type, order and number with the formal parameter list of the *methodname* declared in the class. Note that the invoking statement is an executable one and therefore must end with a semicolon. The values of the actual parameters are assigned to the formal parameters at the time of invocation.

Program 8.1 defines a method named **cube** and is used to compute the cube of a number passed in as a parameter. Note that how an object of **Method** class is created and used to invoke the method **Cube** using the dot operator.

Program 8.1 Defining and invoking a method

```
using System;
class Method  // class containing the method
{
      // Define the Cube method
      int Cube ( int x )
      {
         return ( x * x * x );
      }
}

// Client class to invoke the cube method
class MethodTest
{
      public static void Main( )
      {
         // Create object for invoking cube
         Method M = new Method ( );

         // Invoke the cube method
         int y = M.Cube (5):    //Method call

         // Write the result
         Console.WriteLine( y );
      }
}
```

This program will display an output of 125.

Program 8.2 declares the method **Square ( )** in the same class where the **Main( )** is used and therefore the method **Square ( )** is invoked without using any object. Note how the value 2.5 is passed to the method. The formal parameter **x** is declared **float** and therefore the literal 2.5 which is **double** by default is made float by appending F to it.

Program 8.2 Calling a static method

```
using System;
class StaticMethod
{
      public static void Main( )
      {
         double  y = Square (2.5 F);        //Method call
         Console.WriteLine( y );
      }
      static double Square ( float x)       //Method definition
      {
         return ( x* x );
      }
}
```

Note that the keyword **static** is used to qualify the method **Square**. (More about **static** in Chapter 12.)

Output of this program would be: 6.25

8.5 NESTING OF METHODS

We mentioned earlier that a method of a class can be invoked only by an object of that class if it is invoked outside the class. We have also seen an exception where a method is invoked without using any object and dot operator. That is, a method can be called using only its name by another method of the same class. This is known as *nesting of methods*.

Program 8.3 illustrates nesting of methods inside a class. The class **Nesting** defines two methods, namely **Largest ( )** and **Max ( ).** The method **Largest ( )** calls the method **Max ( )** to determine the largest of the two numbers and then displays the result.

Program 8.3 Nesting of methods

```
using System;
class Nesting
{
        void Largest ( int m, int n )
        {
            int large = Max ( m , n );    //Nesting
            Console.WriteLine( large );
        }
        int  Max (int a, int b)
        {
            int x = ( a > b ) ? a : b ;
            return ( x );
        }
}
class NestTesting
{
        public static void Main( )
        {
            Nesting next = new Nesting ( );
            next.Largest ( 100, 200 ) ;    //Method call
        }
}
```

A method can call any number of methods. It is also possible for a called method to call another method. That is, **Method1** may call **Method2**, which in turn may call **Method3**. Here, the method **Main** calls **Largest** which in turn calls **Max**. This program will produce an output of 200 that is the largest of two values.

8.6 METHOD PARAMETERS

A method invocation creates a copy, specific to that invocation, of the formal parameters and local variables of that method. The actual argument list of the invocation assigns values or variable references to the newly created formal parameters. Within the body of a method, formal parameters can be used like any other variables.

The invocation involves not only passing the values into the method but also getting back the results from the method. For managing the process of passing values and getting back the results, C# employs four kinds of parameters.

- Value parameters
- Reference parameters
- Output parameters
- Parameter arrays

Value parameters are used for passing parameters into methods by *value.* On the other hand, reference parameters are used to pass parameters into methods by *reference.* Output parameters, as the name implies, are used to pass results back from a method. Parameter arrays are used in a method definition to enable it to receive variable number of arguments when called.

8.7 PASS BY VALUE

By default, method parameters are *passed by value*. That is, a parameter declared with no modifier is passed by value and is called a *value parameter*. When a method is invoked, the values of actual parameters are assigned to the corresponding formal parameters. The values of the value parameters can be changed within the method. The value of the actual parameter that is passed by value to a method is not changed by any changes made to the corresponding formal parameter within the body of the method. This is because the methods refer to only copies of those variables when they are passed by value.

Programs 8.1, 8.2 and 8.3 all demonstrated the use of value parameters. Program 8.4 illustrates how the change made to the local copy inside the method does not affect the value of the original argument.

Program 8.4 Illustration of passing by value

```
using System;
class PassByValue
{
        static void Change (int m)
        {
            m = m+10;    //value of m is changed
        }
        public static void Main( )
        {
            int x = 100;
            Change (x);
            Console.WriteLine( "x =" + x );
        }
}
```

This program will produce the output

```
X = 100
```

When the method **change( )** is invoked, the value of **x** is assigned to **m** and a new location for **m** is created in the memory. Therefore, any change in **m** does not affect the value stored in the location **x.**

8.8 PASS BY REFERENCE

We have just seen that pass by value is the default behaviour of methods in C#. We can, however, force the value parameters to be passed by reference. To do this, we use the **ref** keyword. A parameter declared with the **ref** modifier is a reference parameter. *Example:*

```
void Modify ( ref int x )
```

Here, **x** is declared as a reference parameter.

Unlike a value parameter, a reference parameter does not create a new storage location. Instead, it represents the same storage location as the actual parameter used in the method invocation. Remember, when a formal parameter is declared as **ref** , the corresponding argument in the method invocation must also be declared as **ref**. *Example:*

```
void Modify ( ref int x )
{
    x += 10;    // value of m will be changed
}
. . . .
int m = 5; // m is initialized
Modify ( ref m ); // pass by reference
. . . .
```

Note the use of ***ref int*** as the parameter type in the definition and the use of ***ref*** keyword in the method call to tell the compiler *to pass by reference* rather than by value.

Reference parameters are used in situations where we would like to change the values of variables in the calling method. When we pass arguments by reference, the 'formal' arguments in the called method become aliases to the 'actual' arguments in the calling method. This means that when the method is working with its own arguments, it is actually working with the original data. Program 8.5 shows how the contents of two locations can be exchanged using reference parameters.

Program 8.5 Swapping values using ref parameters

```
using System;
class PassByRef
{
        static void Swap ( ref int x, ref int y )
        {
            int temp = x;
            x = y;
```

```
        y = temp;
    }
    public static void Main( )
    {
        int m = 100;
        int n = 200;
        Console.WriteLine("Before Swapping:");
        Console.WriteLine("m = " + m);
        Console.WriteLine("n = " + n);

        Swap( ref m , ref n );

        Console.WriteLine("After Swpaping:");
        Console.WriteLine("m = " + m);
        Console.WriteLine("n = " + n);
    }
}
```

Program 8.5 produces the following output:

```
Before Swapping
m = 100
n = 200
After Swapping
m = 200
n = 100
```

Note that a variable must be definitely assigned a value before it can be passed as a reference parameter.

8.9 THE OUTPUT PARAMETERS

As pointed out earlier, *output* parameters are used to pass results back to the calling method. This is achieved by declaring the parameters with an **out** keyword. Similar to a reference parameter, an output parameter does not create a new storage location. Instead, it becomes an alias to the parameter in the calling method. When a formal parameter is declared as **out**, the corresponding actual parameter in the calling method must also be declared as **out.** *Example:*

```
void Output ( out int x )
{
   x = 100;
}
. . . .
int m; //m is uninitialized
Output ( out m ); //value of m is set
```

Note that the actual parameter **m** is not assigned any values before it is passed as output parameter. Since the parameters **x** and **m** refer to the same storage location, **m** takes the value that is assigned to **x**. Note that every formal output parameter of a method must be definitely assigned a value before the method returns. Program 8.6 illustrates a simple application of **out** parameters.

Program 8.6 Defining an out parameter

```
using System;
class Output
{
        static void Square ( int x, out int y )
        {
            y = x * x;
        }
        public static void Main( )
        {
            int m;    //need not be initialized
            Square ( 10, out m );
            Console.WriteLine("m = " + m);
        }
}
```

Output of Program 8.6 would be:

```
m = 100
```

8.10 VARIABLE ARGUMENT LISTS

In C#, we can define methods that can handle variable number of arguments using what are known as *parameter arrays*. Parameter arrays are declared using the keyword ***params***. *Example:*

```
void Function1 ( Params int [ ] x )

{
    . . . .
    . . . .
}
```

Here, **x** has been declared as a parameter array. Note that parameter arrays must be one-dimensional arrays. A parameter may be a part of a formal parameter list and in such cases, it must be the last parameter.

The method **Function1** defined above can be invoked in two ways:

- Using **int** type array as a value parameter. *Example:* Function1(a);
  Here, **a** is an array of type **int**
- Using zero or more **int** type arguments for the parameter array. *Example:* Function(10, 20);

The second invocation creates an **int** type array with two elements 10 and 20 and passes the newly created array as the actual argument to the method.

Program 8.7 illustrates the concept of applying variable argument lists using the keyword **params**.

Program 8.7 Concept of variable arguments

```
using System;
class Params
{
    static void Parray ( params int [ ] arr )
    {
        Console.Write("Array elements are:");
        foreach (int i in arr)
            Console.Write(" " + i);
        Console.WriteLine( );
    }
    public static void Main( )
    {
        int [ ] x = { 11, 22, 33 } ;
        Parray ( x) ;        // call 1
        Parray ( ) ;         // call 2
        Parray ( 100, 200 ) ;  // call 3
    }
}
```

The output of Program 8.7 would be:

```
Array elements are : 11 22 33
Array elements are :
Array elements are : 100 200
```

Note that the invocations **Parray ( );** and **Parray (100, 200);** are equivalent to:

```
Parray ( new int [ ] { } ) ;
Parray ( new int [ ] { 100, 200 } ) ;
```

We can also use parameter arrays of type **object**. *Example:*

```
public static void Main( )
{
    Oarray ( 10, 20, "abc" ) ;
}
static void Oarray ( params object [ ] x )
{
    foreach ( object i in x )
    {
        Console.WriteLine( i );
    }
}
```

The method **Oarray** is passed for an array of object references. The method call automatically builds an **object** array to hold the arguments, boxing value types as necessary.

Note that while it is permitted to use parameter arrays along with the value parameters, it is not allowed to combine the **params** modifier with the **ref** and **out** modifiers.

Program 8.8 uses the concept of variable number of parameters being passed to a single method. This program calculates the average of any set of numbers passed as parameters to the method **avgVal( )**.

Program 8.8 Params sample

```
using System;
class Avg
{
        public int avgVal(params int[] intData)
        {
            int sum=0,avg=0;
            if(intData.Length == 0)
            {
                Console.WriteLine("Error: no arguments.");
                return 0;
            }
            for(int i=0; i < intData.Length; i++)
            {
            sum=sum+intData[i];
            }
            avg=sum/intData.Length;
            return avg;
        }
}
public class ParamsSample
{
        public static void Main()
        {
            Avg objAvg = new Avg();
            int average;
            int num1 = 50, num2 = 100;
            // Calling the Average function using two paramters
            average = objAvg.avgVal(num1, num2);
            Console.WriteLine("Average of two parameter values is: " + average);
            // Calling the Average function using three paramters
            average = objAvg.avgVal(num1, num2, 150);
            Console.WriteLine("Average of three parameter values is: " + average);
            // Calling the Average function using six paramters
            average = objAvg.avgVal(30, 40, 50, 60, 70);
            Console.WriteLine("Average of six parameter values is: " + average);
            // Calling the Average function using an integer array
            int[ ] args = { 25, 35, 45, 55, 65, 75 };
            average = objAvg.avgVal(args);
            Console.WriteLine("Average of parameter values contained in an array is: " + average);
        }
}
```

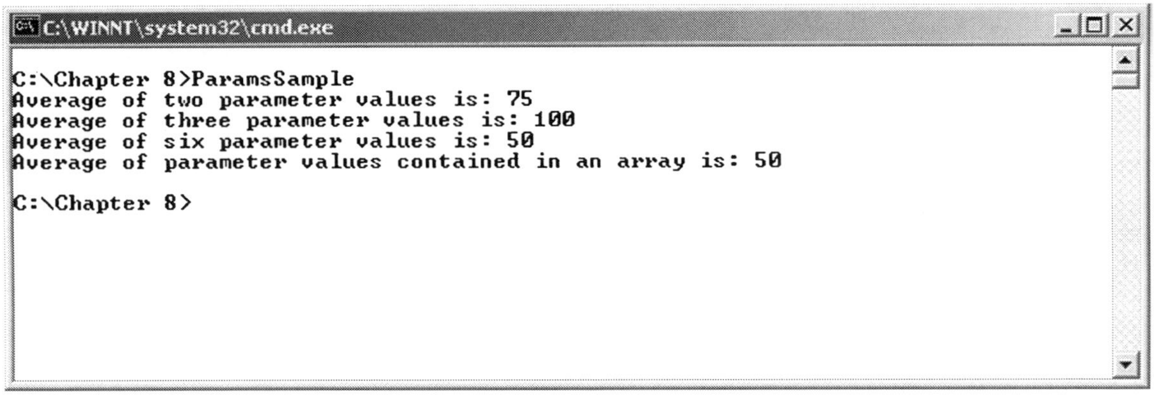

8.11 METHODS OVERLOADING

C# allows us to create more than one method with the same name, but with the different parameter lists and different definitions. This is called *method overloading*. Method overloading is used when methods are required to perform similar tasks but using different input parameters.

Overloaded methods must differ in number and/or type of parameters they take. This enables the compiler to decide which one of the definitions to execute depending on the type and number of arguments in the method call. Note that the method's return type does not play any role in the overload resolution.

Using the concept of method overloading, we can design a family of methods with one name but different argument lists. For example, an overloaded add( ) method handles different types of data as shown below:

```
// Method definitions
int add ( int a, int b ) { … }                    //Method1
int add ( int a, int b, int c ) { … }             //Method2
double add ( float x, float y ) { … }             //Method3
double add ( int p, float q ) { … }               //Method4
double add (float p, int q ) { … }                //Method5

//Method calls
int m = add ( 5, 10 ) ;  //calls method1
double x = add ( 15, 5.0F ) ;                     //calls method4
double x = add ( 1.0F, 2.0 F );                   //calls method3
int m = add ( 5, 10, 15 ) ;                       //calls method2
double x = add ( 2.0F, 10 ) ;                     //calls method5
```

The method selection involves the following steps:

1. The compiler tries to find an exact match in which the types of actual parameters are the same and uses that method.
2. If the exact match is not found, then the compiler tries to use the implicit conversions to the actual arguments and then uses the method whose match is unique. If the conversion creates multiple matches, then the compiler will generate an error message.

Program 8.9 illustrates method overloading.

Program 8.9 Method overloading

```
using System;
class Overloading
{
      public static void Main( )
      {
         Console.WriteLine(volume (10));
         Console.WriteLine(volume(2.5 F, 8));
         Console.WriteLine(volume(100L,75,15));
      }
      static int volume ( int x )                   // cube
      {
         return (  x * x * x  ) ;
      }
      static double volume ( float r , int h )      // cylinder
      {
         return ( 3.14519 * r * r * h ) ;
      }
      static long volume ( long l , int b , int h ) // box
      {
         return ( l * b * h ) ;
      }
}
```

The output of Program 8.9 would be:

```
1000
157.2595
112500
```

Program 8.10 demonstrates another example of method overloading in C#, using a **Multiply** method for both integer and float multiplication.

Program 8.10 Another example of method overloading

```
class Mult
{
      public void Multiply(ref int first, ref int second,out int result)
      {
         result=first*second;
      }
```

```
    public void Multiply(ref float first, ref float second,out float result)
    {
        result=first*second;
    }
}
public class MethodsExample
{
    public static void Main()
    {
        Mult multObj = new Mult();
        int intFirst = 33;
        int intSecond = 5;
        int intResult=0;

        System.Console.WriteLine("Integer Values Before Multiplication:");
        System.Console.WriteLine("\tFirst Integer: {0}\n\tSecond Integer:
{1}\n\tInteger Result: {2}", intFirst,intSecond,intResult);
        multObj.Multiply(ref intFirst, ref intSecond,out intResult);
        System.Console.WriteLine("Integer Values After Multiplication:");
        System.Console.WriteLine("\tFirst Integer: {0}\n\tSecond Integer: {1}\n\tInteger
Result: {2}", intFirst,intSecond,intResult);

        float floatFirst = 33f;
        float floatSecond = 5f;
        float floatResult=0f;

        System.Console.WriteLine("\nFloat Values Before Multiplication:");
        System.Console.WriteLine("\tFirst Float: {0}\n\tSecond Float:
{1}\n\tFloat Result: {2}", floatFirst,floatSecond,floatResult);

        multObj.Multiply(ref floatFirst, ref floatSecond,out floatResult);

    System.Console.WriteLine("Float Values After Multiplication:");
    System.Console.WriteLine("\tFirst Float: {0}\n\tSecond Float:
{1}\n\tFloat Result: {2}", floatFirst,floatSecond,floatResult);
    }
}
```

```
C:\WINNT\system32\cmd.exe

C:\Chapter 8>MethodsExample
Integer Values Before Multiplication:
        First Integer: 33
        Second Integer: 5
        Integer Result: 0
Integer Values After Multiplication:
        First Integer: 33
        Second Integer: 5
        Integer Result: 165

Float Values Before Multiplication:
        First Float: 33
        Second Float: 5
        Float Result: 0
Float Values After Multiplication:
        First Float: 33
        Second Float: 5
        Float Result: 165

C:\Chapter 8>_
```

Case Study

Problem Statement

Mr. Harish Chadda owns a two storied house in B.N. Colony, Ahmedabad. He has to pay the property tax for the house every year to the Municipal Corporation of the city. Currently, Mr. Chadda has been manually calculating the property tax each year using a calculator. As a result, errors occur during calculation and Mr. Chadda is not able to pay the tax on time. Mr. Chadda also needs to spend a large amount of time while manually calculating the property tax for his house. This too delays the payment of property tax to the Municipal Corporation of the city. If the property tax is not paid on time, the Municipal Corporation levies a late fee of Rs. 2000. What must Mr. Chadda do to avoid delay in the payment of property tax?

Solution

After a lot of research, Mr. Chadda comes to the conclusion that he should have a software application installed on his Home PC that helps in automating the task of calculating property tax. He, therefore requests his neighbour, Mr. Gautam who is a software professional, to spare some hours for him, to create an application for him so that the task of calculating property tax is automated. After analyzing the requirements of Mr. Chadda, Gautam decides to use the **method overloading** feature of C# in the application for automating the task of calculating property tax. Method overloading is an important feature of C# that allows you to use the same name for multiple methods. Gautam then creates the following C# application for Mr. Chadda.

```
//
***********************************************************************
// This program demonstrates method overloading in a real world scenario
for tax calculation.
//
```

```
******************************************************************************
using System;
class TaxCalc
{
        // This method takes four arguments: Two amounts values and Two
           rates values and returns the taxable amount
        public double TaxCalculator(double taxAmount1, double taxRate1,
           double taxAmount2, double taxRate2)
        {
        double taxAmount;
        Console.WriteLine("-Using TaxCalculator method containing 4
           arguments.-");
        taxAmount = (taxAmount1 * taxRate1) + (taxAmount2 * taxRate2);
        return taxAmount;
     }
     // This method takes two arguments: One amount value and One rate
        value and returns the taxable amount
     public double TaxCalculator(double taxAmount1, double taxRate1)
     {
        double taxAmount;
        Console.WriteLine("-Using TaxCalculator method containing 2
           arguments.-");
        taxAmount = taxAmount1 * taxRate1;
     return taxAmount;
     }
     // This method takes one argument: One amount value and returns the
        taxable amount
     public double TaxCalculator(double taxAmount1)
     {
        double taxRate = 0.12;
        double taxAmount = 0;
        Console.WriteLine("-Using TaxCalculator method containing 1
           argument only.-");
     taxAmount = taxAmount1 * taxRate;
        return taxAmount;
     }
     // This method also takes one argument: Tax Type (string) and returns
        the tax rate
     public double TaxCalculator(string taxType)
     {
        double taxRate = 0;
     Console.WriteLine("-Using TaxCalculator method containing 1 argument
onlyofStringtype.-");
        if (taxType == "Personal")
           taxRate = 0.10;
              return taxRate;
```

```
      }
}
class MethodOverload
{
      static void Main(string[] args)
      {
            string userResponse;
            bool flagHome = false;
            bool flagBusiness = false;
         double houseRate = 0;
         double houseValue = 0;
         double businessSales = 0;
         double taxRate = 0;
         double earning = 0;
         double totalTax = 0;
            TaxCalc tc=new TaxCalc();
         Console.WriteLine("Do you have a house registered in your name? (y/n)");
         userResponse = Console.ReadLine();
         if (userResponse == "y")
         {
            flagHome = true;
            Console.WriteLine("Enter the worth of the house in Rupees: ");
               userResponse = Console.ReadLine();
               houseValue = Convert.ToDouble(userResponse);
                  Console.WriteLine("Enter the tax rate applicable for a
house: ");
            userResponse = Console.ReadLine();
            houseRate = Convert.ToDouble(userResponse);
            }
                  Console.WriteLine("Are you an owner of a business? (y/n)");
                  userResponse = Console.ReadLine();
                  if (userResponse == "y")
                  {
                     flagBusiness = true;
                     Console.WriteLine("Enter the total sales for the year: ");
                     userResponse = Console.ReadLine();
                     businessSales = Convert.ToDouble(userResponse);
                     Console.WriteLine("Enter the gross sales tax rate: ");
                     userResponse = Console.ReadLine();
                     taxRate = Convert.ToDouble(userResponse);
                  }
                  if (flagHome && !flagBusiness)
                        totalTax = tc.TaxCalculator(houseValue,
houseRate);
                     else if (!flagHome && flagBusiness)
```

```
                        totalTax = tc.TaxCalculator(businessSales,
taxRate);
                    else if (flagHome && flagBusiness)
                        totalTax = tc.TaxCalculator(houseValue,
houseRate, businessSales, taxRate);
                    Console.WriteLine("Enter your total earnings for the
year in Rupees: ");
                userResponse = Console.ReadLine();
                earning = Convert.ToDouble(userResponse);
                    totalTax = totalTax + tc.TaxCalculator(earning);
                Console.WriteLine("Your total tax calculated is {0}",
totalTax);
                Console.WriteLine("The Personal tax rate is
{0}",tc.TaxCalculator("Personal") );
    }
}
```

Remarks

In method overloading, more than one method is defined with the same name in a C# application but different parameters are passed to these methods. Method overloading is useful when in an application a number of methods are required to perform the same task but with different arguments.

Common Programming Errors

- *Compound declaration of method parameters in the header, like (int x, y).*
- *Forgetting to declare return type.*
- *Forgetting to return a value from a method when it is supposed to return a value.*
- *Returning a value from a method that has been declared void.*
- *Defining a local variable with the same name as a method parameter*
- *Invoking a method with incompatible parameter types.*
- *Invoking a method that results in type conversion from a wider data type to a smaller one may cause errors in output.*
- *Defining a method inside another method.*
- *Forgetting to use appropriate namespaces for using certain methods defined in other namespaces.*
- *When a formal parameter is declared as **ref**, forgetting to declare the corresponding argument as **ref** in the method invocation.*
- *When a formal parameter is declared as **out**, forgetting to declare the corresponding actual parameter also as **out**.*
- *Combining params modifier either with the **ref** or with the **out** modifiers.*
- *Not matching the type of value returned by a method with its return type declared.*
- *Placing semicolon immediately after the method header.*
- *Changing only return types for method overloading.*
- *Forgetting to initialize a **ref** parameter in the calling method.*
- *Forgetting to assign a value to an **out** parameter in the called method before it returns.*

Review Questions

8.1 State whether the following statements are true or false.

(a) When a method does not return anything, we can omit the return type in its definition.

(b) When a method does not take any input parameters, we must include void keyword in the place of parameter list.

(c) The body of a method must always be enclosed in curly braces.

(d) The virtual may be used as a method modifier.

(e) While invoking a method, only data values should be used as actual parameters.

(f) While invoking a method, we can pass a float value to a double type parameter.

(g) While invoking a method, we can pass an int type value to byte type parameter.

(h) A method that has been invoked by another method cannot invoke any other method.

(i) Arrays cannot be used as method parameters.

(j) We can overload methods by changing their return types.

(k) By default, all parameters are passed by values.

(l) The values of value parameters that are passed to a method are not modified by any changes made to their formal parameters inside the method.

(m) A method call of a method that returns value can be used in an expression like any other variable.

(n) A variable declared outside a method definition cannot be accessed by the method.

(o) A formal parameter declared as out must be definitely assigned a value before the method returns.

(p) A method has exactly one return statement.

(q) A method has at least one return statement.

(r) Parameter arrays must be single-dimensional arrays.

(s) A method can have all the four kinds of parameters together.

8.2 How do we invoke a method in C#?

8.3 What is nesting of methods? Give an example of typical use of methods nesting.

8.4 We can write the Main method in four different ways. List them. State the significance of each of them.

8.5 What are value parameters? When do we use them?

8.6 What are reference parameters? When do we use them?

8.7 What are output parameters? Why do we need them?

8.8 What are parameter arrays? List three restrictions in declaring parameter arrays.

8.9 What is method overloading? Give two examples where method overloading is applied.

8.10 Distinguish between ref and out parameters.

8.11 Give typical examples of the following situations.

(a) A method with an int type parameter and a string type return value.

(b) A method with two double type parameters and a bool type return value.

(c) A method with no parameters and an int type return value.

(d) A method with string type parameters and an int type return value.

(e) A method with string type parameters and a bool type return value.

(f) A method with two return statements.

8.12 Find errors, if any, in the following method headers:
(a) void Display (int x, y)
(b) public Multiply (float x, float y)
(c) private int Product (int m, int n);
(d) void ArrayOut (params object [ ] x, out int y)
(e) void Fun1 (Params in [ ] x, float y)

8.13 Find errors if any, in the following method invocation statements.
(a) void Display ( );
(b) int y = Product (int a, int b);
(c) int y, Mul (x,y);
(d) float area = Area (10,20)
(e) largest (100,200) = y;

8.14 Given below are pairs of method headers and corresponding invocation statements. Which of them are incorrect? Why?
(a) void Product (int x, int y) { }
int a = Product (10, 20);
(b) float divide (float x, float y) { }
int b = Divide (25.34, 1.56);
(c) double Large (double x, double y) { }
Large (a,b);
(d) void Change (ref int x) { }
Change (y);
(e) void Mul (int x, int y, out int z) { }
Mul (a,b,c);
(f) void Tarry (params int [ ] x) { }
Tarry (10,20, "abc");

8.15 Which of the following are overloading the method int Sum (int x, int y) { }
(a) int Sum (int x, int y, int x) { }
(b) float Sum (int x, int y) { }
(c) int Sum (float x, float y) { }
(d) int Sum (int a, int b) { }
(e) float Sum (int x, int y, float x) { }
(f) int Sum (params int [ ] a) { }

Debugging Exercises

8.1. Given below is a program to print an array using pass by reference. Will the program compile and generate the desired output? If not, why?

```
using System;
class RefCheck
{
```

```
        static void ShowArray(ref int[] intArray)
        {
            if (intArray == null)
            {
                intArray = new int[10];
            }
                intArray[1] = 13;
                intArray[4] = 15;
            }
            static void Main()
            {
                int[] arrObj = { 3, 5, 7, 9, 11 };

                ShowArray(arrObj);

                System.Console.WriteLine("Elements of the Array are:");
                for (int i = 0; i < arrObj.Length; i++)
                {
                    System.Console.Write(arrObj[i] + " ");
                }
        }
}
```

8.2. Debug the program given below.

```
class TrigMaths
{
        public void TrigFunctions(double angle, out double sin, out double
cos,out double tan) {
            sin = System.Math.Sin(angle);
            cos = System.Math.Cos(angle);
            tan = System.Math.Tan(angle);
        }
}
public class OutParamExample
{
        public static void Main()
        {
            double angle;
            double sin;
```

```
        double cos;
        double tan;

        TrigMaths trig = new TrigMaths();
        trig.TrigFunctions(angle, out sin, out cos, out tan);

        System.Console.WriteLine("Sine of the angle is: " + sin);
        System.Console.WriteLine("Cosine of the angle is: " + cos);
        System.Console.WriteLine("Tangent of the angle is: " + tan);
    }
}
```

Programming Exercises

*8.1 Write a PrintLine method that will print a line of 20 character length using * character.*

8.2 Write a program using the PrintLine method defined in Exercise 8.1 to display a message as follows:

```
* * * * * * * * * * * * * * * * * * * *
      C# Programming
* * * * * * * * * * * * * * * * * * * *
```

8.3 Write a method that would calculate the value of money for a given period of years and print the results giving the following details on one line of output:

– Principal amount

– Interest rate

– Period in years

– Final value

The method takes principal amount, interest rate and period as input parameters and does not return any value. The following formula may be used repeatedly:

Value at the end of year = Value at start of year (1 + interest rate)

Write a program to test the method.

8.4 Modify the above program such that the method returns the final value computed to the Main method which will then display the required output.

8.5 Modify the PrintLine method such that it can take the "character" to be used and the "length" of the line as input parameters. Write a program using the new PrintLine method to draw three different types of lines with different sizes, Example:

```
AAAAAAAAA
ZZZZZZZZZZZZZZZZZZ
WWWWWWWWWWWWWWWWWWW
```

8.6 Write a program to calculate the standard deviation of an array of values. Use methods Standard and Mean to calculate standard deviation and mean of the values.

8.7 Develop a program that uses a method to sort an array of integers.

8.8 Write a method Space(int n) that can be used to provide a space of n positions between output of two numbers. Test your method.

8.9 Write a method that will round a floating point number to an indicated decimal places. For example, the number 123.4567 would yield a value 123.46 when it is rounded off to two decimal places.

8.10 Write a method Prime that returns true if its argument is a prime number and returns false otherwise.

8.11 Write a void type method that takes two int type value parameters and one int type out parameter and returns the product of two value parameters through the output parameter. Write a program to test its working.

8.12 Redo the Exercise 8.11 using all the three parameters as ref parameters.

8.13 Write a method that takes three values as input parameters and returns the largest of the three values.

8.14 Modify the method developed in Exercise 8.13 so that the method returns the smallest of the three values.

8.15 Write a method that takes an array as an input parameter and uses two methods, one to find the largest array element and other to compute the average of array elements.

Chapter

9

Handling Arrays

9.1 INTRODUCTION

An array is a group of contiguous or related data items that share a common name. For instance, we can define an array name **marks** to represent a set of marks of a class of students. A particular value is indicated by writing a number called *index* number or *subscript* in brackets after the array name. For example,

```
marks[10]
```

represents the marks obtained by the 10th student. While the complete set of values is referred to as an *array*, the individual values are called *elements*. Arrays can be of any variable type.

The ability to use a single name to represent a collection of items and to refer to an item by specifying the item number enables us to develop concise and efficient programs. For example, a loop with the subscript as the control variable can be used to read the entire array, perform calculations and print out the results.

9.2 ONE-DIMENSIONAL ARRAYS

A list of items can be given one variable name using only one subscript and such a variable is called a *single-subscripted* variable or a *one-dimensional* array. In mathematics, we often deal with variables that are single-subscripted. For instance, we use the equation

$$A = \frac{\sum_{i=1}^{n} X_i}{n}$$

to calculate the average of n values of x. The subscripted variable $\mathbf{x}_i$ refers to the ith element of x. In C#, a single-subscripted variable $\mathbf{x}_i$ can be expressed as

```
x[1], x[2], x[3] .......... x[n]
```

The subscript can begin with number 0. That is

```
x[0]
```

is allowed. For example, if we want to represent a set of five numbers, say (35,40,20,57,19), by an array variable **number**, then we may create the variable **number** as follows

```
int [ ] number = new int[5];
```

and the computer reserves five storage locations as shown below:

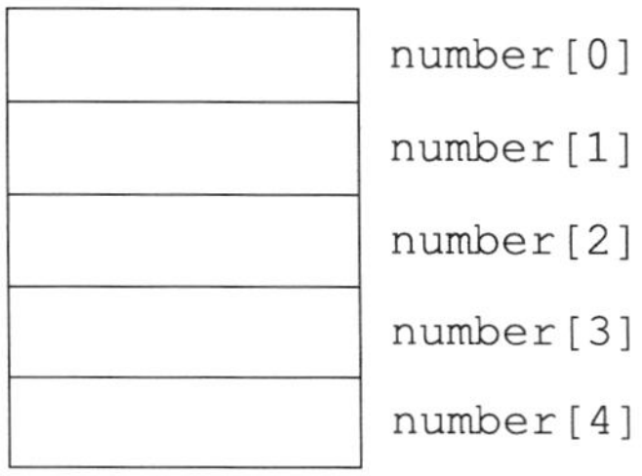

The values to the array elements can be assigned as follows:

```
number[0] = 35;
number[1] = 40;
number[2] = 20;
number[3] = 57;
number[4] = 19;
```

This would cause the array **number** to store the values shown as follows**:**

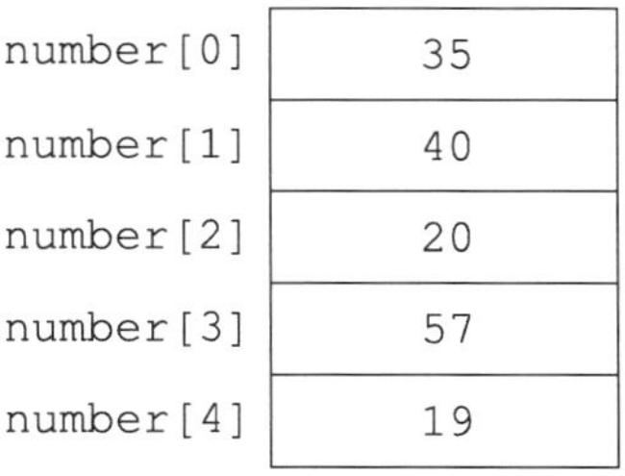

These elements may be used in programs just like any other C# variable. For example, the following are valid statements:

```
aNumber    = number[0] + 10;
number[4]  = number[0] + number[2];
number[2]  = x[5] + y[10];
value[6]   = number[i] * 3;
```

The subscript of an array can be integer constants, integer variables like i or expressions that yield integers.

9.3 CREATING AN ARRAY

Like other variables, arrays must be declared and created in the computer memory before they are used. Creation of an array involves three steps:

1. Declaring the array
2. Creating memory locations
3. Putting values into the memory locations.

9.3.1 Declaration of Arrays

Arrays in C# are declared as follows:

```
type[ ] arrayname;
```

Examples:

```
int[ ] counter;    //declare int array reference
float[ ]  marks;   //declare float array reference
int[ ] x,y;        //declare two int array reference
```

Remember, we do not enter the size of the arrays in the declaration.

9.3.2 Creation of Arrays

After declaring an array, we need to create it in the memory. C# allows us to create arrays using **new** operator only, as shown below:

```
arrayname = new type[size];
```

Examples:

```
number  = new int[5];       //create a 5 element int array
average= new float[10];     // create a 10 element float array
```

These lines create the necessary memory locations for the arrays **number** and **average** and designate them as **int** and **float** respectively. Now, the variable **number** refers to an array of five integers and **average** refers to an array of ten floating-point values.

It is also possible to combine the two steps, declaration and creation, into one as shown below:

```
Int[ ] number = new int[5]; //declare and create 5 element int array
```

Figure 9.1 illustrates the creation of an array in memory.

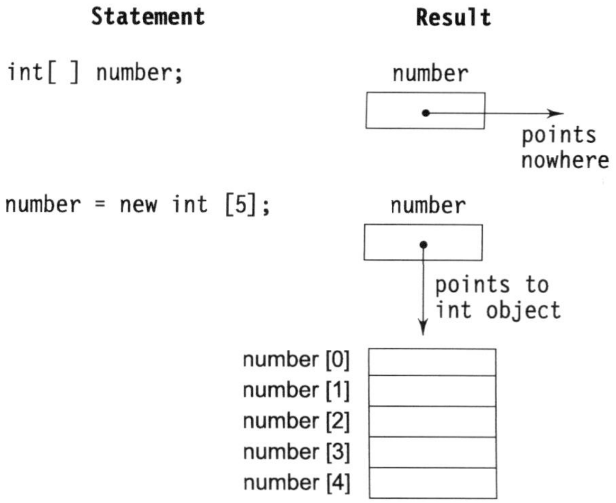

Figure 9.1 *Creation of an array in memory*

9.3.3 Initialization of Arrays

The final step is to put values into the array created. This process is known as *initialization*. This is done using the array subscripts as shown below.

```
arrayname[subscript] = value ;
```

Example:

```
number[0] = 35;
number[1] = 40;
..........
..........
number[4] = 19;
```

Note that C# creates arrays starting with a subscript of 0 and ends with a value one less than the *size* specified.

Unlike C, C# protects arrays from overruns and underruns. Trying to access an array beyond its boundaries will generate an error message. *Example:*

```
number [5] = 100; // error!
```

We can also initialize arrays automatically in the same way as the ordinary variables when they are declared, as shown below:

```
type [ ] arrayname = {list of values};
```

The array initializer is a list of values separated by commas and defined on both ends by curly braces. Note that no size is given. The compiler allocates enough space for all the elements specified in the list.

Example:

```
int[ ] number = {35, 40, 20, 57, 19};
```

The preceding line is equivalent to:

```
int [ ] number = new int [5] { 35, 40, 20, 57, 19 };
```

This combines all the three steps, namely declaration, creation and initialization.

It is possible to assign an array object to another. For instance,

```
int[ ] a = {1,2,3};
int[ ] b;
b = a;
```

are valid in C#. Both the arrays will have the same values.

Every member in an array is automatically set to a default value. For example, if we have an array of numerical type, each element is set to number 0. (See Table 4.6)

A C# array can also hold reference type elements. An array of reference types will contain only references to the elements and not the actual values. Reference types in an array are automatically initialized to null. Any attempt to access an element in an array of reference types before they are initialized will cause an error.

9.3.4 Array Length

In C#, all arrays are class-based and store the allocated size in a variable named **Length.** We can access the length of the array **a** using **a.Length**. *Example:*

```
int aSize = a.Length;
```

This information will be useful in the manipulation of arrays when their sizes are not known.

Loops may be used to initialize large-size arrays. *Example:*

```
...........
..........
for(int i = 0; i < 100; i++)
{
    if(i < 50)
        sum[i] = 0.0;
    else
        sum[i] = 1.0;
}
..........
..........
```

The first fifty elements of the array **sum** are initialized to zero while the remaining are initialized to 1.0.

Consider another example as shown below:

```
for    (int x = 0; x < 10; x++)
        average[x] = (float)x;
```

This loop initializes the array **average** to the values 0.0 to 9.0. Program 9.1 illustrates the use of an array for sorting a list of numbers.

Program 9.1 Sorting a list of numbers

```
using System;
class NumberSorting
{
    public static void Main( )
    {
        int[ ]number = { 55, 40, 80, 65, 71 };
        int n = number.Length;

        Console.Write("Given list : ");

        for (int i = 0; i < n; i++)
        {
            Console.Write("   " + number[i]);
        }

        Console.WriteLine("\n");

        / /   Sorting begins

        for (int i = 0; i < n; i++)
        {
            for (int j = i+1; j < n; j++)
            {
                if (number[i] < number[j])
                {
                    / /   Interchange values

                    int temp = number[i];
                    number[i] = number[j];
                    number[j] = temp;
                }
            }
        }

        Console.Write("Sorted list : ");

        for (int i = 0; i < n; i++)
        {
            Console.Write("   " + number[i]);
        }

        Console.WriteLine("   ");
    }
}
```

Program 9.1 displays the following output:

```
Given list    : 55 40 80 65 71
Sorted list   : 80 71 65 55 40
```

9.4 TWO-DIMENSIONAL ARRAYS

So far we have discussed the array variables that can store a list of values. There will be situations where a table of values will have to be stored. Consider the following data table, which shows the value of sales of three items by four salesgirls:

	Item1	Item2	Item3
Salesgirl #1	310	275	365
Salesgirl #2	210	190	325
Salesgirl #3	405	235	240
Salesgirl #4	260	300	380

The table contains a total of twelve values, three in each line. We can think of this table as a matrix consisting of four *rows* and three *columns*. Each row represents the value of sales made by a particular salesgirl and each column represents the value of sales of a particular item.

In mathematics, we represent a particular value in a matrix by using two subscripts such as v_{ij}. Here v denotes the entire matrix and v_{ij} refers to the value in the ***i***th row and ***j***th column. For example, in the above table v_{23} refers to the value 325. C# allows us to define such tables of items by using *two-dimensional* arrays. The table discussed above can be represented in C# as

```
v[4,3]
```

Two-dimensional arrays are stored in memory as shown in Fig. 9.2. As with the one-dimensional arrays, each dimension of the array is indexed from zero to its maximum size minus one; the first index selects the row and the second index selects the column within that row.

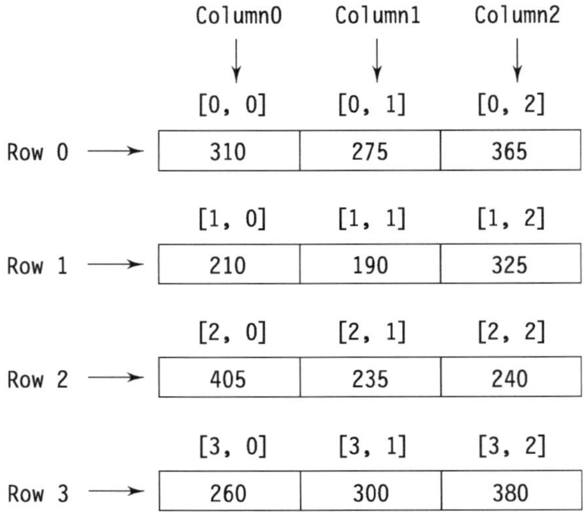

Figure 9.2 *Representation of a two-dimensional array in memory*

For creating two-dimensional arrays, we must follow the same steps as that of simple arrays. We may create a two-dimensional array like this:

```
int[,] myArray;
myArray = new int[3,4];
```

or

```
int[,] myArray = new int[3,4];
```

This creates a table that can store twelve integer values, four across and three down.

Like the one-dimensional arrays, two-dimensional arrays may be initialized by following their declaration with a list of initial values enclosed in braces. For example,

```
int[,] table ={{0,0,0},{1,1,1}};
```

initializes the elements of the first row to zero and the second row to one. The initialization is done row by row. We can also initialize a two-dimensional array in the form of a matrix as shown below:

```
int[,] table = {
                    {0,0,0},
                    {1,1,1}
               };
```

Note the syntax of the above statements. Commas are required after each brace that closes off a row, except in the case of the last row.

We can refer to a value stored in a two-dimensional array by using subscripts for both the column and row of the corresponding element. For example:

```
int value = table[1,2];
```

This retrieves the value stored in the second row and third column of the **table** matrix.

A quick way to initialize a two-dimensional array is to use nested **for** loops as shown below:

```
for (i = 0; i < 5; i++)
{
    for (j = 0; j <5; j++)
    {
        if (i == j)
            table[i,j] = 1;
        else
            table[i,j] = 0;
    }
}
```

This will set all the diagonal elements to 1 and others to zero as given below:

```
1    0    0    0    0
0    1    0    0    0
0    0    1    0    0
0    0    0    1    0
0    0    0    0    1
```

Program 9.2 illustrates the use of two-dimensional arrays in real-life situations.

Program 9.2 Application of two-dimensional arrays

```
using System;
class MulTable
{
    static int ROWS = 20;
    static int COLUMNS = 20;

   public static void Main( )
   {
      int[,] product = new int[ROWS,COLUMNS];

      Console.WriteLine("MULTIPLICATION TABLE");
      Console.WriteLine("  ");

      int i,j;
      for (i=10; i<ROWS; i++)
      {
         for (j=10; j<COLUMNS; j++)
         {
            product[i,j] = i*j;
            Console.Write("   " +product[i,j]);
         }
         Console.WriteLine("  ");
      }
   }
}
```

Program 9.2 produces the following output:

```
MULTIPLICATION TABLE
   100  110  120  130  140  150  160  170  180  190
   110  121  132  143  154  165  176  187  198  209
   120  132  144  156  168  180  192  204  216  228
   130  143  156  169  182  195  208  221  234  247
   140  154  168  182  196  210  224  238  252  266
   150  165  180  195  210  225  240  255  270  285
   160  176  192  208  224  240  256  272  288  304
   170  187  204  221  238  255  272  289  306  323
   180  198  216  234  252  270  288  306  324  342
   190  209  228  247  266  285  304  323  342  361
```

9.5 VARIABLE-SIZE ARRAYS

C# treats multidimensional arrays as 'arrays of arrays'. It is possible to declare a two-dimensional array as follows:

```
int[ ][ ] x= new int[3][];   //three rows array
x[0]   = new int[2];         //first row has two elements
x[1]   = new int[4];         //second row has four elements
x[2]   = new int[3];         //third row has three elements
```

These statements create a two-dimensional array having different lengths for each row as shown in Fig. 9.3. Variable-size arrays are called *jagged* arrays.

The elements can be accessed as follows:

```
x [1] [1] = 10;
int y = x [2[[2];
```

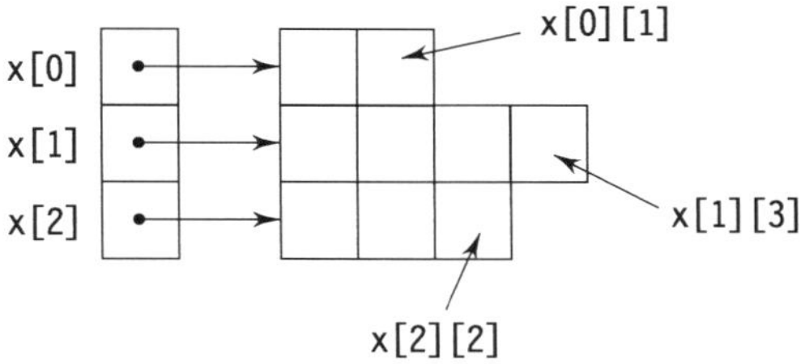

Figure 9.3 *Variable size arrays*

Note the difference in the way we access the two types of arrays. With rectangular arrays, all indices are within one set of square brackets, while for jagged arrays each element is within its own square brackets.

9.6 THE SYSTEM.ARRAY CLASS

In C#, every array we create is automatically derived from the **System.Array** class. This class defines a number of methods and properties that can be used to manipulate arrays more efficiently. Table 9.1 lists some of the commonly used methods and their purpose.

Table 9.1 *Some commonly used methods of System.Array class*

Method/Property	Purpose
Clear ( )	Sets a range of elements to empty values
CopyTo ( )	Copies elements from the source array into the destination array
GetLength ( )	Gives the number of elements in a given dimension of the array
GetValue ( )	Gets the value for a given index in the array
Length	Gives the length of an array
SetValue ( )	Sets the value for a given index in the array
Reverse ( )	Reverses the contents of a one-dimensional array
Sort ( )	Sorts the elements in a one-dimensional array

Program 9.3 illustrates the implementation of **Sort ( )** and **Reverse ( )** methods.

Program 9.3 Sorting and reversing an array

```
using System;
class SortReverse
{
    public static void Main( )
    {
        //creating an array
        int [ ] x = { 30, 10, 80, 90, 20 };

        Console.WriteLine ("Array before sorting");
        foreach (int i in x)
            Console.Write(" " + i );
        Console.WriteLine ( );

        //Sorting and reversing the array elements
        Array.Sort(x); Array.Reverse (x);

        Console.WriteLine("Array after Sorting and Reversing");
        foreach ( int i in x )
            Console.Write(" " + i);
        Console.WriteLine ( );
        }
    }
```

Output of Program 9.3:

```
Array before sorting
30 10 80 90 20
Array after Sorting and Reversing
90  80  30  20  10
```

9.7 ARRAYLIST CLASS

System.Collections namespace defines a class known as **ArrayList** that can store a dynamically sized array of objects. The **ArrayList** class includes a number of methods to support operations such as sorting, removing and enumerating its contents. It also supports a property **Count** that gives the number of objects in an array list and a property **Capacity** to modify or read the capacity.

An array list is very similar to an array, except that it has the ability to grow dynamically. We can create an array list by indicating the initial capacity we want. *Example:*

```
ArrayList cities = new ArrayList (30);
```

It creates **cities** with a capacity to store thirty objects. If we do not specify the size, it defaults to sixteen. That is,

```
ArrayList cities = new ArrayList ( );
```

will create a **cities** list with the capacity to store sixteen objects. We can now add elements to the list using the **Add ( )** method:

```
cities.Add ("Bombay");
cities.Add ("Anand");
```

We can also remove an element:

```
cities.RemoveAt (1);
```

This will remove the object in position1. We can modify the capacity of the list using the property **Capacity:**

```
cities.Capacity = 20;
```

We may obtain the actual number of objects present in the list using the property **Count** as follows:

```
int n = cities.Count;
```

An array list can be really useful if we need to create an array of objects but we do not know in advance how big the array would be. Further, an array list can contain any object reference.

Table 9.2 lists some of the most important methods and properties supported by the **ArrayList** class.

Table 9.2 *Some important ArrayList methods and properties*

Methods/Property	Purpose
Add ( )	Adds an object to a list
Clear ( )	Removes all the elements from the list
Contains ( )	Determines if an element is in the list
CopyTo ( )	Copies a list to another
Insert ( )	Inserts an element into the list
Remove ( )	Removes the first occurrence of an element
RemoveAt ( )	Removes the element at the specified place
RemoveRange ( )	Removes a range of elements
Sort ( )	Sorts the elements
Capacity	Gets or sets the number of elements in the list
Count	Gets the number of elements currently in the list.

Program 9.4 demonstrates how an array list may be created and manipulated.

Program 9.4 Using ArrayList class

```
using System;
using System.Collections;
class City
{
    public static void Main( )
    {
        ArrayList n = new ArrayList ( );
        n.Add ("Madras");
        n.Add ("Bombay");
        n.Add ("Anand");
        n.Add ("Calcutta");
        n.Add   ("Delhi");
        Console.WriteLine("Capacity = "+ n.Capacity);
        Console.WriteLine("Elements present = " + n.Count);

        n.Sort( );
        for (int i = 0, i < n.Count; i ++)
        {
            Console.WriteLine(n[i]);
        }
        Console.WriteLine( );

        n.RemoveAt(4);
        for (int i = 0 ; i < n.Count ; i ++)
        {
            Console.WriteLine(n[i]);
        }
    }
}
```

Program 9.4 produces the following output:

```
capacity = 16
Elements present = 5
Anand
Bombay
Calcutta
Delhi
Madras

Anand
Bombay
Calcutta
Delhi
```

Program 9.5 creates an arraylist of 10 prime numbers and counts the elements and capacities of the arraylist. After this, the program removes two elements from the start, counts elements and again replaces two prime numbers in the beginning of the arraylist.

Program 9.5 Another example of ArrayList class

```
using System;
using System.Collections;
using System.Text;

namespace ArrayListExample
{
   class Program
    {
       static void Main(string[] args)
        {
          //Creating an array list
          ArrayList arr = new ArrayList();

          Console.WriteLine("arr ArrayList having capacity : " + arr.Capacity);
          Console.WriteLine("arr ArrayList having elements : " + arr.Count);

          Console.WriteLine();

          Console.WriteLine("Adding 10 prime number in the ArrayList arr.");

           arr.Add(2);
           arr.Add(3);
           arr.Add(5);
           arr.Add(7);
           arr.Add(11);
           arr.Add(13);
           arr.Add(17);
           arr.Add(19);
           arr.Add(23);
           arr.Add(29);

          Console.WriteLine("Now the capacity of the ArrayList is : " + arr.Capacity);
          Console.WriteLine("The number of elements in the ArrayList are : " +
arr.Count);


          Console.Write("Elements of the ArrayList are : ");
          for (int a = 0; a < arr.Count; a++)
             Console.Write(arr[a] + " ");
          Console.WriteLine("\n");

          Console.WriteLine("Now removing 2 elements of the ArrayList from the starting
position.");

           arr.Remove(2);
           arr.Remove(3);
```

```
        Console.WriteLine("The capacity of the ArrayList is now : " + arr.Capacity);
        Console.WriteLine("Number of elements present in the ArrayList become : " +
arr.Count);

        Console.Write("Elements in the ArrayList after 2 elements are removed : ");
         foreach (int m in arr)
            Console.Write(m + " ");
         Console.WriteLine("\n");

        Console.WriteLine("Replacing first 2 elements of the ArrayList with prime
numbers 31, 37 and replacing last 2 elements with prime numbers 41, 43.");
         arr[0] = 31;
         arr[1] = 37;
         arr[6] = 41;
         arr[7] = 43;

        Console.Write("Now the elements of the ArrayList are : ");
         foreach (int m in arr)
            Console.Write(m + " ");
         Console.WriteLine();

      }
   }
}
```

```
C:\WINNT\system32\cmd.exe
arr ArrayList having capacity : 0
arr ArrayList having elements : 0

Adding 10 prime number in the ArrayList arr.
Now the capacity of the ArrayList is : 16
The number of elements in the ArrayList are : 10
Elements of the ArrayList are : 2 3 5 7 11 13 17 19 23 29

Now removing 2 elements of the ArrayList from the starting position.
The capacity of the ArrayList is now : 16
Number of elements present in the ArrayList become : 8
Elements in the ArrayList after 2 elements are removed : 5 7 11 13 17 19 23 29

Replacing first 2 elements of the ArrayList with prime numbers 31, 37 and replac
ing last 2 elements with prime numbers 41, 43.
Now the elements of the ArrayList are : 31 37 11 13 17 19 41 43
Press any key to continue . . .
```

Program 9.6 displays array items in an array and then reverses the elements of the array. Another array having string elements is used to sort the array elements alphabetically and then displays the reversed data.

Program 9.6 Application of Sort and Reverse methods

```
using System;
using System.Collections.Generic;
using System.Text;

namespace ArraySort_ReverseExample
{
   class Program
    {
      public static void ArrDisplay(string[] arr1)
        {

          foreach (string str in arr1)
            {
             Console.WriteLine("Value: {0}", str);
            }
          Console.WriteLine("\n");
        }

      static void Main(string[] args)
        {
          String[] arr2 =
            {
             "John", "Norman", "Greg", "Steve", "Victor" , "Logan"
            };

          ArrDisplay(arr2);
          Array.Reverse(arr2);
          Console.WriteLine("Here array is reversed.");
          ArrDisplay(arr2);

          String[] arr3 =
            {
             "This", "is", "an", "example",
            "of", "an", "array", "and","how", "to", "sort", "it", "alphabetically"
            };

          ArrDisplay(arr3);
          Console.WriteLine("Now array is sorted alphabetically.");
          Array.Sort(arr3);
          ArrDisplay(arr3);

        }
    }
}
```

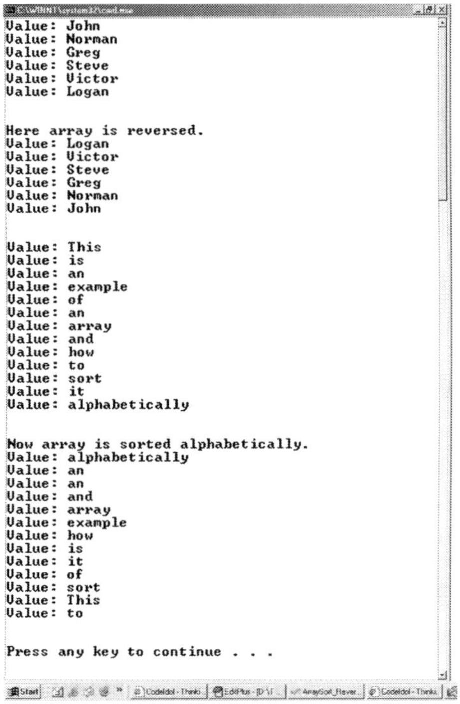

Program 9.7 displays 3 jagged arrays, with partially filled elements in it and displays them. The empty element values are filled with 0.

Program 9.7 Illustration of jagged arrays

```
using System;
using System.Collections.Generic;
using System.Text;

namespace jaggedArrayExample
{
   class Program
   {
      static void Main(string[] args)
      {
         const int rows = 3;

         // declare the jagged array as 3 rows high
         int[][] jagArr = new int[rows][];

         //a row with 2 elements
         jagArr[0] = new int[2];
```

```
        // a row with 3 elements
        jagArr[1] = new int[3];

        // a row with 4 elements
        jagArr[2] = new int[4];


        // Fill some (but not all) elements of the rows
        jagArr[0][1] = 54;
        jagArr[1][0] = 26;
        jagArr[1][1] = 18;
        jagArr[2][0] = 72;
        jagArr[2][3] = 404;

        for (int i = 0; i < 2; i++)
          {
               Console.WriteLine("This is jagged Array 1 having elements : [0][{0}] = {1}",
             i, jagArr[0][i]);
          }

        for (int i = 0; i < 3; i++)
          {
               Console.WriteLine("This is jagged Array 2 having elements : [1][{0}] = {1}",
             i, jagArr[1][i]);
          }

        for (int i = 0; i < 4; i++)
          {
          Console.WriteLine("This is jagged Array 3 having elements : [2][{0}] = {1}",
             i, jagArr[2][i]);
          }

      }
    }
}
```

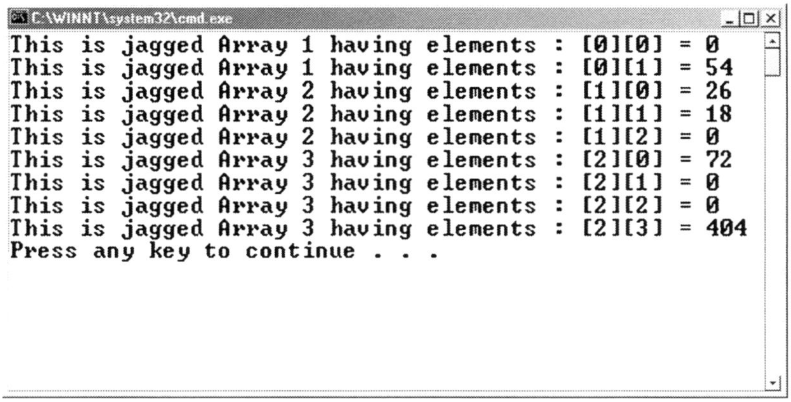

```
C:\WINNT\system32\cmd.exe
This is jagged Array 1 having elements : [0][0] = 0
This is jagged Array 1 having elements : [0][1] = 54
This is jagged Array 2 having elements : [1][0] = 26
This is jagged Array 2 having elements : [1][1] = 18
This is jagged Array 2 having elements : [1][2] = 0
This is jagged Array 3 having elements : [2][0] = 72
This is jagged Array 3 having elements : [2][1] = 0
This is jagged Array 3 having elements : [2][2] = 0
This is jagged Array 3 having elements : [2][3] = 404
Press any key to continue . . .
```

Case Study

Problem Statement

B.G International School is an educational institute that provides primary education to children from K.G class to 10th class. The administrative staff of B.G International School has to maintain the records, which include name, class and roll number of students studying in different classes. Manually maintaining the record of each student is a tedious process for the administrative staff of B.G International School. A large number of errors occur when the administrative staff enters the records manually in registers. How can B.G International School automate the task of entering student records?

Solution

In order to make the process of maintaining student records easy and efficient, B.G International School contacts SoftDev software development organisation to create an application which will help in the task of entering the student records. SoftDev assigns the task of creating the application for B.G International School to Siddharth, who is a senior programmer in the organisation. Siddharth decides to use the C# programming language for creating the application to enter student records. C# provides the **ArrayList** class in the **System.Collections** namespace. This **ArrayList** class allows us to create an array list, which can store array objects. The size of the array list is fixed at run time. Thus the array list created using **ArrayList** class is a dynamic array list. Because a large amount of data has to be entered in case of B.G International School, use of **ArrayList** will be helpful for Siddharth.

After understanding the requirements of B.G International School, Siddharth develops the following C# application.

```
using System;
using System.Collections;
using System.Collections.Generic;
using System.Text;

namespace AddStudentDetails
{
   class StudentDetails
    {


      static void Main(string[] args)
       {
        string studname1, studname2, studname3;
        string classname1, classname2, classname3;
        string rollno1, rollno2, rollno3;
        string getresult;
        ArrayList arrlist = new ArrayList();
        Console.WriteLine("The initial capacity of arraylist =" + arrlist.Capacity);
        Console.WriteLine("The initial elements are =" + arrlist.Count);

        Console.WriteLine("Enter Details of Student 1");
```

```
        Console.WriteLine("________________________________");
        Console.Write("Enter Name of Student1 : ");
        studname1 = Console.ReadLine();
        Console.Write("Enter Class of Student1 : ");
        classname1 = Console.ReadLine();
        Console.Write("Enter Roll No. of Student1 : ");
        rollno1 = Console.ReadLine();
        Console.WriteLine("________________________________");
        Console.Write("Enter Name of Student2 : ");
        studname2 = Console.ReadLine();
        Console.Write("Enter Class of Student2 : ");
        classname2 = Console.ReadLine();
        Console.Write("Enter Roll No. of Student2 : ");
        rollno2 = Console.ReadLine();
        Console.WriteLine("________________________________");
        Console.Write("Enter Name of Student3 : ");
        studname3 = Console.ReadLine();
        Console.Write("Enter Class of Student3 : ");
        classname3 = Console.ReadLine();
        Console.Write("Enter Roll No. of Student3 : ");
        rollno3 = Console.ReadLine();
        Console.WriteLine("________________________________");

        arrlist.Add(studname1);
        arrlist.Add(classname1);
        arrlist.Add(rollno1);
        arrlist.Add(studname2);
        arrlist.Add(classname2);
        arrlist.Add(rollno2);
        arrlist.Add(studname3);
        arrlist.Add(classname3);
        arrlist.Add(rollno3);

        Console.WriteLine("Current Capacity=" + arrlist.Capacity);
        Console.WriteLine("ArrayList having elements = " + arrlist.Count);
        Console.Write("ArrayList is having content:-");
        for (int i = 0; i < arrlist.Count; i++)
          {
           Console.Write(arrlist[i] + " ");

          }

        Console.WriteLine("\n");
        Console.WriteLine("To remove 2nd student details press y or to interchange
student details1 with student details3 press n.");
        getresult = Console.ReadLine();
        if (getresult == "y")
          {
```

```
            Console.WriteLine("Removing 2nd Student details");
            Console.WriteLine("_____________________________");
            arrlist.Remove(studname2);
            arrlist.Remove(classname2);
            arrlist.Remove(rollno2);
            Console.WriteLine("Current Capcity of the arraylist is now =" +
arrlist.Capacity);
            Console.WriteLine("Arraylist is having elements=" + arrlist.Count);
            Console.Write("The elements of Arraylist are now:-");
            for (int m = 0; m < arrlist.Count; m++)
                {
                 Console.Write(arrlist[m] + " ");
                }
            Console.WriteLine("\n");
            }
            else
            {
            Console.WriteLine("\n");
            Console.WriteLine("Changing elements");
            Console.WriteLine("_____________________________");
            arrlist[0] = studname3.ToString();
            arrlist[1] = classname3;
            arrlist[2] = rollno3;
            arrlist[6] = studname1;
            arrlist[7] = classname1;
            arrlist[8] = rollno1;

            Console.Write("Now the elements of the ArrayList are : ");
            for (int m = 0; m < arrlist.Count; m++)
               {
                Console.Write(arrlist[m] + " ");
               }
            Console.WriteLine("\n");

            }
        }

    }
}
```

Remarks

With the use of the C# application created by Siddharth, the administrative staff of B.G International School will be able to save time in entering the details such as names and classes of students studying in different classes.

Common Programming Errors

- *Forgetting to declare an array.*
- *Forgetting to initialize the elements of an array whose elements must be initialized.*
- *Referring an element outside the array bounds (e.g. Off-by-one error).*
- *Referencing the double subscripted array as x [ ] [ ] instead of x [,].*
- *Referencing a jagged array as x [,] instead of x[ ] [ ].*
- *Treating an **Array** type array as an **ArrayList** type array.*
- *Forgetting to use capital letters for the first characters of the methods and properties used form the namespaces.*
- *Forgetting the use of the **System.Collections** namespaces while using **ArrayList** class.*

Review Questions

9.1 State whether the following statements are true or false.
  (a) It is necessary to specify the size of an arry during its declaration.
  (b) In C#, arrays are created in memory using the operator **new**.
  (c) In C#, arrays start with the subscript zero.
  (d) Trying to access an array beyond its boundaries will cause a run time error.
  (e) C# arrays are created on the heap.
  (f) C# arrays cannot store reference type data.
  (g) A jagged array is declared using the same syntax as that of rectangular arrays.
  (h) C# permits the assignment of an array object to another.
  (i) An array of type **System.Array** is of fixed size.
  (j) An array of type **ArrayList** is of fixed size.
  (k) The default size of an **ArrayList** array is 16.
  (l) An array of type **ArrayList** can contain many different types of objects.
  (m) An array declaration creates memory space for the array.
  (n) It is an error if an initilizer list contains more values than the number elements in the array.

9.2 Explain the need for array variables.

9.3 What are the two ways to initialize an array of three values?

9.4 Are arrays reference or value types? And where are they created?

9.5 Identify errors, if any, in the following array declaration statements.
  (a) int score ( );
  (b) float values [ ];
  (c) float table [ ] [ ];
  (d) string name [5];
  (e) int [ ] m, n;
  (f) int [d] x;
  (g) int [ ] counter = int [5];

9.6 Identify errors, if any, in the following initialization statement.
  (a) int [ ] number = (0, 0, 0);
  (b) int [ ] x = new int [3] {1, 2, 3, 4};
  (c) list [3] = {1, 2, 3};

(d) int [ ] code = {1, 2, 3,;}
(e) int [,] table = {0, 0, 0}, {1, 1, 1};
(f) float [ ] MARKS = {5.75, 9.25, 6.0};

9.7 Assume that the arrays **a** and **b** are declared as follows:

```
int [,] a = new int [5,4];
float [ ] b = new float [4];
```

Find errors, if any, in the following program segments:

(a)
```
for (int i =1, i <=5; i++)
    for (int j = 1, j <=4; j++)
        a[ i, j ] = 0;
```
(b)
```
for (int i = 1; i < 4; i++)
    b [i] = 0.0F;
```
(c)
```
for (int i = 0; i <=4; i++)
    b [i] = b [i] + i;
```
(d)
```
for (int i = 4; i >=0, i - - )
    for (int j = 0; j < 4; j++)
        a[i,j] = b[j] + 1.0;
```

9.8 What is a variable size array? How is it different from a rectangular array?
9.9 Give an example of typical use of variable size arrays.
9.10 How does an array of **ArrayList** class differ from an array of **Array** class?
9.11 When do we decide to use the **ArrayList** class over the **Array** class?
9.12 State the class that contains each of the following methods and properties.

(a) Add( )
(b) Insert( )
(c) Length
(d) Remove( )
(e) Reverse( )
(f) Count
(g) Sort( )
(h) RemoeAt( )

Debugging Exercises

9.1 Debug the following program.

```
using System;
using System.Collections.Generic;
using System.Text;

namespace debugApp2_chap9
{
   class Program
    {
      int m=0;int n = 0;

     static void Main(string[] args)
```

```
        {
          const int rowValues = 7;
          const int columnValues = 4;

          int[] recArr =
         {
             {0,14,16,27}, {2,24,36,37}, {3,34,46,47}, {4,44,56,57}, {5,54,66,67},
{6,64,76,77}, {7,74,86,87}
         };
         Console.WriteLine("This is an example of rectangular Array.");
          for (m = 0; m < rowValues; m++)
            {
              for (n = 0; n < columnValues; n++)
                 {
                 Console.WriteLine("[{0},{1}] = {2}",m, n, recArr[]);
                 }
            }

        }
    }
}
```

9.2 Find errors in the following program that displays arrays.

```
using System;
using System.Collections.Generic;
using System.Text;

namespace debugApp1_chap9
{
   class Program
    {
      static void Main(string[] args)
        {
          Program prg = new Program();
          prg.Display(1, 2, 3, 4);
          int exArr[] = new int{"23","33","43","53"};
          Console.WriteLine("Example of array.");
          Console.WriteLine("New array elements added.");
          prg.Display(exArr[]);


        }
      public void Display(params int[] val1)
        {
          foreach (int i in val1)
            {
```

```
            Console.WriteLine("The array is having : {0}", i);
            }
        }
    }
}
```

Programming Exercises

9.1 Write an example of a numeric two-dimensional array and reverse them before printing.

9.2 Write a program that initializes an array of size 5 × 5 as follows:

1	0	1	0	1
0	1	0	1	0
1	0	1	0	1
0	1	0	1	0
1	0	1	0	1

9.3 Write a program for fitting a straight line through a set of points (x_i, y_i), $i = 1, \ldots n$. *The straight line equation is*

$$y = mx + c$$

and the values of m and c are given by

$$m = \frac{n\sum(xiyi) - (\sum xi)(\sum yi)}{n(\sum x_i^2) - (\sum x_i)^2}$$

$$c = \frac{1}{n}(\Sigma y_i - m\Sigma x_i)$$

All summations are from 1 to n.

9.4 The daily maximum temperature recorded in 10 cities during the month of January (for all 31 days) have been tablulated as follows:

Day \ City	1	2	3	10
1				
2				
3				
.	.			
.	.			
.	.			
31				

Write a program to read the table elements into a two-dimensional array temperature, and to find the city and day corresponding to (a) the highest temperature and (b) the lowest temperature.

9.5 *An election is contested by 5 candidates. The candidates are numbered 1 to 5 and the voting is done by marking the candidate number on the ballot paper. Write a program to read the ballots and count the votes cast for each candidate using an array variable count. In case, a number read is outside the range 1 to 5, the ballot should be considered as a 'spoilt ballot' and the program should also count the number of spoilt ballots.*

9.6 *The following set of numbers is popularly known as Pascal's triangle.*

1					
1	1				
1	2	1			
1	3	3	1		
1	4	6	4	1	
1	5	10	10	5	1
.	.	.	.	.	.
.	.	.	.	.	.

If we denote rows by i and columns by j, then any element (except the boundry elements) in the triangle is given by

$$P_{ij} = P_{i-1, j-1} + P_{i-1, j}$$

Write a program to calculate the elements of the Pascal triangle for 10 rows and print the results.

9.7 *The annual examination results of 100 students are tabulated as follows:*

Roll No.	Subject 1	Subject 2	Subject 3

Write a program to read the data and determine the following:

(a) Total marks obtained by each student.

(b) The highest marks in each subject and the Roll No. of the student who secured it.

(c) The student who obtained the highest total marks.

9.8 *Given are two one-dimensional arrays A and B which are sorted in ascending order. Write a program to merge them into a single sorted array C that contains every items from arrays A and B, in ascending order.*

9.9 *Two matrices that have the same number of rows and columns can be multiplied to produce a third matrix. Consider the following two matrices.*

$$A = \begin{bmatrix} a_{11} & a_{12} \ldots\ldots\ldots\ldots a_{1n} \\ a_{21} & a_{22} \ldots\ldots\ldots\ldots a_{2n} \\ \vdots & \\ a_{n1} \ldots\ldots\ldots\ldots\ldots & a_{nn} \end{bmatrix}$$

$$B = \begin{bmatrix} b_{11} & b_{12} \ldots\ldots\ldots\ldots b_{1n} \\ b_{21} & b_{22} \ldots\ldots\ldots\ldots b_{2n} \\ \vdots & \\ b_{n1} & \ldots\ldots\ldots\ldots\ldots b_{nn} \end{bmatrix}$$

The product of A and B is a third matrix C of size n × n where each element of C is given by the following equation.

$$c_{ij} = \sum_{K=1}^{n} a_{ik} b_{kj}$$

Write a program that will read the values of elements of A and B and produce the product matrix C.

9.10 *Write a program that accepts a shopping list of five items from the command line, stores them in string type array and then prints the list in alphabetical order.*

9.11 *Modify the program of Exercise 9.9 so that we may accomplish the following.*

- *To delete an item in the list.*
- *To add an item in the list at a specified location*
- *To add an item at the end of the list*
- *To print the contents of the list*

9.12 *Write a method that would take an array as input and check whether a given element is present in the list or not. The method must return true if it is present, otherwise return false.*

9.13 *Write a program that stores a list of numbers in an array and computes the maximum and minimum values in the list.*

9.14 *Write your own method reverse that would reverse the sequence of elements in an array. Test your method with a suitable program.*

9.15 *Write a method that would print the alternate elements in an array. Write a program using this method two separate list of alternate elements of a given list. For example, if the given list is say,*

1 2 3 4 5 6 7 8 9

then the program should print out two lists as follows:

1 3 5 7 9
2 4 6 8

9.16 *Write a method that would take two arrays as input and produces a third array by appending one to the other.*

Chapter

10

Manipulating Strings

10.1 INTRODUCTION

String manipulation is the most common part of many C# programs. Strings represent a sequence of characters. The easiest way to represent a sequence of characters in C# is by using a character array. For example:

```
char [ ] charArray = new char [4];
charArray [0] = 'N';
charArray [1] = 'a';
charArray [2] = 'm';
charArray [3] = 'e';
```

Although character arrays have the advantage of being able to query their length, by themselves they are not good enough to support the range of operations we may like to perform on strings. For example, copying one character array into another might require a lot of book-keeping effort. Fortunately, C# is equipped with features that could handle these situations more efficiently.

C# supports two types of strings, namely, *immutable* strings and *mutable* strings (Fig. 10.1). We shall consider in this chapter the most common operations performed on both these types of strings.

C# also supports a feature known as *regular expressions* that can be used for complex string manipulations and pattern matching. We shall also see how regular expressions are used for manipulating strings.

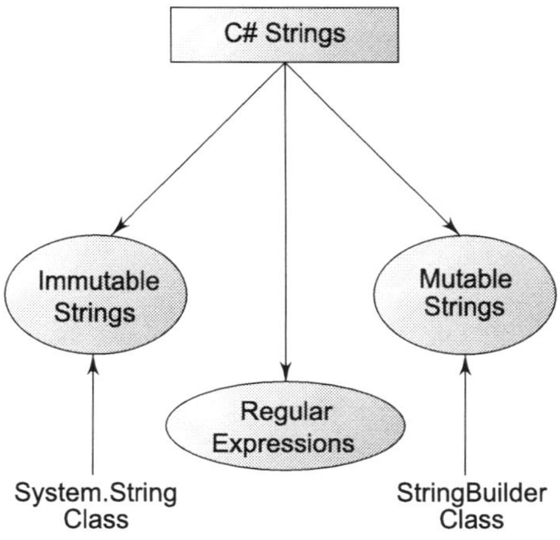

Figure 10.1 *Types of C# strings*

10.2 CREATING STRINGS

C# supports a predefined reference type known as **string**. We can use **string** to declare **string** type objects. Remember, when we declare a string using **string** type, we are in fact declaring the object to be of type **System.String**, one of the built-in types provided by the .NET Framework. A string type is therefore a **System.String** type as well.

We can create immutable strings using **string** or **String** objects in a number of ways.

- Assigning string literals
- Copying from one object to another
- Concatenating two objects
- Reading from the keyboard
- Using **ToString** method

10.2.1 Assigning String Literals

The most common way to create a string is to assign a quoted string of characters known as *string literal* to a string object. For example:

```
string s1; //declaring a string object
s1 = "abc"; //assigning string literal
```

Both these statements may be combined into one as follows:

```
string s1 = "abc"; //declaring and assigning
```

10.2.2 Copying Strings

We can also create new copies of existing strings. This can be accomplished in two ways:

- Using the overloaded = operator
- Using the static **Copy** method

Example:

```
string s2 = s1;                //assigning
string s2 = string.Copy(s1);   //copying
```

Both these statements would accomplish the same thing, namely, copying the contents of **s1** into **s2.**

10.2.3 Concatenating Strings

We may also create new strings by concatenating existing strings. There are a couple of ways to accomplish this.

- Using the overloaded + operator
- Using the static **Concat** method

Examples:

```
string s3 = s1 + s2; //s1 and s2 exist already
string s3 = string.Concat(s1, s2)
```

If s1 = 'abc' and s2 = 'xyz', then both the statements will store the string 'abcxyz' in s3. Note that the contents of s2 is simply appended to the contents of s2 and the result is stored in s3.

10.2.4 Reading from the Keyboard

It is possible to read a string value interactively from the keyboard and assign it to a string object.

```
string s = Console.ReadLine( );
```

On reaching this statement, the computer will wait for a string of characters to be entered from the keyboard. When the 'return key' is pressed, the string will be read and assigned to the string object **s.**

10.2.5 The ToString Method

Another way of creating a string is to call the **ToString** method on an object and assign the result to a string variable.

```
int number = 123;
string numStr = number.ToString( );
```

This statement converts the number 123 to a string '123' and then assigns the string value to the string variable **numStr.**

10.2.6 Verbatim Strings

Strings can also be created using what are known as v*erbatim strings.* Verbatim strings are those that start with the @ symbol. This symbol tells the compiler that the string should be used verbatim even if it includes escape characters.

```
string s1 = @ "\EBG\Csharp\string.cs";
```

In order to obtain the same output without using the symbol @, the input string should be written as follows:

```
string s1 = "\\EBG\\Csharp\\string.cs";
```

We may also use the escape characters such as \n and \t in @ - quoted strings. They will not be processed during output and appear in the text as such.

Note that if an ordinary string contains an embedded \n, the string that follows \n will be displayed on the next line when the string is processed for output.

10.3 STRING METHODS

String objects are ***immutable*,** meaning that we cannot modify the characters contained in them. However, since the string is an alias for the predefined **System.String** class in the Common Language Runtime (CLR), there are many built-in operations available that work with strings. All operations produce a modified version of the string rather than modifying the string on which the method is called.

Table 10.1 lists various methods that could be used for various operations such as comparison, appending, inserting, conversion, copying, indexing, joining, splitting, padding, trimming, removing, replacing and searching. We have already used some methods like **Concat** and **Copy** in the previous section.

Table 10.1 *String class methods*

Method	Operation
Compare ( )	Compares two strings
CompareTo ( )	Compares the current instance with another instance
ConCat ( )	Concatenates two or more strings
Copy( )	Creates a new string by copying another
CopyTo ( )	Copies a specified number of characters to an array of Unicode characters
EndsWith ( )	Determines whether a substring exists at the end of the string
Equals( )	Determines if two strings are equal
IndexOf ( )	Returns the position of the first occurrence of a substring
Insert ( )	Returns a new string with a substring inserted at a specified location
Join ( )	Joins an array of strings together
LastIndexOf ( )	Returns the position of the last occurrence of a substring
PadLeft ( )	Left-aligns the strings in a field
PadRight ( )	Right-aligns the string in a field
Remove ( )	Deletes characters from the string
Replace ( )	Replaces all instances of a character with a new character
Split ( )	Creates an array of strings by splitting the string at any occurrence of one or more characters
StartsWith ( )	Determines whether a substring exists at the beginning of the string
Substring ( )	Extracts a substring
ToLower ( )	Returns a lower-case version of the string
ToUpper ( )	Returns an upper-case version of the string
Trim ( )	Removes white space from the string
TrimEnd ( )	Removes a string of characters from the end of the string
TrimStart ( )	Removes a string of characters from the beginning of the string

10.4 INSERTING STRINGS

String methods are called using the string object on which we want to work. Program 10.1 illustrates the use of the **Insert ( )** method and the indexer property supported by the **System.String** class.

Program 10.1 Using the insert ( ) method

```
using System;
class StringMethod
{
      public static void Main( )
      {
         string s1 = "Lean";
      string s2 = s1.Insert (3,"r");          // s2 = Learn
      string s3 = s2.Insert (5,"er");         // s3 = Learner
         for (int i = 0; i < s3.Length; i++)
            Console.Write(s3[i]);
         Console.Writeline( );
      }
}
```

When the statement

```
string s2 = s1.Insert(3, "r");
```

is executed, the string variable s2 contains the string "Learn". The string "r" is inserted in s1 after 3 characters. Similarly, the string "er" is inserted at the end of the string. Finally, the variable s3 contains the value "Learner".

Note that we are not modifying the contents of a given string variable. Rather, we are assigning the modified value to a new string variable. For instance:

```
s1 = s1.Insert(3, "r");
```

is illegal. We are trying to modify the immutable string object s1.

Program 10.2 takes two inputs as string from the users and copies the input of string2 to string5 and checks for the string3 ends with 'IDE' or not. The program shows true if string3 ends with 'IDE'. Searches char 'a' from the string1. Inserts 'hello' in string6 at position 6 and shows the string6.

Program 10.2 Copying, searching and inserting strings

```
using System;
using System.Collections.Generic;
using System.Text;

namespace SearchString
{
   class Program
   {
      public void Display()
      {
```

```
            string str1 = "";
            Console.Write("Enter a string : ");
            str1 = Console.ReadLine();

            string str2 = "";
            Console.Write("Enter another string string : ");
            str2 = Console.ReadLine();

            string str3 = "C# 2005 is developed in Visual Studio 2005 IDE";
            Console.WriteLine("String str3 is : {0}", str3);
            // the string copy method
            string str5 = string.Copy(str2);
            Console.WriteLine("String str5 is copied from str2: {0}", str5);

            Console.WriteLine("\nString str5 is {0} characters long. ",
str5.Length);

            Console.WriteLine("The 10th character of string str3 is : {0}",
str3[9]);

            // check if a string ends with a set of characters
            Console.WriteLine("String str3:{0}\nEnds with IDE?: {1}\n",
            str3,
            str3.EndsWith("IDE"));
            Console.WriteLine("Ends with Studio?: {0}",
            str3.EndsWith("Studio"));

            // return the index of the substring
            Console.WriteLine("\nThe first time character 'a' occured in
string str1 at position : {0}", str1.IndexOf("a")+1);

            string str6 = str2.Insert(6, "hello");
            Console.WriteLine("'hello' is inserted in string str6. String s6
is now : {0}\n", str6);

        }
        static void Main(string[] args)
        {
            Program prg = new Program();
            prg.Display();
        }
    }
}
```

```
C:\WINNT\system32\cmd.exe
Enter a string : hello sumantto
Enter another string string : India is a great country
String str3 is : C# 2005 is developed in Visual Studio 2005 IDE
String str5 is copied from str2: India is a great country

String str5 is 24 characters long.
The 10th character of string str3 is : s
String str3:C# 2005 is developed in Visual Studio 2005 IDE
Ends with IDE?: True

Ends with Studio?: False

The first time character 'a' occured in string str1 at position : 10
'hello' is inserted in string str6. String s6 is now : India hellois a great cou
ntry

Press any key to continue . . .
```

10.5 COMPARING STRINGS

String class supports overloaded methods and operators to compare whether two strings are equal or not. They are:

- Overloaded **Compare( )** method
- Overloaded **Equals( )** method
- Overloaded == operator

10.5.1 Compare Method

There are two versions of overloaded static **Compare** method. The first one takes two strings as parameters and compares them. *Example:*

```
int n = string.Compare(s1,s2);
```

This performs a case-sensitive comparison and returns different integer values for different conditions as under:

- Zero integer, if s1 is equal to s2
- A positive integer (1), if s1 is greater than s2
- A negative integer (–1), if s1 is less than s2

For example, if s1 = "abc" and s2 = "ABC", then n will be assigned a value of –1. Remember, a lowercase letter has a smaller ASCII value than an uppercase letter.

We can use such comparison statements in **if** statements like:

```
if ( string.Compare(s1, s2) == 0)
     Console.WriteLine("They are equal");
```

The second version of **Compare** takes an additional **bool** type parameter to decide whether case should be ignored or not. If the bool parameter is **true,** case is ignored. *Example:*

```
int n = string.Compare(s1, s2, true);
```

This statement compares the strings s1 and s2 ignoring the case and therefore returns zero when s1 = "abc" and s2 = "ABC", meaning they are equal.

10.5.2 Equals Method

The **string** class supports an overloaded **Equals** method for testing the equality of strings. There are again two versions of **Equals** method. They are implemented as follows:

bool b1 = s2.Equals(s1);

bool b2 = string.Equals (s2, s1);

These methods return a Boolean value **true** if s1 and s2 are equal, otherwise **false**.

10.5.3 The == Operator

A simple and natural way of testing the equality of strings is by using the overloaded = = operator. *Example:*

```
bool b3 = (s1 == s2); //b3 is true if they are equal
```

We very often use such statements in decision statements, like:

```
if (s1 == s2)
    Console.WriteLine("They are qual");
```

10.6 FINDING SUBSTRINGS

It is possible to extract substrings from a given string using the overloaded **Substring** method available in **String** class. There are two version of **Substring:**

- `s.Substring(n)`
- `s.Substring(n1, n2)`

The first one extracts a substring starting from the nth position to the last character of the string contained in **s**. The second one extracts a substring from **s** beginning at n1 position and ending at n2 position. *Examples:*

```
string s1 = "NEW YORK";
string s2 = s1.Substring(5);
string s3 = s1.Substring(0,3);
string s4 = s1.Substring(5,8);
```

When executed, the string variables will contain the following substrings:

```
s2:   YORK
s3:   NEW
s4:   YORK
```

10.7 MUTABLE STRINGS

Mutable strings that are modifiable can be created using the **StringBuilder** class. *Examples:*

```
  StringBuilder str1 = new StringBuilder("abc");
v StringBuilder str2 = new StringBuilder ( );
```

The string object **str1** is created with an initial size of three characters and **str2** is created as an empty string. They can grow dynamically as more characters are added to them. They can grow either unbounded or up to a configurable maximum. Mutable strings are also known as *dynamic strings.*

The **StringBuilder** class supports many methods that are useful for manipulating dynamic strings. Table 10.2 lists some of the most commonly used methods.

Table 10.2 *Some useful stringBuilder methods*

Method	Operation
Append ( )	Appends a string
AppendFormat ( )	Appends strings using a specific format
EnsureCapacity ( )	Ensures sufficient size
Insert ( )	Inserts a string at a specified position
Remove ( )	Removes the specified characters
Replace ( )	Replaces all instances of a character with a specified one

C# also supports some special functions known as *properties*. They are listed in Table 10.3

Table 10.3 *StringBuilder properties*

Property	Purpose
Capacity	To retrieve or set the number of characters the object can hold
Length	To retrieve or set the length
MaxCapacity	To retrieve the maximum capacity of the object
[ ]	To get or set a character at a specified position

Program 10.3 shows how some of the methods are used for manipulating strings. The program implements the following methods and properties:

- Length
- Indexer [ ]
- Append ( )
- Insert ( )

The **System.Text** namespace contains the **StringBuilder** class and therefore we must include the **using System.Text** directive for creating and manipulating mutable strings.

Program 10.3 Using stringbuilder methods

```
using System.Text;    //For using StringBuilder
using System;
class StringBuilderMethod
{
    public static void Main( )
    {
        StringBuilder s = new StringBuilder ("Object ");
        Console.WriteLine("Original string   : " + s);
        Console.WriteLine("Length            : " + s.Length);

        //Appending a string
        s.Append("language ");
        Console.WriteLine("String now        : " + s);

        //Inserting a string
        s.Insert (7,"oriented ");
        Console.WriteLine("Modified string   : " + s);

        //Setting a character
        int n = s.Length;
        s[n-1] = '!' ;
        Console.WriteLine("Final string      : " + s);
    }
}
```

Look at the output produced by Program 10.3:

```
Original string  :  Object
Length           :  7
String now       :  Object language
Modified string  :  Object oriented language
Final string     :  Object oriented language!
```

Program 10.4 accepts two string inputs from users and appends the first string to a predefined value. Using **StringBuilder** class, a string value is inserted in the string. Then, all spaces in first string are replaced with * and the program calculates the length of the two strings after appending.

Program 10.4 Using stringbuilder class

```
using System;
using System.Collections.Generic;
using System.Text;

namespace StringExample
{
   class Program
   {
       public void Show()
       {
          string str1="";
          string str2="";
          Console.Write("Please enter a string : ");
          str1 = Console.ReadLine();

          Console.Write("Please enter another string : ");
          str2 = Console.ReadLine();

          StringBuilder builder1 = new StringBuilder(str1, 4);
          StringBuilder builder2 = new StringBuilder(str2, 4);
          int cap = builder1.EnsureCapacity(55);
          Console.Write("string str1 appended to : ");
          builder1.Append(". This is a class test.");
          Console.WriteLine(builder1);

          Console.Write("string str2 inserted with : ");
          builder2.Insert(2, " String Builder");
          Console.WriteLine(builder2);

          Console.WriteLine("\nSecond character of string2 removed.\n");
          builder2.Remove(2, 4);
          Console.WriteLine(builder2);

          builder1.Replace(' ', '*');
          Console.WriteLine("\nSpaces removed from string1\n");
          Console.WriteLine(builder1);

          Console.WriteLine("\nLength of string1 is:" + builder1.Length.ToString());
          Console.WriteLine("\nLength of string2 is:" +
builder2.Length.ToString());
       }
       static void Main(string[] args)
       {
          Program prg = new Program();
          prg.Show();
       }
   }
}
```

```
C:\WINNT\system32\cmd.exe
Enter a small string : TEST
Enter a small string : test
s5 is having the value of string str2: test

String str5 is 4 characters long.
The value of 5th character is 0

str3:C# 2005 is one of the core Languages of .NET. It provides a custom .NET dev
elopment platform for developing applications.
Ends with development?: False

Ends with applications?: False

The first occurrence of development
in str3 is at position 72

str10: C# 2005 is one of the core Languages of .NET. It provides a custom .NET d
nice evelopment platform for developing applications.

str11: C# 2005 is one of the core Languages of .NET. It provides a custom .NET d
evelopment platform for developing nice applications.

Press any key to continue . . .
```

10.8 ARRAYS OF STRINGS

We can also create and use arrays that contain strings. The statement

```
string [ ] itemArray = new string [3];
```

will create an **itemArray** of size 3 to hold three strings. We can assign the strings to the **itemArray** element by element using three different statements, or more efficiently using a **for** loop. We could also provide an array with a list of initial values in curly braces:

```
string [ ] itemArray = {"Java", "C++", "Csharp"};
```

The size of the array is determined by the number of elements in the initialization list. The size of the array, once created, cannot be changed.

If we want an array whose length is determined dynamically or an array which can be extended at run time, we have to use the **ArrayList** class to create a list.

Once an array of strings is created, we can sort them into ascending order or reverse their order using the methods of **Array** class. Program 10.5 stores a list of countries in a string array, sorts the array into reverse alphabetical order, and then prints the list.

Program 10.5 Manipulating string arrays

```
using System;
class Strings
{
   public static void Main( )
   {
```

```
            string[ ]countries = {"India", "Germany",
                                              "America","France"};
         int n = countries.Length;

         //Sort alphabetically
         Array.Sort(countries);
         for (int i = 0; i < n; i++)
         {
            Console.WriteLine(countries[i]);
         }
         Console.WriteLine( );

         //Reverse the array elements
         Array.Reverse(countries);
         For (int i = 0; i < n; i++)
         {
            Console.WriteLine(countries[i]);
         }
      }
   }
```

Program 10.6 takes two inputs as string from the users and copies the input of string2 to string5 by replacing its value and checks for the length of string5, checks for a word 'development' in string3 and shows its position. It inserts 'nice' in string10 at position 73 and then places 'nice' word after 'applications word in string11.

Program 10.6 Working with strings

```
using System;
using System.Collections.Generic;
using System.Text;

namespace WorkingwithStrings
{
   class Program
   {
      public void Display()
      {
         string str1 = "";
         Console.Write("Enter a small string : ");
         str1 = Console.ReadLine();

         string str2 = "";
         Console.Write("Enter a small string : ");
         str2 = Console.ReadLine();
```

```
        string str3 = @"C# 2005 is one of the core Languages of .NET. It
provides a custom .NET development platform for developing applications.";

        // the string copy method
        string s5 = string.Copy(str2);
        Console.WriteLine(
        "s5 is having the value of string str2: {0}", s5);

        // Two useful properties: the index and the length
        Console.WriteLine(
        "\nString str5 is {0} characters long. ",
        s5.Length);

        Console.WriteLine(
        "The value of 5th character is {0}\n", str3[4]);

        // test whether a string ends with a set of characters
        Console.WriteLine("str3:{0}\nEnds with development?: {1}\n",
        str3,
        str3.EndsWith("platform"));
        Console.WriteLine(
        "Ends with applications?: {0}",
        str3.EndsWith("applications"));

        // return the index of the substring
        Console.WriteLine(
        "\nThe first occurrence of development");
        Console.WriteLine("in str3 is at position {0}\n",
        str3.IndexOf("development"));

        // insert the word nice in development"
        string str10 = str3.Insert(73, "nice ");
        Console.WriteLine("str10: {0}\n", str10);

        // Combining the two as follows:
        string str11 = str3.Insert(str3.IndexOf("applications"),
        "nice ");
        Console.WriteLine("str11: {0}\n", str11);
        }

    static void Main(string[] args)
    {
        Program prg = new Program();
        prg.Display();
    }
  }
}
```

```
C:\WINNT\system32\cmd.exe
Please enter a string : test1
Please enter another string : string
string str1 appended to : test1. This is a class test.
string str2 inserted with : st String Builderring

4 characters from second character position of string2 removed.

sting Builderring

Spaces removed from string1

test1.*This*is*a*class*test.

Length of string1 is:28

Length of string2 is:17
Press any key to continue . . .
```

10.9 REGULAR EXPRESSIONS

Regular expressions provide a powerful tool for searching and manipulating a large text. A regular expression may be applied to a text to accomplish tasks such as:

- To locate substrings and return them
- To modify one or more substrings and return them
- To identify substrings that begin with or end with a pattern of characters
- To find all words that begin with a group of characters and end with some other characters
- To find all the occurrences of a substring pattern, and
- Many more

A detailed discussion of regular expressions is beyond the scope of this book. We give here a simple example of using regular expressions.

A regular expression (also known as a *pattern string)* is a string containing two types of characters.

- Literals
- Metacharacters

Literals are characters that we wish to search and match in the text. *Metacharacters* are special characters that give commands to the regular expression parser. Examples of regular expressions are:

Expression	Meaning
"\bm"	Any word beginning with m
"er\b"	Any word ending with er.
"\BX\B"	Any X in the middle of a word
"\bm\S*er\b"	Any word beginning with m and ending with er.
" |,"	Any word separated by a space or a comma

In these expressions, \b, \B, \S* and | are metacharacters and m, er, X, space and comma are literals.

The .NET Framework provides support for regular expression matching and replacement. The namespace **System.Text.RegularExpressions** supports a number of classes that can be used for searching, matching and modifying a text document. The important classes are:

- Regex
- MatchCollection
- Match

These classes provide a number of useful overloaded instance and static methods that could be exploited for various situations. Program 10.7 illustrates the use of **Regex** class and regular expressions for splitting a string into separate words.

Program 10.7 Use of regular expressions

```
using System;
using System.Text;    //for StringBuilder class
using System.Text.RegularExpressions;  //for Regex class

Class RegexTest

{
  public static void Main ( )
  {

     string str;
     str = "Amar, Akbar, Antony are friends!";

     Regex reg = new Regex (" |, ");
     StringBuilder sb = new StringBuilder( );
     int count = 1;

  foreach(string sub in reg.Split(str))
  {
     sb.AppendFormat("{0}: {1}\n", count++, sub);
  }
  Console.WriteLine(sb);
}
}
```

Program 10.7 would produce the following output:

```
1: Amar
2: Akbar
3: Antony
4: are
5: friends!
```

The program creates a regular expression object **reg** using the class **Regex** and stores the regular expression

" |,"

in it. This pattern is used to search the string

"Amar, Akbar, Antony are friends!"

with the help of **Split( )** method of **Regex** class. The call

```
reg.Split(str)
```

in the **foreach** statement splits the string wherever a comma or space appears and returns the substring to **sub**. These substrings are appended to the **StringBuilder** string **sb** using the **AppendFormat** method of **StringBuilder.** The output shows that the given string has been split into separate words using the separators 'space' and 'comma'.

Case Study

Problem Statement

QTest is an advertising company that is approached by big organisations such as ABCL and SoftCom to conduct quizes for advertising their products. For conducting quizes, QTest first asks different individuals to register by specifying details such as their name and surname. After individuals get registered, QTest selects some individuals from a group of individuals and invites them to participate in the quiz. The task of obtaining details from different individuals is so far performed manually by a group of QTest staff. However, manually performing the task of obtaining details from different individuals is time consuming. QTest has to employ additional staff for performing the task of getting details from individuals. As a result, QTest is facing financial loss in the advertisement business. How can QTest automate the task of receiving details such as the name and surname from individuals for registration?

Solution

QTest needs to use an application created in a programming language such as C# to automate the task of obtaining details from individuals for registration. After having decided that it needs an application for automating the task, QTest contacts ADev Software Solutions Private Limited, which is a small software development company, to create the application for automating the task of obtaining details from individuals. After an analysis of the requirement, ADev creates a C# application using the string methods such as **Insert** and **Append** available in C#. These string methods allow us to perform different manipulation tasks such as appending a string to another string and inserting a string. The following C# application is ultimately created by ADev Software Solutions for QTest.

```
using System;
using System.Collections.Generic;
using System.Text;
```

```
namespace AboutStrings
{
    class Program
    {
        static void Main(string[] args)
        {
            string str1 = "", str2, str3, str4;
            Console.WriteLine("This is an example of how strings can be
manipulated/validated in C#.");
Console.WriteLine("==============================================================");
            int n = 0;

            Console.Write("Please enter your Name : ");
            str1 = Console.ReadLine();

            if (str1.Length > 3)
            {

                if (string.IsNullOrEmpty(str1))
                {
                    Console.WriteLine("string cannot be empty");
                }

                else if (char.IsNumber(str1, 0) || char.IsNumber(str1, 1) ||
char.IsNumber(str1, 2) || char.IsNumber(str1, 3))
                {
                    Console.WriteLine("Name is a number type");
                }
                else if (char.IsLower(str1, 0))
                {
                    Console.WriteLine("First character of the Name should be
capital.");
                }
                else if (char.IsUpper(str1, 1) || char.IsUpper(str1, 2) ||
char.IsUpper(str1, 3))
                {
                    Console.WriteLine("Only the first character should be in
Capital.");
                }
                else
                {
                Console.WriteLine("Correct Name inserted.");
                Console.Write("Enter your surname : ");
                str2 = Console.ReadLine();
```

```
                if (str2.Length > 3)
                {

                   if (string.IsNullOrEmpty(str2))
                   {
                      Console.WriteLine("Surname cannot be empty");
                   }

                   else if (char.IsNumber(str2, 0) || char.IsNumber(str2,
1) || char.IsNumber(str2, 2) || char.IsNumber (str2, 3))
                   {
                      Console.WriteLine("Surname is a number type");
                   }
                   else if (char.IsLower(str2, 0))
                   {
                      Console.WriteLine("First character of the Surname
should be capital.");
                   }
                   else if (char.IsUpper(str2, 1) || char.IsUpper(str2, 2)
|| char.IsUpper(str2, 3))
                   {
                      Console.WriteLine("Only the first character of
Surname should be in Capital.");
                   }
                   else
                   {
                      str3 = str1 + " " + str2;
                      Console.WriteLine("You have entered your name as :"
+ str3);
                      if (str3.Contains("o"))
                      {
                         Console.WriteLine("Your name contains 'o'
character");
                      }
                      else
                      {
                         Console.WriteLine("Your name does not contain 'o'
character");
                      }
                      n = string.Compare(str1, str2);
                      if (n == 0)
                      {
                         Console.WriteLine("Your firstname and surname is
equal in all respect.");
```

```
                    }
                    else if (n > 0)
                    {
                        Console.WriteLine("Your firstname is greater than
your surname");
                    }
                    else
                    {
                        Console.WriteLine("Your surname is greater than
your firstname");
                    }

                    StringBuilder sb1 = new StringBuilder("Sir/Madam ");

                    Console.WriteLine("Added before your name : " +
sb1.Append(str3));

                    Console.WriteLine("Added in your Name: " +
sb1.Insert(10, " Mr./Ms. "));

                    str4 = string.Copy(str3);
                    Console.WriteLine("String str4 is now having your
full Name : " + str4);
                    Console.WriteLine("Removing Sir/Madam from your
current Name : " + sb1.Remove(0, 9));
                    Console.WriteLine("Replacing Mr. with Mister we
have: " + sb1.Replace("Mr.", "Mister"));

                }// End of checks for str2
            }
            else
            {
                Console.WriteLine("Surname should be at least 4
characters.");
            }//End for check of str2 length is at least 4

        }//End for check of str1 after proper output
    }
    else
    {
        Console.WriteLine("Name should be at least 4 characters.");
    }//End for check of str1 length is at least 4
  }
 }
}
```

Remarks

C# supports the String class, which contains methods such as Compare() and Insert() that help in comparing two strings and inserting a string. String manipulation is important for performing tasks such as removing and replacing a string in a group of string.

Common Programming Errors

- *Attempting to modify a basic string type object.*
- *Forgetting to use uppercase characters in method names as defined in Framework Class Library.*
- *Attempting to invoke a static method using a class instance.*
- *Attempting to call an instance method using its class name.*
- *Improper choice of overloaded methods.*
- *Forgetting to use appropriate namespaces.*
- *Improper choice of metacharacters in regular expressions.*
- *Using StringBuilder methods that are not members of String class to manipulate string objects.*
- *Attempting to access a character that is outside the bounds of a string.*

Review Questions

10.1 Distinguish between String and StringBuilder classes.

10.2 Distinguish between a string object and string literal.

10.3 What is concatenation? How is it achieved in C#?

10.4 What is the typical use of the ToString method?

10.5 How would you use a ReadLine method to create a string?

10.6 What is a verbatim string? Where and when do you use it?

10.7 What is an immutable string? How can we modify such a string?

10.8 Compare the functioning of two versions of the Compare method. Give typical use of these methods.

10.9 How do you invoke the overloaded Equals method?

10.10 String class supports two Substring methods. How do they differ? Give an example where both the methods will return the same substring.

10.11 When do you use the System.Text namespace?

10.12 How would you create an array of strings? Give an example of its typical use.

10.13 How would you create a mutable string? How does it differ from creating an immutable string?

10.14 What does an Append method achieve?

10.15 What is AppendFormat method? How is it different from the Append method?

10.16 What are regular expressions? When do you use them?

10.17 Lists atleast three properties provided by the StringBuilder class. State their typical use in string manipulations.

10.18 Find errors, if any, in the following statements:

(a) string s1 = (“abc + xyz”);

(b) String s2 = s1 + "abc";
(c) string s1.Copy ("abc");
(d) string s3 = "@\ABC\n\Csharp;"

10.19 State what the following statements would accomplish?

(a) int n = s1.Length;
(b) int m = s1.IndexOf ("are");
(c) string s2 = s1.Insert (s1.IndexOf("OK"), "not");
(d) int n = string.Compare (s1, s2, false);
(e) Console.WriteLine (s1[5]);

10.20 State whether the following statements are true or false:

(a) All methods of System.String class can work on the objects of string type.
(b) The ToString method is used to convert a System.String object to a string type string.
(c) Verbatim strings may contain escape characters.
(d) All the methods of StringBuilder class can work on the objects of System.String class
(e) We can obtain the length of string s1 using the expression s1.length( ).

Debugging Exercises

10.1. Correct the following program which manipulates strings.

```
using System;
using System.Collections.Generic;
using System.Text;

namespace debugAppl_chap10
{
   class Program
   {
       public void Display()
           {
           // create some strings to work with
           string strl = "";
           Console.Write("Please enter a string : ");
           strl = Console.ReadLine();

           // constants for the space and comma characters
           const char Space = ' ';
           const char Comma = ',';

           // array of delimiters to split the sentence with
           char[] spacers = new char()
           {
              Space;
```

```
            Comma;
        };

        // use a StringBuilder class to build the output string
        StringBuilder sb = new StringBuilder();
        int increment = 1;

        // splitting the string
        Console.WriteLine("The whole string is displayed below.");
        foreach (string subString in str1.Split(spacers))
        {
            // AppendFormat appends a formatted string

            sb.AppendFormat("{0}: word {1} is at position {0}\n",
+increment++, +subString);
        }
        Console.WriteLine(sb);

    }

    static void Main(string[] args)
    {
        Program prg = new Program();
        prg.Display();

    }
  }
}
```

10.2. Find errors in the following program for inserting and copying strings.

```
using System;
using System.Collections.Generic;
using System.Text;

namespace debugApp2_chap10
{
  class Program
  {
    public void Display()
    {
        string str1 = "";
        Console.Write("Please enter a string : ");
```

```
            str1 = Console.ReadLine();

            string str2 = "";
            Console.Write("Please enter another string : ");
            str2 = Console.ReadLine();

            Console.WriteLine(" string str1: {0}", str1);
            Console.WriteLine(" inserting value of string str1 in string str2:
");
            str2 == str1;
            Console.WriteLine("str1: {0} \nstr2: {1}", str1, str2);
            Console.WriteLine("str1 == str2? {0}", str1 == str2);

            Console.WriteLine(" \nstring str2 = string.Copy( str1 ); ");
            Console.WriteLine(" Copying string str1 in string str3");
            str3 = string(copy(str1,str2));
            Console.WriteLine("str1: {0} \nstr3: {1}", str1, str3);
            Console.WriteLine("str1 == str3? {0}", str1 = str3);

            Console.WriteLine(" \nstr2 = \"New string\"; ");
            str1 = "New string";
            Console.WriteLine("str1: {0} \nstr2: {1}", str1, str2);
            Console.WriteLine("str1 == str2? {0}", str1 == str2);

        }

        static void Main(string[] args)
        {
            Program prg = new Program();
            prg.Display();

        }
    }
}
```

10.3. Debug the program given below.

```
using System;
using System.Collections.Generic;
using System.Text;
using System.Text.RegularExpressions;

namespace debugApp3_chap10
{
```

```
class Program
{
    static void Main(string[] args)
    {
        Program prg = new Program();
        Show();

    }

    public void Show()
    {
        string str1 = "";
        Console.Write("Please enter a string : ");
        str1 = Console.ReadLine();

        Regex reg = new Regex(" |, |.");
        StringBuilder sb = new StringBuilder();
        int i = 1;

        for(string subString as reg.Split(str1))
        {
            sb.Appendformat("{0}: {1}\n", i++, subString);
        }
        Console.WriteLine("{0}", +sb);
    }

  }
}
```

Programming Exercises

10.1 Write a program that takes a string from the command line input and takes another input of a character from the user and searches if the character exists in the string. If yes, counts the no. of times the characters occurred in the string.

10.2 Write a program to demonstrate the working of Copy and Concat methods.

10.3 Write a program that would illustrates the use of ReadLine method to read a string interactively.

10.4 Write a program that reads the name
INDIRA GANDHI
into two separate string objects and then concatenates them into a new string using:
(a) + operator
(b) Append method

10.5 Write a program using while construct and indexer property to display the contents of a string object.

10.6 Write a program that reads a line of text containing three words and then replaces all the blank spaces with an undercore (_)

10.7 Write a program that will read a string and rewrite it in alphabetical order. For example, the word STRING should be written an GINRST.

10.8 Write a program that counts the number of occurrences of a particular character in a line of text.

10.9 Write a program that counts the number of words in a text document.

10.10 Write a program that reads a list of names in random order using an array and displays them in alphabetical order.

(a) using Compare method

(b) using Equals method

10.11 Store a string "123456789" in a string variable and use it to display the following pattern.

```
1
2 3 2
3 4 5 4 3
4 5 6 7 6 5 4
5 6 7 8 9 8 7 6 5
. . .
```

Chapter

11

Structures and Enumerations

11.1 INTRODUCTION

We have already seen that C# supports two kinds of value types, namely, *predefined* types and *user-defined* types. We have defined and used the predefined data types such as **int** and **double** throughout our earlier discussions. C# allows us to define our own complex value types (known as user-defined value types) based on these simple data types. There are two sorts of value types we can define in C#:

- Structures
- Enumerations

As we know, value type variables store their data on the stack and therefore the structures and enumerations are stored on the stack.

We shall discuss here briefly how these data types are defined and used in programming.

11.2 STRUCTURES

Structures (often referred to as *structs*) are similar to classes in C#. Although classes will be used to implement most objects, it is desirable to use structs where simple composite data types are required. Because they are value types stored on the stack, they have the following advantages compared to class objects stored on the heap:

- They are created much more quickly than heap-allocated types.
- They are instantly and automatically deallocated once they go out of scope.
- It is easy to copy value type variables on the stack.

The performance of programs may be enhanced by judicious use of structs.

11.2.1 Defining a Struct

A struct in C# provides a unique way of packing together data of different types. It is a convenient tool for handling a group of logically related data items. It creates a *template* that may be used to define its data properties. Once the structure type has been defined, we can create variables of that type using declarations that are similar to the built-in type declarations.

Structs are declared using the **struct** keyword. The simple form of a struct definition is as follows:

```
struct struct-name
{
    data member1;
    data member2;
    . . .
    . . .
}
```

Example:

```
struct Student
{
    public string Name;
    public int RollNumber;
    public double TotalMarks;
}
```

The keyword **struct** declares **Student** as a new data type that can hold three variables of different data Types. These variables are known as *members* or *fields* or *elements*. The identifier **Student** can now be used to create variables of type **Student**. *Example:*

```
Student s1 ; //declare a student
```

s1 is a variable of type **Student** and has three member variables as defined by the template.

11.2.2 Assigning Values to Members

Member variables can be accessed using the simple dot notation as follows:

```
s1.Name       =  "John";
s1.RollNumber =  999 ;
s1.TotalMarks =  575.50 ;
```

We may also use the member variables in expressions on the right-hand side. *Example:*

```
FinalMarks  =  s1.TotalMarks + 5.0 ;
```

11.2.3 Copying Structs

We can also copy values from one struct to another. *Example:*

```
Student s2; // s2 is declared
s2 = s1 ;
```

This will copy all those values from s1 to s2.

Note that we can also use the operator **new** to create **struct** variables.

```
Student s3 = new Student ( );
```

A **struct** variable is initialized to the default values of its members as soon as it is declared.

Note that struct's data members are 'private' by default and therefore cannot be accessed outside the struct definition. The use of access modifier **public** means that anyone can use the member. A member not declared public cannot be accessed using the dot operator. *Example:*

```
struct ABC
{
   int a ;              //private by default
   public int b;
   private int c;       //explicitly declared as private
}
. . . .
. . . .
ABC   abc ;
abc.a = 10;    // Error, a is private
abc.b = 20;  // OK, b is  public
. . . .
. . . .
```

Program 11.1 illustrates a simple application of **struct** type objects.

Program 11.1 A simple application of structs

```
using System;
struct Item
{
   public string name;
   public int code;
   public double price;
}

class StructTest
{

   public static void Main( )
   {
      Item fan;      //create an item

     //Assign values to members
     fan.name  = "Bajaj";
      fan.code  =  123;
      fan.price =  1576.50;
```

```
        //Display item details
        Console.WriteLine("Fan name: " + fan.name);
        Console.WriteLine("Fan code: " + fan.code);
        Console.WriteLine("Fan cost: " + fan.price);
    }
}
```

Output of Program 11.1:

```
Fan name : Bajaj
Fan code : 123
Fan cost : 1576.5
```

Note that we cannot initialize the data fields inside the struct definition.

Structs may also contain methods, properties, operator functions, indexers and nested types. While the use of methods and nested types are discussed in the sections that follow, other members are considered in Chapter 12.

11.3 STRUCTS WITH METHODS

We have seen that values may be assigned to the data members using **struct** objects and the dot operator. We can also assign values to the data members using what are known as *constructors*.

A constructor is a method which is used to set values of data members at the time of declaration. Consider the code below:

```
sruct Number
{
    int number;                     // data member
    public Number ( int value )     // constructor
    {
        number = value;
    }
}
```

The constructor method has the same name as **struct** and declared as public. The constructor is invoked as follows:

Number n1 = new Number(100);

This statement creates a struct object **n1** and assigns the value 100 to its only data member **number**. C# does not support default (i.e., parameterless) constructors.

Structs can also have other methods as members. These methods may be designed to perform certain operations on the data stored in struct objects. Note that a struct is not permitted to declare a destructor.

Program 11.2 shows how constructors and methods are used in a struct implementation.

Program 11.2 Using methods in structs

```
using System;
struct Rectangle
{
    int a, b;
```

```
    public Rectangle ( int x, int y )    //constructor
    {
       a =  x;
       b = y;
    }
    public int Area( )                   //a method
    {
       return ( a * b);
    }
    public void Display ( )              //another method

    {
       Console.WriteLine("Area = " + Area( ) );
    }
}
class TestRectangle
{
    public static void Main( )
    {
       Rectangle rect = new Rectangle ( 10, 20 );
       rect.Display ( );   // invoking Display ( ) method
    }
}
```

Program 11.2 produces the following output:

```
Area = 200
```

This code contains one constructor method to give values to the data members, another method **Area ( )** to compute the area of the rectangle and the third method **Display ( )** to display the area computed. Note that how the method **Area ( )** is nested inside the method **Display ( )** . The statement

```
Rect.Display ( );
```

invokes the method **Display ( )** which in turn invokes the method **Area ( ).**

11.4 NESTED STRUCTS

C# permits declaration of structs nested inside other structs. The following code is valid:

```
struct Employee
{
   public string name;
   public int code;
   public struct Salary
   {
```

```
        public double basic;
        public double allowance;
    }
}
```

We can also use struct variables as members of another struct. This implies nesting of references to structs.

```
struct M

{
    public int x;
}
struct N
{
    public M m;    // object of M
    public int y;
}
. . .
. . .
N n;
n.m.x = 100;    // x is a member of m, a member of n
n.y = 200;      // y is a member of n
```

Program 11.3 is an example of nesting of structures in C#, one structure **teacher** is declared within structure **student** and their values are displayed using functions **show(), show details(),** etc. The values of structure and method overriding is done using **getvalues()** method.

Program 11.3 Nested structures

```
using System;
using System.Collections.Generic;
using System.Text;

namespace NestedStructures
{

    struct student
    {
        public string studentname;
        public string rollno;
        public static string grade;

        //Setting the property with struct student
        public string RollNo
        {
```

```
    set
    {
       rollno = "S000018_ClassV";
    }
    get
    {
       return rollno;
    }
 }

 //Declaring a class school with in the structure student
 public class school
 {
    public string schoolname;
    public string classname;

    //Setting property for schoolproperty

    public string schoolproperty
    {
       set
       {
          schoolname = value;
       }
       get
       {
          return schoolname;
       }
    }
    public school()
    {
       schoolname = "St. Johns Schhol, Janakpuri, Delhi";
       classname = "Class Vth";
    }
    public void insert_details(string str1, string str2)
    {
       schoolname = str1;
       classname = str2;
    }
    public void show_details()
    {
       Console.WriteLine("School Name is {0} ", schoolname);
```

```
            Console.WriteLine("Class is {0} ", classname);

        }
    }//End of inner class school

    public struct teachers
    {
        public string tcode;
        public string tname;

        public void getValues(string st, string str)
        {
            grade = "PGT";
            tcode = st;
            tname = str;
        }

        public void show()
        {
            Console.WriteLine("Teacher Code is {0} ", tcode);
            Console.WriteLine("Teacher Name is {0} ", tname);
        }
    }

    //End of nested structure
    //Here method overloading applicable for the function addStudents()
and addStudents(string str3, string str4)

    public void addStudents()
    {
        studentname = "Suresh Goyal";
        rollno = "S0011_class4";
    }

    public void addStudents(string str3, string str4)
    {

        studentname = str3;
        rollno = str4;
    }

    public void show()
```

```
        {
            Console.WriteLine("Student Name {0} ", studentname);
            Console.WriteLine("Roll Number {0} ", rollno);
        }
    }; // End of Structure student

    class Program
    {
        static void Main(string[] args)
    {
    //creating an instance of student
    student st=new student();
    st.RollNo="S0007_Class3";
    Console.WriteLine("Roll Number after overriding the property {0}",
st.RollNo);

    //Creating an instance of inner structure teachers
    student.teachers te=new student.teachers();

    //Directly Refering the fields of inner structure teachers
    te.tcode="T00014";
    te.tname="Kishan Singh";

    //Accessing the getValues() of inner structure teachers
    te.getValues("T00015","Naresh Kumar");

    //calling the show() of inner structure of teachers
    te.show();

    //overriding the instance of student.teachers
    student.teachers intfl=new student.teachers();

    intfl.getValues("T00014","Sunil Mohan");
    intfl.show();

    //Creating the instance of the nested class examination
    student.school ee=new student.school();
    ee.schoolproperty="St. Johns School, Janakpuri, Delhi";
    ee.show_details();
        }
    }
}
```

```
C:\WINNT\system32\cmd.exe
Roll Number after overriding the property S000018_ClassV
Teacher Code is T00015
Teacher Name is Naresh Kumar
Teacher Code is T00014
Teacher Name is Sunil Mohan
School Name is St. Johns School, Janakpuri, Delhi
Class is Class Vth
Press any key to continue . . .
```

11.5 DIFFERENCES BETWEEN CLASSES AND STRUCTS

A struct is a simple lightweight alternative to a class. Although structs are basically the same as classes, structs are designed more for the situations where we simply want to group some data together. Structs, therefore, differ significantly from classes in design and implementation. (Classes are discussed in Chapters 12 and 13). Important differences are listed in Table 11.1.

Table 11.1 *Class and struct differences*

Category	Classes	Structs
Data Type	Reference type and therefore stored on the heap	Value type and therefore stored on the stack. Behave like simple data types
Inheritance	Support inheritance	Do not support inheritance
Default values	Default value of a class type is null	Default value is the value produced by 'zeroing out' the fields of the struct
Field initialization	Permit initialization of instance fields	Do not permit initialization of instance fields
Constructors	Permit declaration of parameterless constructors	Do not permit declaration of parameterless constructors
Destructors	Supported	Not supported
Assignment	Assignment copies the reference	Assignment copies the values.

The fact that structs are value types will affect performance when they are passed as parameters to methods. We can avoid this performance loss by passing structs as **ref** parameters to methods. In this case, only the address in memory will be passed in, which is as fast as passing in a class.

11.6 ENUMERATIONS

An enumeration is a user-defined integer type which provides a way for attaching names to numbers, thereby increasing the comprehensibility of the code. The **enum** keyword automatically enumerates a list of words by assigning them values 0, 1, 2 and so on. This facility provides an alternative means of creating 'constant' variable names. The syntax of an **enum** statement is illustrated below:

```
enum Shape
{
   Circle,        //ends with comma
   Square,        //ends with comma
   Triangle       //no comma
}
```

This can be written in one line as follows:

```
enum Shape {Circle, Square, Triangle}
```

Here, **Circle** has the value 0, **Square** has the value 1 and **Triangle** has the value 2. Other examples of **enum** are:

```
enum Colour { Red, Blue, Green, Yellow }
enum Position { Off, On }
enum Day { Mon, Tue, Wed, Thu, Fri, Sat, Sun }
```

Program 11.4 illustrates the use of **enum** type dots.

Program 11.4 Implementing enum type

```
using System;
class Area
{
   public enum Shape
   {
      Circle,
      Square
   }
   public void AreaShape ( int x, Shape shape)
   {
      double area;
      switch (shape)
      {
         case  Shape.Circle:
               area = Math.PI * x * x;
               Console.WriteLine("Circle Area = "+area);
               break;
         case  Shape.Square:
               area = x * x ;
```

```
                    Console.WriteLine("Square Area = "+area);
                           break;
                         default:
                      Console.WriteLine("Invalid Input"); break;
            }
        }
}

class EnumTest:
{
    public static void Main( )
    {

        Area area = new Area ( );
        area.AreaShape ( 15, Area.Shape.Circle);
        area.AreaShape ( 15, Area.Shape.Square);
        area.AreaShape ( 15, (Area.Shape) 1 );
        area.AreaShape ( 15, (Area.Shape) 10 );
    }
}
```

The **Shape** enum defines the values that can be specified for the enum, and the same enum is used in the method call to specify the shape whose area is required. Note the way the enum members are accessed.

```
Shape.Circle // Access to Circle inside Area class
Area.Shape.Circle // Access to Circle outside Area class
```

The declaration;

```
Shape shape
```

in the method definition declares **shape** as an enum type object.

In method calls, the values that can be specified for an enum are not limited to the identifiers specified in the enum definition. The third and fourth calls

```
area.AreaShape ( 15, ( Area.Shape) 1 );
area.AreaShape ( 15, ( Area.Shape) 10 );
```

are legal calls. However, the values passed must be in the range of valid values. Since the value of enum members range from 0 to 1, the third call refers to the case **Shape.Square** while fourth call passes a value that is outside the range. Note that when **int** type values are passed, they are **cast** into **enum** type values.

Program 11.4 will produce the following output:

```
Circle Area = 706.8583
Square Area = 225.00000
Square Area = 225.00000
Invalid Input
```

Program 11.5 stores values in two enumerations, **Staff** and **Company**. It uses two functions to display the data contained in **Staff** and **Company** enumerations.

Program 11.5 Enumeration illustrated further

```
using System;
using System.Collections.Generic;
using System.Text;

namespace EnumerationExample
{
   enum Staff
   {
      Directors,
      Managers,
      Executives
   }

   enum Company
   {
      NewSoftLtd,
      TechnologiesInc,
      HillRockLtd
   }

   class Program
   {
      public static void Show(Staff st)
      {

         switch (st)
         {
            case Staff.Directors:
               Console.WriteLine("You are a Director.");
               break;

            case Staff.Managers:
               Console.WriteLine("You are a Manager.");
               break;

            case Staff.Executives:
               Console.WriteLine("You are an Executive.");
               break;
```

```
            default: break;
        }

    }
    public static void CompDisplay(Company com)
    {

        switch (com)
        {
            case Company.NewSoftLtd:
                Console.WriteLine("NewSoftLtd.");
                break;

            case Company.TechnologiesInc:
                Console.WriteLine("TechnologiesInc.");
                break;

            case Company.HillRockLtd:
                Console.WriteLine("HillRockLtd.");
                break;

            default: break;
        }

    }

        static void Main(string[] args)
        {
            Staff st;
            st = Staff.Directors;
            Console.WriteLine("This is an example of Enumeration in C#.\n");
            Show(st);
            Company com;
            com = Company.NewSoftLtd;
            Console.Write("Director belongs to Company : ");
            CompDisplay(com);
        }
    }
}
```

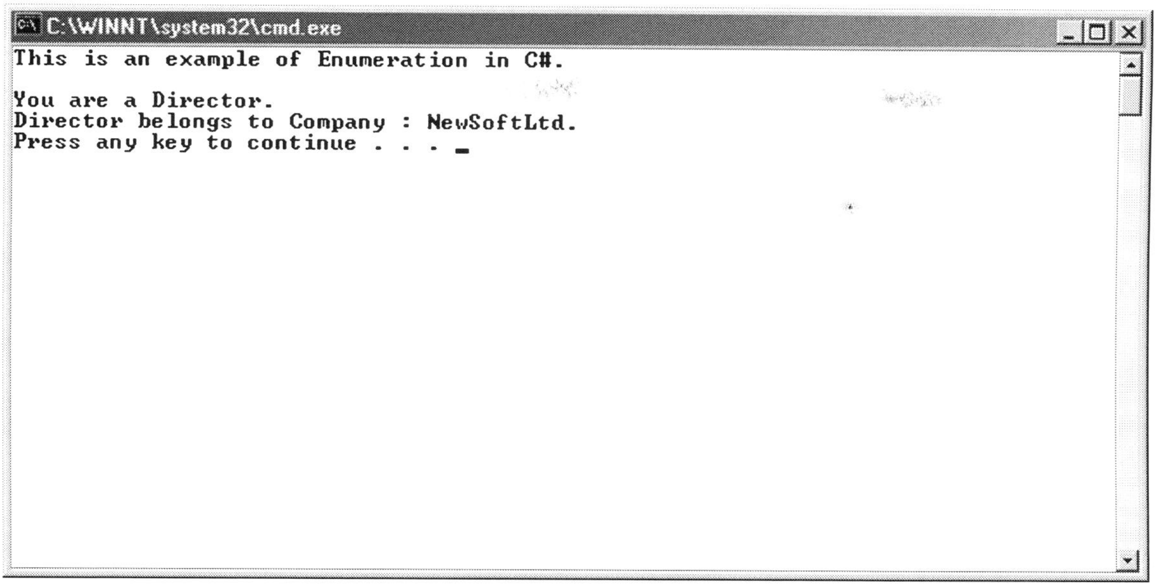

11.7 ENUMERATOR INITIALIZATION

As mentioned earlier, by default, the value of the first enum member is set to 0, and that of each subsequent member is incremented by one. However, we may assign specific values for different members, if we so desire. *Example:*

```
enum Colour
{
   Red    = 1,
   Blue   = 3,
   Green  = 7,
   Yellow = 5
}
```

We can also have expressions, as long as they use the already defined enum members. *Example:*

```
enum Colour
{
   Red     = 1,
   Blue    = Red + 2,
   Green   = Red + Blue + 3,
   Yellow  = Blue + 2
}
```

If the declaration of an enum member has no initializer, then its value is set implicitly as follows:

- If it is the first member, its value is zero.

- Otherwise, its value is obtained by adding one to the value of the previous member

Consider the following enum declaration:

```
enum Alphabet
{
    A,
    B = 5,
    C,
    D = 20,
    E
}
```

The member A is set to zero. Since the member B is explicitly given the value 5, the value of C is set to 6 (i.e, 5 + 1). Similarly, E is set to 21.

The following enum declaration is not valid because the declarations are circular.

```
enum ABC
{
    A,
    B = C,
    C
}
```

11.8 ENUMERATOR BASE TYPES

By default, the type of an enum is **int**. However, we can declare explicitly a base type for each enum. The valid base types are:

```
byte, sbyte, short, ushort, int, uint, long and ulong
```

Examples:

```
enum Position : byte
{
    off,
    on
}

enum Shape : long
{
    Circle,
    Square = 100,
    Triangle
}
```

The values assigned to the members must be within the range of values that can be represented by the base type. For example, if the base type is **byte**, assigning a value 300 is illegal.

11.9 ENUMERATOR TYPE CONVERSION

Enum types can be converted to their base type and back again with an explicit conversion using a cast. *Example:*

```
enum Values
{
   Value0,
   Value1,
   Value2,
   Value3
}
. . . .
. . . .
Values u1 = (Values) 1;
int a = (int ) u1;
. . . .
. . . .
```

The exception to this is that the literal 0 can be converted to an enum type without a cast. That is

```
Values u0 = 0;
```

permitted. Program 11.6 illustrates how enumerator types are converted.

Program 11.6 Enumerator type conversion

```
using System;
class Enumtype
{
   enum Direction
   {
      North ,
      East = 10 ,
      West ,
      South
   }
   public static void Main( )
   {
      Direction d1 = 0,              //implicit conversion
      Direction d2 = Direction.East;
      Direction d3 = Direction.West;
      Direction d4 = (Direction)12;   //explicit conversion

      Console.WriteLine("d1 = " + d1);
```

```
        Console.WriteLine("d2 = " + (int) d2);
        Console.WriteLine("d3 = " + d3);
        Console.WriteLine("d4 = " + d4);
    }
}
```

Program 11.6 will output the following:

```
d1 = 0
d2 = 10
d3 = 11
d4 = 12
```

Case Study

Problem Statement

CompTech is a medium sized organisation that is involved in the sale of computer hardware. It also has a small software development department, which develops customised software and Web sites for different organisations. CompTech has been till now manually maintaining the records of its employees, which is time consuming and decreases the efficiency of the HR department staff. How can CompTech solve the problem?

Solution

CompTech requests one of the programmers, Harsh, who has been working for the past two years in the software development department, to create an application, which will enable the HR department to automate the task of entering employee data. In order to judiciously utilise memory space, Harsh decides to use the **structure** feature of C#. A **structure** is a value type that allows us to organise related data of different data types. After much research, Harsh creates the following C# application.

```
using System;
using System.Collections.Generic;
using System.Text;

namespace StructureExample
{
    public struct Employee
    {
        private string emp_details;

        public string Write
        {
            get { return emp_details; }
            set { emp_details = value; }
        }

        public string Read()
```

```
    {
       return Console.ReadLine();
    }
}

public struct EmployeeDetails
{
   Employee name;
   Employee address;
   Employee telephone;
   Employee dept;
   Employee sal;

   public Employee Name
   {
      get { return name; }
      set { name = value; }
   }

   public Employee Address
   {
      get { return address; }
      set { address = value; }
   }

   public Employee Telephone
   {
      get { return telephone; }
      set { telephone = value; }
   }

   public Employee Department
   {
      get { return dept; }
      set { dept = value; }
   }

   public Employee Salary
   {
      get { return sal; }
      set { sal = value; }
   }

   public void ShowAllDetails()
```

```
    {
        Employee emp = new Employee();
        Console.WriteLine("Employee Details Registration System.");
        Console.WriteLine("----------------");
        Console.Write("Enter the Name of the Employee: ");
        name.Write = emp.Read();
        Console.Write("Enter the Address of the Employee: ");
        address.Write = emp.Read();
        Console.Write("Enter the Telephone of the Employee: ");
        telephone.Write = emp.Read();
        Console.Write("Enter the Department of the Employee: ");
        dept.Write = emp.Read();
        Console.Write("Enter the Salary of the Employee: ");
        sal.Write = emp.Read();

    }
}

class ShowEmployeeDetails
{
    static void Main(string[] args)
    {
        string input = "";
        EmployeeDetails empl = new EmployeeDetails();

        empl.ShowAllDetails();
        Console.WriteLine();

        Console.WriteLine("Showing Employee Details");
        Console.WriteLine("---------------");
        Console.WriteLine("Name: {0}", empl.Name.Write);
        Console.WriteLine("Address: {0}", empl.Address.Write);
        Console.WriteLine("Telephone: {0}", empl.Telephone.Write);
        Console.WriteLine("Department: {0}", empl.Department.Write);
        Console.WriteLine("Salary: {0}", empl.Salary.Write);
        Console.WriteLine("----------------");
        Console.WriteLine("Press y to save the record....");
        input = Console.ReadLine();
        if (input == "y")
        {
            Console.WriteLine("Record Saved.");
        }
        else
```

```
            {
                Console.WriteLine("Record not saved.");
            }

        }
    }
}
```

Remarks

The use of structure in a C# application allows you to manage the memory of a computer efficiently as all related data is organised in a structure. It also increases the performance of a C# application.

Common Programming Errors

- *Placing the semicolon at the end of structure definition.*
- *Using the upper case S in the keyword struct.*
- *Giving initial values to data fields inside struct definition.*
- *Trying to access a private member outside the struct definition.*
- *Assigning a structure of one type to a structure different type.*
- *Attempting to access a struct member by using only the member name.*
- *Placing the semicolon at the end of enum definition.*
- *Placing a comma after the last item in the enum list.*
- *Forgetting to use the dot operator to access enum members.*
- *In method calls, forgetting to cast **int** type values to **enum** type*
- *In method calls, passing the values outside the range of valid enum values.*

Review Questions

11.1 State whether the following statements are true or false. If false, correct the statement to make it true.

(a) Struct type data are stored on heap.
(b) It is faster to manipulate struct type data compared class type data.
(c) We cannot initialize a data field inside a struct definition.
(d) Structs are implicitly sealed and therefore no class or struct can derive from the struct.
(e) Like classes, structs support inheritance.
(f) By default, the type of an enum is byte.
(g) Integer values can be converted to enum types using casts.
(h) The values of enum type members may be modified during the execution of a program.
(i) Structs can implement multiple interfaces.
(j) C# does not allow structure variables to be members of structures.

11.2 State two significant differences between classes and structs. In what ways are they similar?

11.3 When do we prefer the use of structs over classes?

11.4 Can we initialise variables in a struct?

11.5 Do structs have constructors and destructors? Does a struct support inheritance?

11.6 What is the purpose of a constructor in a structure?

11.7 Why do we use methods as members of a struct?

11.8 What is nesting of structures? Give an example of typical use of nested structures.

11.9 What is enumeration? How is it useful in C# programming?

11.10 Give two examples of typical use of enum type values.

11.11 Find errors, if any, in the following struct definitions and statements.

```
(a) structure Employee
    {
        string name;
        float age;
    };
(b) struct Point
    {
        int x = 10;
        int y = 20;
    }
(c) struct A
    {
        int a;
        A ( int x ) { a = x; }
    }
(d) struct vector
    {
        public int x, y;
        public Vector (int a, int b)
        { x = a, y = b; }
        }
    Vector V1 = Vector (10, 20);
    Vector V2 = V1;
    . . . .
    . . . .
```

11.12 Find errors, if any, in the following code segments:

```
(a) enum XYZ
    {
        X,
        Y,
        Z = 100;
    }
```

```
(b) enum Colour : byte
    {
        Red,
        Blue = 300,
        Green
    }
(c) enum Colour
    {
        Red,
        Blue,
        Green = Red
    }
(d) enum Type
    {
        Type 1,
        Type 2
    }
    Type t1 = 10
        int n = t1
        . . . .
```

11.13 Given these structure definitions and declarations

```
    struct Abc
    {
        public int a1;
        public double d1;
    }
    struct Xyz
    {
        public int x1;
        public int x2;
    }
        Abc abc;
        Xyz xyz;
```

find errors, if any, in the following statements:

(a) abc = xyz;

(b) Abc.a1 = 10;

(c) int m = a1 + x1;

(d) int n = xyz.x2+10;

Debugging Exercises

11.1. Debug the program given below.

```
using System;
using System.Collections.Generic;
using System.Text;

namespace debugAppl_chap11
{
      enum Students : byte
      {
          passed,
          failed
      }

      struct Stud
      {
          Students stu;
          int id;
      }

      class Program
      {
          static void Main(string[] args)
          {
              Stud st = new Stud();
              st.stu = Students.passed;
              st.id = "S00017_Class5";
              Console.Write("You have ");
              Console.WriteLine(st.stu);
              Console.Write("Your roll number is : ");
              Console.WriteLine(st.id);
          }
      }
}
```

11.2. The program given below demonstrates the use of **public** keyword. Debug the errors so that the program compiles successfully.

```
using System;
using System.Collections.Generic;
using System.Text;
```

```
namespace debugApp2_chap11
{
struct StructExample {
   public int m;
   public StructExample(int n) {
   m=n*n;
   }

   void Show() {
        System.Console.WriteLine("This is from Show method in a Struct.");
      }
   }

   class Program
   {
      static void Main(string[] args)
      {
         StructExample st;
         System.Console.WriteLine(st.i);
         st.show();

      }
   }
}
```

11.3. Find errors in the following program for displaying the months using **enum**.

```
using System;
using System.Collections.Generic;
using System.Text;

namespace debugApp3_chap11
{
   enum DateMonth : long
   {
      int oct = 10,
      int nov = 20,
      int dec
   }

   class Program
   {
      static void Main(string[] args)
      {
```

```
            long a = DateMonth.oct;
            long b = DateMonth.nov;
            long c = DateMonth.dec;
            Console.WriteLine("October = {0}, November = {1},
December = ", a, b, c);

        }
    }
}
```

Programming Exercises

11.1 *Design a structure data type named Date Of Birth to contain date, month and year of birth. Develop a C# program using this data structure that would assign your date of birth to the individual members and display the date of birth in the following format:*
My date of birth is 15/06/75
Do not use any methods in your program.

11.2 *Modify the above program using*
(a) *a constructor to input values to the individual members, and*
(b) *a method to display the date of birth.*

11.3 *Design a structure type data using a suitable name for each of the following records:*
(a) *A student record consisting of name, date of birth, and total marks obtained.*
(b) *A mailing list consisting of name, door number, street, city and pincode.*
(c) *An inventory, record consisting of item code, item name, item cost and the total items available.*
(d) *A book record consisting of the author, title, year of publication and cost.*

11.4 *Modify the data structure defined in the Exercise 11.3(a) as a nested data structure using the date of birth as a structure type data. Develop a suitable C# program to test the nested data structure.*

11.5 *Develop a program that prompts the user to input the name, door number, street, city and pincode and stores them in the address record designed in the Exercise 11.3(b) and display them in an appropriate manner.*

11.6 *Write a C# program that uses an array of five items of data type designed in the Exercise 11.3(c), accept interactively the data of all the five items into the array, and display them in the tabular form as shown below:*

Code	Name	Cost	Total items
—	—	—	—
—	—	—	—
—	—	—	—

Chapter

12

Classes and Objects

12.1 INTRODUCTION

C# is a true object-oriented language and therefore the underlying structure of all C# programs is classes. Anything we wish to represent in a C# program must be encapsulated in a class that defines the *state* and *behaviour* of the basic program components known as *objects*. Classes create objects and objects use methods to communicate between them as shown in Fig. 12.1. This is object-oriented programming (OOP). We have already used classes and objects in a number of examples.

Classes provide a convenient approach for packing together a group of logically related data items and functions that work on them. In C#, the data items are called *fields* and the functions are called *methods*. Calling a specific method in an object is described as sending the object a message.

A class is essentially a description of how to construct an object that contains fields and methods. It provides a sort of *template* for an object and behaves like a basic data type such as **int**. It is therefore important to understand how the fields and methods are defined in a class and how they are used to build a C# program that incorporates the basic OOP concepts.

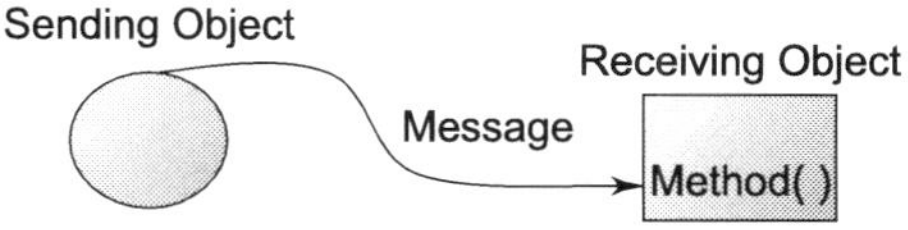

Figure 12.1 *Message passing between objects*

12.2 BASIC PRINCIPLES OF OOP

All object-oriented languages employ three core principles, namely,

- encapsulation,
- inheritance, and
- polymorphism.

These are often referred to as three 'pillars' of OOP.

Encapsulation provides the ability to hide the internal details of an object from its users. The outside user may not be able to change the state of an object directly. However, the state of an object may be altered indirectly using what are known *accessor* and *mutator* methods. In C#, encapsulation is implemented using the access modifier keywords ***public, private***, and ***protected*** which are discussed in Section 12.6.

The concept of encapsulation is also known as *data hiding* or *information hiding*. When done properly, we can create software 'black boxes' that can be independently tested and used.

Inheritance is the concept we use to build new classes using the existing class definitions. Through inheritance we can modify a class the way we want to create new objects. The original class is known as *base* or *parent* class and the modified one is known as *derived class* or *subclass* or *child class*.

The concept of inheritance facilitates the reusability of existing code and thus improves the integrity of programs and productivity of programmers.

Polymorphism is the third concept of OOP. It is the ability to take more than one form. For example, an operation may exhibit different behaviour in different situations. The behaviour depends upon the types of data used in the operation. For example, an addition operation involving two numeric values will produce a sum and the same addition operation will produce a string if the operands are string values instead of numbers. Similarly, a method when called with one set of parameters may draw a circle but when called with another set of parameters may draw a triangle.

Polymorphism is extensively used while implementing inheritance. More about polymorphism and how it is achieved will be discussed later.

12.3 DEFINING A CLASS

As stated earlier, a class is a user-defined data type with a template that serves to define its properties. Once the class type has been defined, we can create 'variables' of that type using declarations that are similar to the basic type declarations. In C#, these variables are termed as *instances* of classes, which are the actual *objects*. The basic form of a class definition is :

```
class classname
{
    [ variables declaration; ]
    [ methods declaration; ]
}
```

class is a keyword and *classname* is any valid C# identifier. Everything inside the square brackets is optional. This means that the following would be a valid class definition:

```
class Empty //class name is Empty
{
}
```

Because the body is empty, this class does not contain any properties and therefore cannot do anything. We can, however, compile it and even create objects using it. *C++ programmers may note that there is no semicolon after the closing brace.*

Note:

- The above form really represents a simplified class definition. A C# class may include many more items such as properties, indexers, constructors, destructors and operators as shown in Fig. 12.2. These items which behave like methods will be discussed later. We often use the term *member* to denote any item that is part of a class.

- Classes and structs are very similar, but class is a reference type. Also, classes include more features compared to structs.
- Some of the categories of class members shown in Fig.12.2 will be discussed later in this chapter.

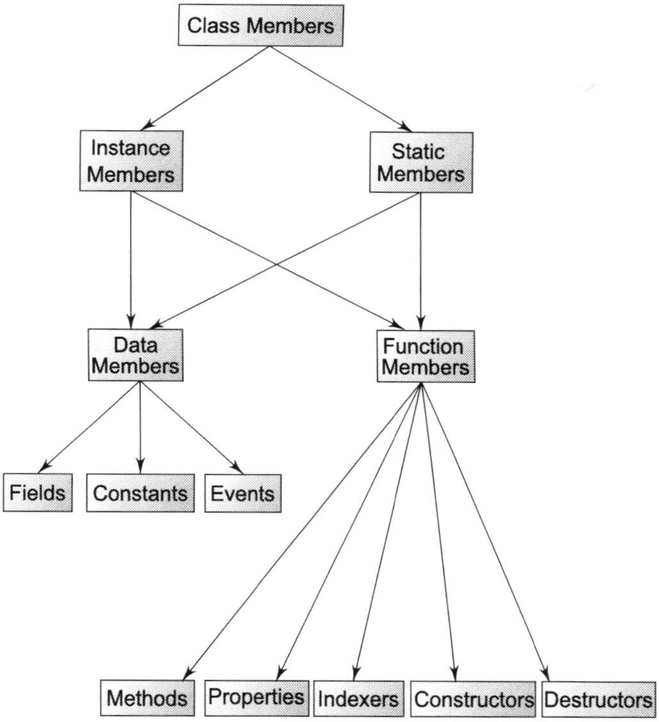

Figure 12.2 *Categories of class members*

12.4 ADDING VARIABLES

Data is encapsulated in a class by placing data fields inside the body of the class definition. These variables are called *instance variables* because they are created whenever an object of the class is instantiated. We can declare the instance variables exactly the same way as we declare local variables.
Example:

```
class Rectangle
{
   int length;    //instance variable
   int width;     //instance variable
}
```

The class **Rectangle** contains two integer type instance variables. It is allowed to declare them in one line as:

```
   int length, width;
```

Remember that these variables are only declared and therefore no storage space has been created in the memory. Instance variables are also known as *member variables*.

12.5 ADDING METHODS

A class with only data fields and without methods that operate on that data has no life. The objects created by such a class cannot respond to any messages. We must therefore add methods that are necessary for manipulating the data contained in the class. Methods are declared inside the body of the class, usually after the declaration of instance variables. We know that the general form of a method declaration is:

```
type methodname (parameter-list)
{
    method-body;
}
```

The *body* actually describes the operations to be performed on the data. Let us consider the **Rectangle** class again and add a method **GetData ( )** to it.

```
class Rectangle
{
    int length;
    int width;
    public void GetData(int x , int y)//mutator method
    {
        length = x ;
        width = y ;
    }
}
```

Note that the method has a return type **void** because it does not return any value. We pass two integer values to the method which are then assigned to the instance variables **length** and **width**. The **GetData** method is basically added to provide values to the instance variables.

Notice that we are able to use directly **length** and **width** inside the method. We have used the keyword **public** in defining the method. We shall discuss the reason for this in Section 12.6. Note that the declaration of instance variables may be placed after the method definition as well. C# permits this.

Let us add some more behaviour to the class. Assume that we want to compute the area of the rectangle defined by the class. This can be done as follows:

```
class Rectangle
{
    int length, width;
    public void GetData(int x , int y)
    {
        length = x ;
        width = y ;
    }
```

```
        public int RectArea( )
        {
           int area = length * width; // area is local variable
           return(area);
        }
    }
```

The new method **RectArea( )** computes the area of the rectangle and *returns* the result. Since the result would be an integer, the return type of the method has been specified as **int.** Also note that the parameter list is empty.

Remember that while the declaration of instance variables (and also local variables) can be combined as

```
    int length, width;
```

the parameter list used in the method header should always be declared independently separated by commas. That is,

```
    void GetData(int x,y)
```

is illegal.

Now, our class **Rectangle** contains two instance variables and two methods. We can add more variables and methods, if necessary.

Most of the times when we use classes, we will have many methods and variables within the class. Instance variables and methods in classes are accessible by all the methods in the class, but a method cannot access the variables declared in other methods. *Example:*

```
class Access
{
   int x ;                          //instance variable
   public void Method1( )          //a method
   {
      int  y ;
      x = 10 ;     // legal
      y =  x   ;   // legal
   }

   public void Method2( )    //another method
   {
      int z ;
      x = 5 ;  // legal
      z = 10;  // legal
      y = 1 ;  // illegal, y defined in Method1
   }
}
```

12.6 MEMBER ACCESS MODIFIERS

One of the goals of object-oriented programming is 'data hiding'. That is, a class may be designed to hide its members from outside accessibility. C# provides a set of 'access modifiers' that can be used with the members of a class to control their visibility to outside users. Table 12.1 lists various access modifiers provided by C# and their visibility control. These modifiers are a part of C# keywords.

Table 12.1 *C# access modifiers*

Modifier	Accessibility Control
private	Member is accessible only within the class containing the member.
public	Member is accessible from anywhere outside the class as well. It is also accessible in derived classes.
protected	Member is visible only to its own class and its derived classes.
internal	Member is available within the assembly or component that is being created but not to the clients of that component.
protected internal	Available in the containing program or assembly and in the derived classes.

In C#, all members have **private** access by default. If we want a member to have any other visibility range, then we must specify a suitable access modifier to it individually. *Example:*

```
class Visibility
{
    public int x;
    internal int y;
    protected double d;
    float p; //private by default
}
```

Note we cannot declare more than one member under a visibility modifier. For instance, the code

```
    public: //allowed in C++
    int x;
    int y;
```

is illegal in C#.

Methods and data fields that are declared **public** are directly accessible from an object instance. Private members cannot be accessed by an object instance but can be used directly by the methods inside the class.

We shall be using the **public** modifier extensively in this chapter and examine the use of **protected** and **internal** later while discussing inheritance.

12.7 CREATING OBJECTS

As pointed out earlier, an object in C# is essentially a block of memory that contains space to store all the instance variables. Creating an object is also referred to as *instantiating* an object.

Objects in C# are created using the **new** operator. The **new** operator creates an object of the specified class and returns a reference to that object. Here is an example of creating an object of type **Rectangle**.

```
Rectangle rect1 ;              // declare
rect1 = new Rectangle( );     // instantiate
```

The first statement declares a variable to hold the object reference and the second one actually assigns the object reference to the variable. The variable **rect1** is now an object of the **Rectangle** class. (See Fig. 12.3).

Both statements can be combined into one as shown below:

```
Rectangle rect1 = new Rectangle( );
```

Note that the object **rect1** does not contain the value for the **Rectangle** object; it contains only the reference (i.e., address) to the object.

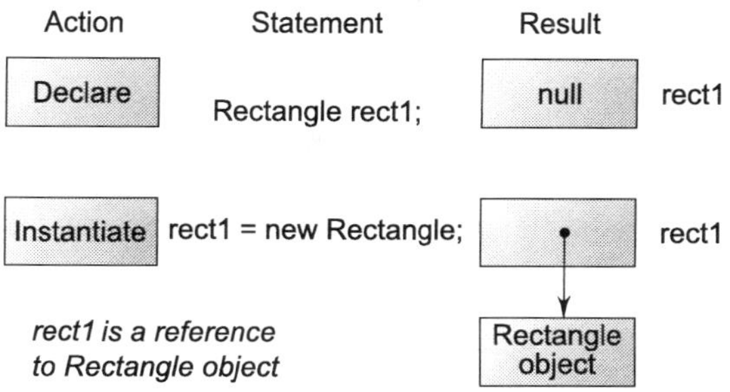

Figure 12.3 *Creating object references*

The method **Rectangle( )** is the default constructor of the class. We can create any number of objects of **Rectangle**. *Example:*

```
Rectangle rect1 = new Rectangle( );
Rectangle rect2 = new Rectangle( );
and so on.
```

It is important to understand that each object has its own copy of the instance variables of its class. This means that any changes to the variables of one object will have no effect on the variables of another. It is also possible to create two or more references to the same object. (See Fig. 12.4).

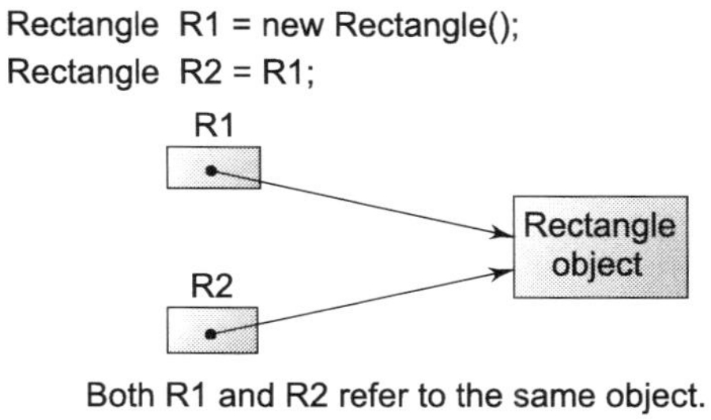

Figure 12.4 *Assigning one object reference variable to another*

12.8 ACCESSING CLASS MEMBERS

Now that we have created objects, each containing its own set of variables, we should assign values to these variables in order to use them in our program. Remember, all variables must be assigned values before they are used. Since we are outside the class, we cannot access the instance variables and the methods directly. To do this, we must use the concerned object and the *dot* operator as shown below:

```
objectname.variable name;
objectname.methodname (parameter-list);
```

Here *objectname* is the name of the object, *variablename* is the name of the instance variable inside the object that we wish to access, *methodname* is the method that we wish to call, and *parameter-list* is a comma separated list of 'actual values' (or expressions) that must match in type and number with the parameter list of the *methodname* declared in the class. The instance variables of the **Rectangle** class may be accessed and assigned values as follows:

```
rect1.length   = 15;
rect1.width    = 10;
rect2.length   = 20;
rect2.width    = 12;
```

Note that the two objects **rect1** and **rect2** store different values as shown below:

	rect1		rect2
rect1.length	15	rect2.length	20
rect1.width	10	rect2.width	12

This is one way of assigning values to the variables in the objects. Another and more convenient way of assigning values to the instance variables is to use a method that is declared inside the class.

In our case, the method **GetData** can be used to do this work. We can call the **GetData** method on any **Rectangle** object to set the values of both **length** and **width.** Here is the code segment to achieve this.

```
Rectangle rect1 = new Rectangle( );
rect1.GetData(15,10);    // calling the method
```

This code creates **rect1** object and then passes in the values 15 and 10 for the **x** and **y** parameters of the method **GetData**. This method then assigns these values to **length** and **width** variables respectively. For the sake of convenience, the method is again shown below:

```
void GetData(int x, int y)
{
   length = x ;
   width = y ;
}
```

Now that the object **rect1** contains values for its variables, we can compute the area of the rectangle represented by **rect1**. This again can be done in two ways.

- First approach is to access the instance variables using the dot operator and compute the area. That is,

```
int area1 = rect1.length * rect1.width ;
```

- The second approach is to call the method **RectArea** declared inside the class. That is,

```
int area1 = rect1.RectArea( );    // calling the method
```

Program 12.1 illustrates the concepts discussed so far.

Program 12.1 Application of classes and objects

```
using System;
class Rectangle
{
  public int length, width;          //  Declaration of variables

  public void GetData(int x, int y) //  Definition of method
  {
    length = x;
    width  = y;
```

```
    }

    public int RectArea()               // Definition of another method
    {
      int area = length * width;
      return (area);
    }
}

class RectArea                          //   class with main method
{
    public static void Main( )
    {
      int area1,area2;                  // Local variables
      Rectangle rect1 = new Rectangle(); // Creating objects
      Rectangle rect2 = new Rectangle();

      rect1.length = 15;              //   Accessing variables
      rect1.width  = 10;
      area1 = rect1.length * rect1.width;

      rect2.GetData(20,12);           //   Accessing methods
      area2 = rect2.RectArea();

      Console.WriteLine("Area1 = " + area1);
      Console.WriteLine("Area2 = " + area2);
    }
}
```

Program 12.1 would output the following:

```
Area1 = 150
Area2 = 240
```

12.9 CONSTRUCTORS

We know that all objects that are created must be given initial values. We have done this earlier using two approaches. The first approach uses the dot operator to access the instance variables and then assigns values to them individually. It can be a tedious approach to initialize all the variables of all the objects.

The second approach takes the help of a function like **GetData** to initialize each object individually using statements like,

```
rect1.GetData(15,10);
```

It would be simpler and more concise to initialize an object when it is first created. C# supports a special type of method, called a *constructor*, that enables an object to initialize itself when it is created.

Constructors have the same name as the class itself. Secondly, they do not specify a return type, not even **void**. This is because they do not return any value.

Let us consider our **Rectangle** class again. We can now replace the **GetData** method by a constructor method as shown below:

```
class Rectangle
{
    public int length ;
    public int width ;

    public Rectangle(int x, int y) // Constructor method
    {
          length = x ;
          width  = y ;
    }
    public int RectArea( )
    {
        return(length * width);
    }
}
```

Program 12.2 illustrates the use of a constructor method to initialize an object at the time of its creation.

Program 12.2 Application of constructors

```
using System;
class Rectangle
{
   public int length, width ;
   public Rectangle(int x, int y) // Defining constructor
   {
      length = x ;
      width = y ;
   }

   public int RectArea( )
   {
      return (length * width);
   }
}

class RectangleArea
{
   public static void Main( )
```

```
    {
        Rectangle rect1 = new Rectangle(15,10); // Calling constructor
        int area1 = rect1.RectArea( ) ;
        Console.WriteLine("Area1 = "+ area1) ;
    }
}
```

Output of Program 12.2:

```
Area1 = 150
```

Constructors are usually **public** because they are provided to create objects. However, they can also be declared as **private** or **protected.** In such cases, the objects of that class cannot be created and also the class cannot be used as a base class for inheritance. This property may be used in the implementation of inheritance, if necessary.

12.10 OVERLOADED CONSTRUCTORS

As discussed earlier, it is possible to create methods that have the same name, but different parameter lists and different definitions. This is called *method overloading*. Method overloading is used when objects are required to perform similar tasks but using different input parameters. When we call a method in an object, C# matches up the method name first and then the number and type of parameters to decide which one of the definitions to execute. This process is known as *polymorphism*. We can extend the concept of method overloading to provide more than one constructor to a class.

To create an overloaded constructor method, all we have to do is to provide several different constructor definitions with different parameter lists. The difference may be in either the number or type of arguments. That is, each parameter list should be unique. Here is an example of creating an overloaded constructor:

```
class Room
{
    public double length ;
    public double breadth ;

    public Room(double x, double y)  // constructor1
    {
        length = x ;
        breadth = y ;
    }

    public Room(double x)            // constructor2
    {
        length = breadth = x ;
    }
```

```
    public int Area( )
    {
        return (length * breadth) ;
    }
}
```

Here, we are overloading the constructor method **Room( )**. An object representing a rectangular room will be created as:

```
Room room1 = new Room(25.0,15.0);    //using constructor1
```

On the other hand, if the room is square, then we may create the corresponding object as:

```
Room room2 = new Room(20.0);  // using constructor2
```

12.11 STATIC MEMBERS

We have seen that a simple class contains two sections. One declares variables and the other declares methods. These variables and methods are called *instance variables* and *instance methods*. This is because every time the class is instantiated, a new copy of each is created. They are accessed using the objects (with dot operator).

Let us assume that we want to define a member that is common to all the objects and accessed without using a particular object. That is, the member belongs to the class as a whole rather than the objects created from the class. Such members can be defined as follows:

```
static int count ;
static int max(int x, int y);
```

The members that are declared **static** as shown above are called *static members*. Since these members are associated with the class itself rather than with individual objects, the static variables and static methods are often referred to as *class variables* and *class methods* in order to distinguish them from their counterparts, instance variables and instance methods.

Static variables are used when we want to have a variable common to all instances of a class. One of the most common examples is to have a variable that can keep a count of how many objects of a class have been created. Remember, C# creates only one copy for a static variable which can be used even if the class is never actually instantiated.

Like static variables, static methods can be called without using the objects. They are also available for use by other classes. Methods that are of general utility but do not directly affect an instance of that class are usually declared as class methods. C# class libraries contain a large number of class methods. For example, the **Math** class of C# System namespace defines many static methods to perform math operations that can be used in any program. We have used earlier statements of the types:

```
double x = Math.Sqrt(25.0);
```

The method **Sqrt** is a class method (or static method) defined in **Math** class.

We can define our own static methods as shown in Program 12.3.

Program 12.3 Defining and using static members

```
using System;
class Mathoperation
{
   public static float mul(float x, float y);
   {
      return x*y;
   }
   public static float divide(float x, float y)
   {
      return x/y ;
   }
}

class MathApplication
{
   public void static Main( )
   {
         float a = MathOperation.mul(4.0F,5.0F) ;
         float b = MathOperation.divide(a,2.0F) ;
         Console.WriteLine("b = "+ b) ;
   }
}
```

Output of Program 12.3:

```
b = 10.0
```

Note that the static methods are called using class names. In fact, no objects have been created for use. Static methods have several restrictions:

- They can only call other **static** methods.
- They can only access **static** data.
- They cannot refer to **this** or **base** in any way. (Refer Section 12.17)

12.12 STATIC CONSTRUCTORS

Like any other static members, we can also have static constructors. A static constructor is called before any object of the class is created. This is useful to do any housekeeping work that needs to be done once. It is usually used to assign initial values to static data members.

A static constructor is declared by prefixing a **static** keyword to the constructor definition. It cannot have any parameters. *Example:*

```
class Abc
{
```

```
    static Abc ( ) //No parameters
    {
            . . . .  //set values for static members here
    }
    . . . .
}
```

Note that there is no access modifier on static constructors. It cannot take any. A class can have only one static constructor.

12.13 PRIVATE CONSTRUCTORS

C# does not have global variables or constants. All declarations must be contained in a class. In many situations, we may wish to define some utility classes that contain only static members. Such classes are never required to instantiate objects. Creating objects using such classes may be prevented by adding a **private** constructor to the class.

12.14 COPY CONSTRUCTORS

A copy constructor creates an object by copying variables from another object. For example, we may wish to pass an **Item** object to the **Item** constructor so that the new **Item** object has the same values as the old one.

Since C# does not provide a copy constructor, we must provide it ourselves if we wish to add this feature to the class. A copy constructor is defined as follows:

```
public Item (Item item)
{
        code = item.code;
        price = item.price;
}
```

The copy constructor is invoked by instantiating an object of type **Item** and passing it the object to be copied. *Example:*

```
Item item2 = new Item (item1);
```

Now, **item2** is a copy of **item1**.

12.15 DESTRUCTORS

A destructor is opposite to a constructor. It is a method called when an object is no more required. The name of the destructor is the same as the class name and is preceded by a tilde (~). Like constructors, a destructor has no return type. *Example:*

```
class Fun
{
        . . . .
        . . . .
```

```
        ~ Fun ( ) //No arguments
        {
        . . . .
        }
}
```

Note that the destructor takes no arguments.

C# manages the memory dynamically and uses a *garbage collector,* running on a separate thread, to execute all destructors on exit. The process of calling a destructor when an object is reclaimed by the garbage collector is called *finalization*.

12.16 MEMBER INITIALIZATION

In addition to using constructors and methods to provide initial values to the objects, C# also allows us to assign initial values to individual data members at the time of declaration. *Example:*

```
class Initialization
{
    int number = 100;
    static double x = 1.0;
    string name = "John";
    Vehicle car = new vehicle (800, "maruti");
    . . . .
    . . . .
}
```

Note that other object-oriented languages like C++ and Java do not support this feature. When the variables are provided with the initial values at the time of declaration, the values are assigned as follows:

- Static variables are assigned when the class is loaded.
- Instance variables are assigned when an instance is created.

If the variables are not provided with the initial values as above, then they are assigned default values as dictated by their types. This is done as follows:

- Static variables are initialized to their default values when the class is loaded.
- Instance variables are initialized to their default values when an instance is created.

Thus a variable is never "uninitialized" in C#.

12.17 THE THIS REFERENCE

C# supports the keyword **this** which is a reference to the object that called the method. The **this** reference is available within all the member methods and always refers to the current instance. It is normally used to distinguish between local and instance variables that have the same name. Consider the code segment shown below:

```
class Integers
{
   int x;
   int y;
   public void SetXY(int x, int y(
   {
      this.x = x;
      this.y = y;
   }
   . . . .
   . . . .
}
```

In the assignment statements, **this.x** and **this.y** refer to the class members named **x** and **y** whereas simple **x** and **y** refer to the parameters of the **SetXY( )** method.

12.18 NESTING OF CLASSES

In C#, it is possible to define a type within the scope of another type. C# allows classes, structs, interfaces and enums to nest others. *Example:*

```
public class Outer
{
   . . . .          //Members of Outer class
   . . . .
   . . . .
   public class Inner
   {
      . . . .       //Members of Inner class
      . . . .
      . . . .
   }
}
```

A class can nest not only complete classes but also objects of other classes. For example:

```
public class Room
{
   . . .
   . . .
}
//A house has a room
public class House
{
   . . . .
   . . . .
```

```
    //The contained room
    private Room room1;
    . . . .
}
```

The **House** class had declared the **Room** object **room1** as one of its members.This relationship between **House** and **Room** is termed as 'has-a' relationship. (See Chapter 13)

Program 12.4 demonstrates the concept of nested classes and the way the methods of the **Main** class and the inner classes are called and displayed.

Program 12.4 Illustration of nested classes

```
using System;
using System.Collections.Generic;
using System.Text;

namespace InnerClassExample
{
    public class nestedClass
     {
        public static void Main()
         {
               outerClass obj2 = new outerClass();
               obj2.show();
               outerClass.innerClass obj1 = new outerClass.innerClass();
               obj1.display();

        }
    }
    public class outerClass
    {
        public void show()
         {
           System.Console.WriteLine("This message is from the class
outerClass containing the\n nested class innerClass.\n");
         }
        public innerClass abc()
         {
           return new innerClass();
         }
        public class innerClass
         {
           public void display(){
           System.Console.WriteLine("This is a nested class example in
```

```
C#.\n This info is in the innermost class innerClass.");
        }
    }
}
```

```
C:\WINNT\system32\cmd.exe
This message is from the class outerClass containing the
 nested class innerClass.

This is a nested class example in C#.
 This info is in the innermost class innerClass.
Press any key to continue . . .
```

12.19 CONSTANT MEMBERS

C# permits declaration of data fields of a class as constants. This can be done using the modifier **const**. *Example:*

```
public const int size = 100;
```

The member **size** is assigned the value 100 at compile time and cannot be changed later. Any attempt to assign a value to it will result in compilation error. It also means its value must be set when it is defined. For instance:

```
public const int size;
```

is wrong and will cause a compilation error. The **const** members are implicitly **static**. The value is by definition constant and therefore only one copy of it is stored which is common for all objects of the class. This implies that we must access the **const** members using the class name as we do with the **static** members.

Although **const** members are implicitly static, we cannot declare them so explicitly using **static.** For example, the statement

```
public static const int size = 100;
```

is wrong and will produce compile-time error.

12.20 READ-ONLY MEMBERS

There are situations where we would like to decide the value of a constant member at run-time. We may also like to have different constant values for different objects of the class. To overcome these shortcomings, C# provides another modifier known as **readonly** to be used with data members. This modifier is designed to set the value of the member using a constructor method, but cannot be modified later. The **readonly** members may be declared as either static fields or instance fields. When they are declared as instance fields, they can take different values with different objects. Consider the code below:

```
class Numbers
{
   public readonly int m;
   public static readonly int n;
   public Numbers ( int x )
   {
      m = x;
   }
   static Numbers ( )
   {
      n = 100;
   }
}
```

The value for **m** is provided at the time of creation of an object using the constructor with parameter **x.** This value will remain constant for that object. Remember, the variable **n** is assigned a value of 100, even before the creation of any objects of **Numbers.**

12.21 PROPERTIES

One of the design goals of object-oriented systems is not to permit any direct access to data members, because of the implications of integrity. It is normal practice to provide special methods known as *accessor methods* to have access to data members. We must use only these methods to set or retrieve the values of these members. Recall that we have used a method **GetData( )** in Program 12.1 to provide values to the data members **length** and **breadth** of **Rectangle** class. Similarly, we could use another method to read the values of these members. Program 12.5 shows how accessor methods can be used to set and get the value of a private data member.

Program 12.5 Accessing private data using accessor methods

```
using System;
class Number
{
```

```
    private int number;
    public void SetNumber( int x )    //accessor method
    {
       number = x;                        //private number accessible
    }
    public int GetNumber( )           //accessor method
    {
       return number;
    }
}

class NumberTest
{
    public static void Main ( )
    {
       Number n = new Number ( );
       n.SetNumber (100);  // set value
       Console.WriteLine("Number = " + n.GerNumber( ));  // get value

       // n.number; //Error! Cannot access private data
    }
}
```

Output of Program 12.5:

```
Number = 100
```

The **SetNumbers** method is also known as the *mutator* method. Using accessor methods works well and is a technique used by several OOP languages, including C++ and Java. However, it suffers from the following drawbacks:

- We have to code the accessor methods manually.
- Users have to remember that they have to use accessor methods to work with data members.

In order to overcome these problems, C# provides a mechanism known as *properties* that has the same capabilities as accessor methods, but is much more elegant and simple to use. Using a property, a programmer can get access to data members as though they are public fields. (Properties are sometimes referred to as 'smart fields' as they add smartness to data fields.)

We could rewrite the code in Program 12.5 using a property as shown in Program 12.6.

Program 12.6 Implementing a property

```
using System;
class Number
{
    private int number;
    public int Anumber  // property
```

```
    {
        get
        {
            return number;
        }
        set
        {
            number = value;
        }
    }
}
class  PropertyTest
{
    public void static Main ( )
    {
    Number n = new Number ( );
    n.Anumber = 100;
    int m = n.Anumber;
    Console.WriteLine("Number = " + m);
    }
}
```

The class now declares a property called **Anumber** of type **int** and defines a *get accessor* method (also known as *getter*) and a *set accessor* method (also known as *setter*). The getter method used the keyword **return** to return the field's value to the caller. The setter method uses the keyword **value** to receive the value being passed in from the user. The type of **value** is determined by the type of property.

As the names imply, getter method is used to get (or read) the value and the setter method is used to set (or write) the value. In Program 12.6, the statement

```
n.Anumber = 100;
```

invokes the setter method and places an integer value 100 in a variable named **value** which in turn is assigned to the field **number**.

Similarly, the statement

```
int m = n.Anumber;
```

invokes the getter method and assigns the value of the property to **m**.

A property can omit either a **get** clause or the **set** clause. A property that has only a getter is called a *read-only* property, and a property that has only a setter is called a *write-only* property. A write-only property is very rarely used. There are other powerful features of properties. They include:

- Other than fetching the value of a variable, a **get** clause uses code to calculate the value of the property using other fields and returns the results. This means that properties are not simply tied to data members and they can also represent dynamic data.

- Like methods, properties are inheritable. We can use the modifiers **abstract, virtual, new** and **override** with them appropriately, so that the derived classes can implement their own versions of properties.
- The **static** modifier can be used to declare properties that belong to the whole class rather than to a specific instance of the class. (Like static methods, static properties cannot be declared with the **virtual, abstract** or **override** modifiers.)

An important point to note here is that we can specify any modifier only at the property level and this will affect both the accessors equally. For instance, we cannot override only one, leaving the other unaffected.

12.22 INDEXERS

Indexers are location indicators and are used to access class objects, just like accessing elements in an array. They are useful in cases where a class is a container for other objects.

An indexer looks like a property and is written the same way a property is written, but with two differences:

- The indexer takes an *index* argument and looks like an array.
- The indexer is declared using the name **this**.

The indexer is implemented through **get** and **set** accessors for the [ ] operator. *Example:*

```
public double this[ int idx ]
{
    get
    {
        //Return desired data
    }
    set
    {
        //Set desired data
    }
}
```

The implementation rules for **get** and **set** accessors are the same as for properties. The return type determines what will be returned, in this case a **double**. The parameter inside the square brackets is used as the *index*. In this example, we have used an **int**, but it can be any object type.

Program 12.7 shows simple implementation using an integer index.

Program 12.7 Implementation of an indexer

```
using System;
using System.Collections;
class List
{
```

```
    ArrayList array = new ArrayList ( );
    public object this[ int index ]
    {
        get
        {
            if(index < 0 || index >= array.Count)
            {
                return null;
            }
            else
            {
                return (array [index] );
            }
        }
        set
        {
            array[index] = value;
        }
    }
}

class IndexerTest
{
    public static void Main ( )
    {
        List list = new List ( );
        list [0] = "123";
        list [1] = "abc";
        list [2] = "xyz";
        for (int i = 0, i < list.Count; i++)
            Console.WriteLine( list[i]);
    }
}
```

Indexers are sometimes referred to as 'smart arrays'. As pointed out earlier, indexers and properties are very similar in concept, but differ in the following ways:

- A property can be static member, whereas an indexer is always an instance member.
- A **get** accessor of a property corresponds to a method with no parameters, whereas a **get** accessor of an indexer corresponds to a method with the same formal parameter list as the indexer.
- A **set** accessor of a property corresponds to a method with a single parameter named **value**, whereas a **set** accessor of an indexer corresponds to a method with the same formal parameter list as the indexer, plus the parameter named **value.**
- It is an error for an indexer to declare a local variable with the same name as an indexer parameter.

Program 12.8 uses indexers and makes an array of 15 values, stores a few values in the array and the null values of array, from 12th position are stored with the word **'empty'**.

Program 12.8 Indexers illustrated further

```
using System;
using System.Collections.Generic;
using System.Text;

namespace IndexersExample
{
    class IndexersEx
    {
         private string[] TechCompanies = new string[14];

         public string this[int indexVal]
         {

            get
            {
               if(indexVal <0 || indexVal >= 11)
                  return "empty";
               else
                  return TechCompanies[indexVal];

         }
         set
         {
            if(!(indexVal <0 || indexVal >= 11))
               TechCompanies[indexVal] = value;
         }

         }
      }
 class theMainClass
 {
  public static void Main()
  {
   IndexersEx Obj = new IndexersEx();

   Obj[0] = "Infosys";
   Obj[3] = "Wipro Infotech";
   Obj[5] = "Satyam Computers";
   Obj[6] = "Patni Computers";
   Obj[8] = "Hexaware";
   Obj[10] = "Polaris Software";
```

```
    Obj[12] = "Mastek Computers";
    Obj[14] = "HCL";

    Console.WriteLine("This is a List of IT Companies in India.\n");
    for(int i=0; i<=14; i++)
    {

        Console.WriteLine("The name of IT Company at index {0} :
  {1} ",i,Obj[i]);
    }

  }
 }
}
```

```
C:\WINNT\system32\cmd.exe
This is a List of IT Companies in India.

The name of IT Company at index 0 : Infosys
The name of IT Company at index 1 :
The name of IT Company at index 2 :
The name of IT Company at index 3 : Wipro Infotech
The name of IT Company at index 4 :
The name of IT Company at index 5 : Satyam Computers
The name of IT Company at index 6 : Patni Computers
The name of IT Company at index 7 :
The name of IT Company at index 8 : Hexaware
The name of IT Company at index 9 :
The name of IT Company at index 10 : Polaris Software
The name of IT Company at index 11 : empty
The name of IT Company at index 12 : empty
The name of IT Company at index 13 : empty
The name of IT Company at index 14 : empty
Press any key to continue . . .
```

Case Study

Problem Statement

Dheeraj, who is working for AB SoftDev Corporation as a senior programmer, has been asked by his manager to create a C# application for CompHard Enterprises that should allow the marketing staff of CompHard Enterprises to obtain information about different varieties of branded computers such as HP and Apple. CompHard Enterprises is involved in the sale of computer hardware. What feature of C# programming language should Dheeraj use in the application which needs to be created for CompHard Enterprises?

Solution

Dheeraj first visits CompHard Enterprises and find out about the varieties of branded computers that are sold by the marketing staff. After finding the information related to branded computers, Dheeraj then decides to create the application using the **class** and **object** features of C#. A class contains a group of related data and methods to perform specific operations on data. The object of a class helps access data and methods contained in a class. Classes and objects are useful for efficient organisation of data related to real-world objects such as students and computers. The following C# application has been created by Dheeraj for the marketing staff of CompHard Enterprises:

```
 using System;
 using System.Collections.Generic;
 using System.Text;

namespace CaseStudy12
 {
     public class AppleComputer
       {

        private string applevariety = "";

        public AppleComputer(string appleComputerVariety)
           {
           this.applevariety = appleComputerVariety;
           }

        public void showAppleComputers()
           {
           System.Console.WriteLine(applevariety);
           }
      }

     public class HPComputer
     {

         private string hpvariety = "";

         public HPComputer(string hpComputerVariety)
            {
             this.hpvariety = hpComputerVariety;
            }

         public void showHPComputers()
            {
             System.Console.WriteLine(hpvariety);
            }
```

```
        }

        public class AcerComputer
          {

            private string acervariety = "";

            public AcerComputer(string acerComputerVariety)
                {
                 this.acervariety = acerComputerVariety;
                }

            public void showAcerComputers()
                {
                 System.Console.WriteLine(acervariety);
                }
        }

    class DisplayComputers
      {

        static void Main()
         {
        Console.WriteLine("This is an example of Classes and objects and
how they can be used in real world scenarios.");
        AppleComputer mac = new AppleComputer("Apple Power Mac G5");
        AppleComputer gra = new AppleComputer("Apple iMac 2 G4");

        Console.WriteLine("---------");
        Console.WriteLine("Apple Computer varities are : ");
        mac.showAppleComputers();
        gra.showAppleComputers();

        HPComputer obj1 = new HPComputer("HP Pavilion Media Center
m8000n");
        HPComputer obj2 = new HPComputer("HP Compaq DC5750 PC ");
        Console.WriteLine("---------");
        Console.WriteLine("HP Computer varities are : ");
        obj1.showHPComputers();
        obj2.showHPComputers();

        AcerComputer acObj1 = new AcerComputer("Acer Aspire E380 PC");
        AcerComputer acObj2 = new AcerComputer("Acer Aspire E560 PC");
        AcerComputer acObj3 = new AcerComputer("Acer Aspire T180 PC");
        AcerComputer acObj4 = new AcerComputer("Acer Aspire T660 PC");
```

```
        Console.WriteLine("---------");
      Console.WriteLine("Acer Computer varities are : ");
      acObj1.showAcerComputers();
      acObj2.showAcerComputers();
      acObj3.showAcerComputers();
      acObj4.showAcerComputers();
      Console.WriteLine("---------");

        }
    }


}
```

Common Programming Errors

- *Forgetting to declare an array.*
- *Placing the semicolon at the end of a class definition.*
- *Specifying a return type for a constructor.*
- *Returning a value from a constructor.*
- *Not declaring a non static constructor as public.*
- *Using the same name for both data member and method member.*
- *Failing to include return type for a method.*
- *Forgetting to call the constructor with the* ***new*** *operator to create an object.*
- *Using a constructor to reset an existing object.*
- *Using a* ***null*** *object to invoke a method.*
- *Forgetting to initialize instance variables using a constructor.*
- *Attempting to invoke a method without using either an object or a class.*
- *Invoking a* ***static*** *method using an object.*
- *Invoking an instance method using its class.*
- *Referring to* ***this*** *in a* ***static*** *method.*
- *Using an access modifier on a* ***static*** *constructor.*
- *Attempting to assign a value to a* ***readonly*** *field after declaration.*
- *Attempting to pass a* ***readonly*** *variable as an* ***out*** *or* ***ref*** *parameter.*
- *Declaring* ***const*** *members as* ***static.***
- *Attempting to assign a value to a* ***const*** *data field at run time.*
- *Forgetting to provide initial value to a* ***const*** *data field at the time of declaration.*
- *Attempting to access a* ***const*** *data using an instance of the class (instead of class name).*

Review Questions

12.1 State whether the following statements are true or false.

(a) We cannot use any access modifiers to a class.

(b) Defining a class without any members is an error.

(c) In C#, all the members have private access by default.

(d) An internal member is available to all the classes in a program.
(e) A protected member is not visible in a derived class.
(f) Data members of class can be accessed only with the help of member methods.
(g) Constructor methods cannot return any values.
(h) A constructor is always declared as public.
(i) A class can implement any number of constructors.
(j) A copy of a static variable is created everytime the class is instantiated.
(k) A static method can access only other static members.
(l) A static constructor is called before any object of the class is created.
(m) A static constructor can be overloaded.
(n) C# does not permit to assign values to data members at the time of declaration.
(o) The reference this can be used only within those member methods that have been declared public.
(p) A class can nest not only complete classes but also objects of other classes.
(q) A const member is implicitly static.
(r) A const data member may be assigned a value using a static constructor.
(s) A read-only member cannot be declared static.
(t) A property that omits get clause is referred to as write-only property.

12.2 What do you mean by a class template?
12.3 What are known as three pillars of object-oriented programming?
12.4 What is encapsulation? What is its role in object-oriented programming.?
12.5 What are the two major categories of class members?
12.6 Describe three different ways of providing values to instance variables of a class.
12.7 How are the static variables given initial values?
12.8 Why do we declare the data fields as public?
12.9 When do we use a private modifier to a data field?
12.10 What is the implication of declaring a member protected?
12.11 Describe two ways of accessing the values of data fields? Compare them.
12.12 What is a constructor?
12.13 State three important characteristics of constructors?
12.14 Why do we usually declare constructors as public?
12.15 Why do we need to use overloaded constructors?
12.16 When do we use static variables?
12.17 When do we use static methods?
12.18 What is the difference between a static method and an instance method?
12.19 What does the keyword **this** refers to?
12.20 What is a static constructor? How is it different from a non-static constructor?
12.21 What are the restrictions of static methods?
12.22 What is a private constructor? When do we use a private constructor?
12.23 What is a copy constructor? Why do we need such a constructor?
12.24 What is finalization?
12.25 Give an example of typical use of this reference.
12.26 What is class nesting? Give an example of typical use of class nesting.
12.27 Distinguish between constant members and read-only members.
12.28 What is a property? Why are they referred to as smart fields?

12.29 What is read_only property? How is it achieved?
12.30 What is an indexer? What is it used for?
12.31 How does an indexer differ from a property in terms of implementation?
12.32 List at least three important features of properties.
12.33 Find errors, if any in the following class definitions:

(a)
```
class Vector
(int x, y};
```
(b)
```
class ABC
{
    public int x;
    protected private y;
}
```
(c)
```
class XYZ
{
    public const int x;
    public static int y;
}
```
(d)
```
class ABC
{
    int x = 1;
    int y = x + 1;
}
```
(e)
```
class ABC
{
static int x = 1, y, z = 2;
new double p = 2.5;
}
```

12.34 What is wrong with the following code? If it is correct, what does it produce?

```
class A
{
 static bool b1;
 int m1;
 public static void Main ( )
 {

      A a = new A( );
      Console.WriteLine (“b1 = ” + b1);
      Console.WriteLine (“m1 = ” + m1);
 }
}
```

12.35 Find errors in the following class definition:

```
class Alpha
{
 int x;
```

```
static int y;
static void F ( )
{
        x = 10;
        y = 20;
}
static void Main ( )
{
        Alpha a1 = new Alpha( );
        a1.x = 100;
        a1.y = 200;
        Alpha.x = 0;
        Alpha .y = 0;
        . . . .
        . . . .
}
}
```

12.36 State whether the following code is legal. If yes, what would be the output?

```
Class A
{
        static double x = y + 1.0;
        static double y = x + 1.0;
        Console.WriteLine (" x = " + x);
        Console.WriteLine ("y = " + y);
}
```

12.37 What is wrong with the following code:

```
class xy
{
        int x = 1;
        int y = x + 1;
}
```

12.38 Find errors in the following class construct:

```
class AB
{
        public int A ( ) { }
        public int B ( )
        { return 0;}
}
```

12.39 What would be the output of the following code?

```
class AB
{
```

```
        int x = 0;
        public void F( )
        {

                int x = 10;
                Console.WriteLine ( x);
        }
}
```

12.40 Comment on the following code:

```
class AB
{
        private int number;
        public int Number
        {
                get { return number;}
        }
        public int Number
        {
                set {number = value;}
        }
}
```

12.41 Consider the following code:

```
class Product
{
        public static void Main ( )
        {
                int x = 10, y = 20;
                Console.WriteLine (mul (x, y));
        }
        int mul (int a, int b)
        {
                return (a*b);
        }
}
```

Will it compile? If not, why?

12.42 What is the use of the following code?

```
class Student
{

        static int m = 0;
        public Student ( )
        { m ++;}
        . . . .
        . . . .
}
```

12.43 Will the following code compile? If no, why?

```
class A
{
      public static void Main ( )
      {
            Test t;
            T.Print( );
      }

}
class Test
{
      public void Print ( ) { }
}
```

Debugging Exercises

12.1 Find syntax errors in the following program.

```
using System;
using System.Collections.Generic;
using System.Text;

namespace DebugApp1_chap12
{
    class Program
    {
        static void Main(string[] args)
        {
            Name nm = new Name();
            nm.Name1 = "Steve";
            nm.Address = "Washington";
            nm.ShowInfo();
            Console.ReadLine();
        }
    }
   class Name
    {
        public string Name1;
        public string Address;

        public void ShowInfo()
        {
            Console.WriteLine("{0} is in city {1}", + Name1, + Address);
        }
    }
}
```

12.2 Debug the program given below.

```
using System;
using System.Collections.Generic;
using System.Text;

namespace debugApp2_chap12
{
    class Program
    {
        static void Main(string[] args)
        {
                Name obj = new Name();
                obj.showinfo();
        }
    }

    class Name
    {
        public showinfo()
        {
                Console.WriteLine("Hello");
        }
    }
}
```

12.3 The program below displays the current date and time. Will the program generate an error? If yes, what type of error?

```
using System;
using System.Collections.Generic;
using System.Text;

namespace debugApp3_chap12
{
    public class TimeDate
    {

        void DisplayCurrentTime_Date()
        {
            Console.Write("The current Date is ");
            Console.WriteLine(DateTime.Now);
        }
    }

    class Program
```

```
    {
       static void Main(string args)
         {
             TimeDate timeObj = new TimeDate();
             timeObj.DisplayCurrentTime_Date();

         }
    }
}
```

12.4 Debug the program below for any errors.

```
using System;
using System.Collections.Generic;
using System.Text;

namespace debugApp4_chap12
{
   class Class2
    {
         int a;
         int b;
        Class2(i, j)
         {
              a = int i;
              b = int j;
         }
        void Display()
         {
              Console.WriteLine(a + b);
         }
    }

   class Program
    {
      static void Main(string[] args)
         {
              Class2 cl = new Class2();
              Console.Write("The output is : ");
               cl.Display();

         }
    }
}
```

Programming Exercises

12.1 Create a class named Number to hold one integer value. Include a constructor to display the message "object is created" and a destructor to display the message "object is destroyed". Write a program to demonstrate the creation and destruction of Number objects.

12.2 Rewrite the program in Exercise 12.1 by adding a Set method to provide a value to the data member and a Get method to display the value stored in the object. Demonstrate your program by storing an integer value, say 2002, and display the same.

12.3 Rewrite the program in Exercise 12.2 by replacing the Set method by a constructor.

12.4 Rewrite the program in Exercise 12.2 by replacing the Set and Get methods by a property.

12.5 Create a class named Integer to hold two integer values. Design a method Swap that could take two integer objects as parameters and swap their contents. Write a Main program to implement the class Integer and the method Swap.

12.6 Design a class named Date with the following members:
- *Data members day, month and year*
- *A constructor to provide values to the data members (using three parameters)*
- *A method to display the date in the format day/month/year (Example: 15/8/1945 for August 15, 1945)*

Write a program to implement the class Date.

12.7 Add to the class designed in Exercise 12.6 another constructor which takes only one parameter that represents the date in the form of a long integer, like 19450815 for the date 15/8/1945 and assigns suitable values for the year, month and day members.
Also modify the Main program to demonstrate the use of both the constructors.
Use a suitable algorithm to convert the long integer 19450815 into year, month and day.

12.8 Define a Time class containing the following members:
- *Two integer data members minutes and hours*
- *Two overloaded constructors*
- *One method to display the class data members*

One of the constructors take two integer parameters to assign values to two data members and the other constructor takes one integer parameter representing total number of minutes and converts it into hours and minutes
Write a complete program to verify the operation of the constructors and the method.

12.9 Add the following two methods to the Time class defined in Exercise 12.8
- *Increment ( ) method that increments time by 1 minute*
- *Decrement ( ) method that decrements time by 1 minute*

Test your methods to ensure that they perform correctly the intended operations.

12.10 Design a class to represent a bank account. Include the following members:
Data members
- *Name of the depositor*
- *Account number*
- *Type of account*
- *Balance amount in the account*

Methods
- *To assign initial values*
- *To deposit an amount*

- *To withdraw an amount after checking balance*
- *To display the name and balance*

Write a program to demonstrate the working of the various class members.

12.11 *Write a class to represent a vector (a series of integer values). Include constructors and methods to perform the following tasks.*

- *To create the vector*
- *To modify the value of a given element*
- *To multiply by a scalar value*
- *To display the vector in the form (10, 20.....)*

Write a program to test your class.

12.12 *Modify the class and the program of Exercise 12.10 for handling 10 customers.*

12.13 *Modify the class and the program of Exercise 12.11 such that the program would be able to add two vectors and display the resultant vector.*

12.14 *A common use of static member variable is to keep track of objects that currently exist for a class. Write a program to illustrate this feature.*

12.15 *Create two classes DM and DB which store the values of distances. DM stores the distances in metres and centimeters and DB in feet and inches. Write a program that can read values for the class objects and add one object of DM with the another object of DB. The program should display the result in the format of feet and inches or metres and centimeters as required.*

12.16 *A book shop maintains the inventory of books that are being sold at the shop. The list includes details such as author, title, price, publisher and stock position. Whenever a customer wants a book, the sales person inputs the title and the author and the system searches the list and displays whether it is available or not. If it is not, an appropriate message is displayed. If it is, then the system displays the book details and requests for the number of copies required. If the requested copies are available, the total cost is displayed; otherwise the message "Insufficient Stock" is displayed.*

Design a system using a class called Books with suitable members.

12.17 *Improve the system design of Exercise 12.16 to incorporate the following features:*

- *The price of the book should be updated as and when required.*
- *The stock value of each book should be automatically updated as soon as a transaction is completed.*
- *The number of successful and unsuccessful transactions should be recorded for the purpose of statistical analysis. Use static data members to keep count of transactions.*

Chapter

13

Inheritance and Polymorphism

13.1 INTRODUCTION

It is always nice if we could reuse something that already exists rather than creating the same over again. C#, being a pure object-oriented language, supports this feature. In fact, the ability to create new classes from the existing classes is the major motivation and power behind using OOP languages.

C# classes can be reused in several ways. Reusability is achieved by designing new classes, reusing all or some of the properties of existing ones. The mechanism of designing or constructing one class from another is called *inheritance*. This may be achieved in two different forms.

- Classical form
- Containment form

Related to inheritance is an equally important feature known as *polymorphism.* This feature permits the same method name to be used for different operations in different derived classes. We shall discuss all these features in this chapter.

13.2 CLASSICAL INHERITANCE

Inheritance represents a kind of relationship between two classes. Let us consider two classes **A** and **B**. We can create a class hierarchy such that **B** is derived from **A** as shown in Fig. 13.1.

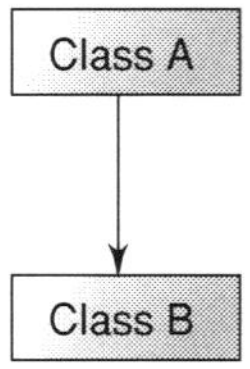

Figure 13.1 *Simple inheritance*

Class **A**, the initial class that is used as the basis for the derived class is referred to as the *base class*, *parent class* or *superclass.* Class **B**, the derived class, is referred to as *derived class, child class* or *sub class.* A derived class is a completely new class that incorporates all the data and methods of its base class. It can also have its own data and method members that are unique to itself. That is, it can enhance the content and behaviour of the base class.

We can now create objects of classes A and B independently. *Example:*

```
A a; // a is object of A
B b; // b is object of B
```

In such cases, we say that the object **b** is a type of **a.** Such relationship between **a** and **b** is referred to as 'is-a' relationship. Examples of is-a relationship are:

- Dog is-a type of animal
- Manager is-a type of employee
- Ford is-a type of car

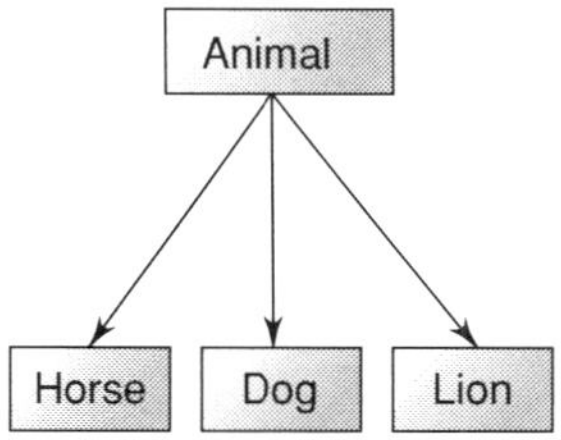

Figure 13.2 *The is-a relationship*

The is-a relationship is illustrated in Fig. 13.2.

The classical inheritance may be implemented in different combinations as illustrated in Fig.13.3. They include:

- Single inheritance (only one base class)
- Multiple inheritance (several base classes)
- Hierarchical inheritance (one base class, many subclasses)
- Multilevel inheritance (derived from a derived class)

C# does not directly implement multiple inheritance. However, this concept is implemented using a secondary inheritance path in the form of interfaces. Interfaces are discussed in Chapter 14.

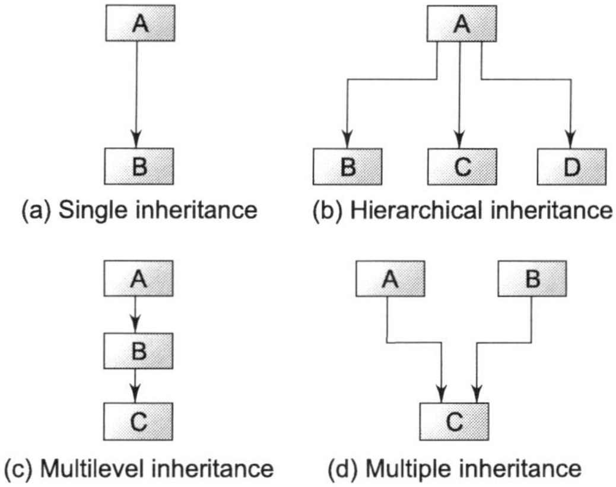

Figure 13.3 *Implementation of inheritance*

13.3 CONTAINMENT INHERITANCE

We can also define another form of inheritance relationship known as *containership* between class A and B. *Example:*

```
class A
{
    . . . .
}
class B
{
    . . . .
    A a; // a is contained in b
}
B b;
. . . .
```

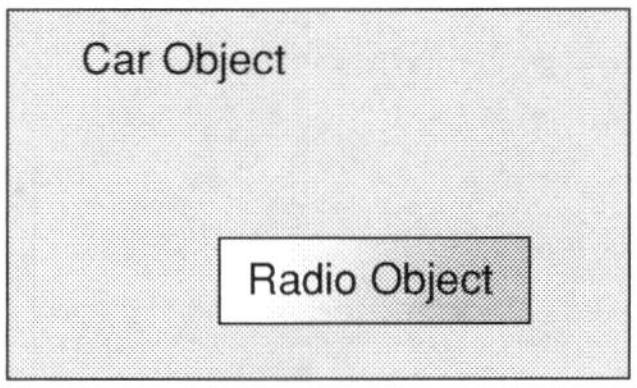

Figure 13.4 *The has-a relationship*

In such cases, we say that the object **a** is contained in the object **b.** This relationship between **a** and **b** is referred to as 'has–a'relationship. The outer class **B** which contains the inner class **A** is termed the 'parent' class and the contained class A is termed a 'child' class. Examples are:

- Car has-a radio
- House has-a store room
- City has-a road

The has-a relationship is illustrated in Fig.13.4.

13.4 DEFINING A SUBCLASS

A subclass is defined as follows:

```
Class subclass-name : baseclass-name

{
    variables declaration ;
    methods declaration ;

 }
```

The definition is very similar to a normal class definition except for the use of colon**:** and *baseclass-name.* The colon signifies that the properties of the baseclass are extended to the *subclass-name.* When implemented the subclass will contain its own members as well those of the baseclass. This kind of situation occurs when we want to add more properties to an existing class without actually modifying it. For example:

```
class Cylinder : Circle
{
    // Add additional fields
    // and methods here
}
```

Circle is the name of an existing class containing value of the radius and **Cylinder** is the name of derived class which could declare the length of the cylinder as an additional data member and a method to compute its volume.

Program 13.1 illustrates the concept of simple inheritance.

Program 13.1 Illustration of a simple inheritance

```
using System;
Class Item
{
    public void Company ( )     // base class
    {
        Console.WriteLine("Item Code = XXX");
    }
}
```

```
class Fan : Item                    // derived class
{
    public void Model ( )
    {
        Console.WriteLine("Fan Model : Classic");
    }
}

class SimpleInheritance
{
    public static void Main( )
    {
        Item item = new Item( ) ;
        Fan fan = new Fan( ) ;
        item.Company( ) ;
        fan.Company( ) ;
        fan.Model( ) ;
    }
}
```

The output of Program 13.1 would be:

```
Item Code = XXX
Item Code = XXX
Fan Model : Classic
```

Note that we are able to invoke the method **Company ( ),** a base class member using the objects of both the base class and derived class. This implies that the class **Fan** has inherited the method **Company ( )** to its own membership. That is, the class **Fan** has now two method members.

Note that the inheritance is transitive. If C is derived from B, and B is derived from A, then C inherits the members declared in B as well as the members declared in A. The inheritance relationship

```
class A : B
{
}
class B : C
{
}
class C : A
{
}
```

is wrong because the classes circularly depend on themselves. However, the example

```
class A
{
    class B : A
    {
    }
}
```

is valid. Note that a class does not depend on the classes that are enclosed. In this example, B depends on A but A does not depend on B.

Some important characteristics of inheritance are:

- A derived class extends its direct base class. It can add new members to those it inherits. However, it cannot change or remove the definition of an inherited member.
- Constructors and destructors are not inherited. All other members, regardless of their declared accessibility in base class, are inherited. However, their accessibility in the derived class depends on their declared accessibility in the base class. (Section 13.5).
- An instance of a class contains a copy of all instance fields declared in the class and its base classes.
- A derived class can hide an inherited member (Section 13.10).
- A derived class can override an inherited member (Section 13.9).

13.5 VISIBILITY CONTROL

When implementing inheritance, it is important to understand how to establish visibility levels for our classes and their members. Recall that we have discussed four types of accessibility modifiers which may be applied to classes and members to specify their level of visibility. They include **public, private, protected** or **internal.**

13.5.1 Class Visibility

Each class needs to specify its level of visibility. Class visibility is used to decide which parts of the system can create class objects.

A C# class can have one of the two visibility modifiers: **public** or **internal.** If we do not explicitly mark the visibility modifier of a class, it is implicitly set to 'internal'; that is, by default all classes are **internal.** Internal classes are accessible within the same program assembly and are not accessible from outside the assembly.

Classes marked **public** are accessible everywhere, both within and outside the program assembly. All the classes contained in the .NET Class Framework are marked **public** so as to enable all developers to have access to them and use them for configuring applications and services.

Although classes are normally marked either **public** or **internal**, a class may also be marked **private** when it is declared as a member of another class. In such cases, the class behaves like a member of the enclosing class and therefore the modifiers applicable to class members are also applicable to it.

13.5.2 Class Members Visibility

As mentioned in the previous chapter, a class member can have any one of the five visibility modifiers:

- public
- protected
- private
- internal
- protected internal

Except for the **protected internal** combination, it is an error to specify more than one access modifier. When no modifier is specified, it defaults to **private** accessibility. Table 13.1 shows how the visibility modifiers control the visibility of members in a program.

Table 13.1 *Visibility of class members*

Keyword	Visibility			
	Containing classes	Derived classes	Containing program	Anywhere outside the containing program
Private	✓			
protected	✓	✓		
Internal	✓		✓	
protected internal	✓	✓	✓	
Public	✓	✓	✓	✓

It is important to remember that the accessibility domain of a member is never larger than that of the class containing it. Table 13.2 gives the visibility domain of members under different combinations of class access specifiers.

Table 13.2 *Accessibility domain of class members*

Member modifier	Modifier of the containing class		
	public	internal	private
public	Everywhere	only program	only class
internal	only program	only program	only class
private	only class	only class	only class

13.5.3 Accessibility of Baseclass Members

When a class inherits from a base class, all members of the base class, except constructor and destructors, are inherited and become members of the derived class. The declared accessibility of a base class member has no control over its inheritability. However, an inherited member may not be accessible in a derived class, either because of its declared accessibility or because it is hidden by a declaration in the class itself.

Let us consider the code shown below:

```
class A
{
   private int x ;
   protected int y ;
   public int z ;
}
class B : A
{
   public void SetXYZ ( )
```

```
    {
        x = 10; // Error; x is not accessible
        y = 20; // ok
        z = 30; // ok
    }
}
```

Note that although B inherits the private member x from A, it is not accessible in B. However, the **protected** and **public** members of A are accessible in the subclass B. But as far as the users of the classes are concerned, **protected** is private. Therefore, the following is illegal:

```
A a = new A ( ); //object of A
a.y = 5; //error! an object cannot access protected data
a.z = 10; // ok
```

13.5.4 Accessibility Constraints

C# imposes certain constraints on the accessibility of members and classes when they are used in the process of inheritance.

- The direct base class of a derived class must be at least as accessible as the derived class itself.
- Accessibility domain of a member is never larger than that of the class containing it.
- The return type of method must be atleast as accessible as the method itself.

For example, the inheritance relationship

```
class A
{
. . . .
}
public class B : A
{
. . . .
}
```

is illegal because **A** is 'internal' by default and **B** is public. **A** should be at least as accessible as **B**. Consider another example:

```
class A
{
    private class B
    {
        public int x;
    }
}
```

Here, the public data **x** is not accessible outside the class **B**. It is due to the constraint imposed by the accessibility of class **B**. Because **B** is private, the public on **x** is reduced to **private.**

In many situations we use methods to handle class objects and therefore they take classes as their return types. In such cases, the return type of a method must be at least as accessible as the method itself. For instance, consider the following code:

```
class A
{
. . . .
}
public class B
{
   A Method1( ) { }             //OK
   internal A Method2( ) { }    //OK
   public A Method3( ) { }      //Error!
}
```

All the three methods declared in class **B** specify **A** as their return type. Since the class **A**, by default, assumes **internal** as the accessibility level, the method

```
public A Method3( ) //public is higher than internal
{
. . . .
}
```

is in error. The method cannot have an accessibility level higher than that of its return type.

13.6 DEFINING SUBCLASS CONSTRUCTORS

We have seen that an object is created when constructor is called. The same principle may be applied for constructing the derived class objects as well. We can define an appropriate constructor for the derived class that may be invoked when a derived class object is created. Remember, the purpose of a constructor is to provide values to the data fields of the class. What if the base class constructor takes arguments for constructing base class objects? How do we pass values to the base class constructor? Program 13.2 answers these questions by implementing a single level inheritance.

Program 13.2 Application of single inheritance

```
using System;
class Room          // base class
{
        public int length;
        public int breadth;

        public Room (int x , int y)  // base constructor
        {
            length  =   x;
            breadth =   y;
        }
        public int Area ( )
        {
            return (length * breadth );
        }
```

```
}

class BedRoom : Room   //Inheriting Room
{
        int height;
                //subclass constructor
        public Bedroom (int x, int y, int z):base (x,y)
        {

            height = z;
        }
        public int Volume ( )
        {
            return (length * breadth * height);
        }
}
class InherTest
{
        public static void Main( )
        {
            BedRoom room1 = new BedRoom (14, 12, 10);

            int area1 = room1.Area ( );     // superclass method
        int volume1 = room1.Volume ( );     // subclass method

            Console.WriteLine("Area1 =   " + area11);
            Console.WriteLine("Volume1 = " + volume1);
    }
}
```

The output of Program 13.2 would be:

```
Area1  = 168
Volume1= 1680
```

The program defines a class **Room** and extends it to another class **BedRoom.** Note that the class **BedRoom** defines its own data members and methods. The subclass Bedroom now includes three instance variables, namely**, length, breadth** and **height** and two methods, **Area** and **Volume**.

The constructor in the derived class uses the **base** keyword to pass values that are required by the base constructor. The statement

```
BedRoom room1 = new BedRoom (14,12,10);
```

calls first the **BedRoom** constructor method which in turn calls the room constructor method by using the **base** keyword.

Finally, the object **room1** of the subclass **BedRoom** calls the method **Area** defined in the super class as well as the method **Volume** defined in the subclass itself.

13.6.1 Subclass Constructor

A subclass constructor is used to construct the instance variables of both the subclass and the superclass. The subclass constructor uses the keyword **base** to invoke the constructor method of the superclass.

```
public BedRoom (int x, int y, int z) : base (x, y)
{
    height = z;
}
```

Here, the derived constructor takes three arguments, the first two to provide values to the base constructor and the third one to provide value to its own class member. Note that **base (x,y)** behaves like a method call and therefore arguments are specified without types. The type and order of these arguments must match the type and order of base constructor arguments.

When the compiler encounters **base (x,y),** it passes the values **x** and **y** to the base class constructor which takes two **int** arguments. The base class constructor is called before the derived class constructor is executed. That is, the data members **length** and **breadth** are assigned values before the member **height** is assigned its value.

Note that the use of keyword **base** is not limited to the constructor logic alone. We may use the **base** in any subclass to access a **public** or **protected** member defined in a parent class. For instance, statements such as:

```
base.length = 100;
int area = base.Area( );
```

are valid inside the derived class **Bedroom.**

13.7 MULTILEVEL INHERITANCE

A common requirement in object-oriented programming is the use of a derived class as a superclass. C# supports this concept and uses it extensively in building its class library. This concept allows us to build a chain of classes as shown in Fig. 13.5.

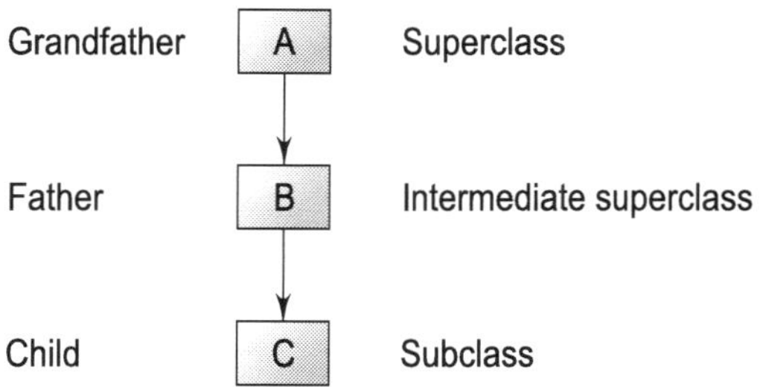

Figure 13.5 *Multilevel inheritance*

The class **A** serves as a base class for the derived class **B** which in turn serves as a base class for the derived class **C**. The chain **ABC** is known as *inheritance path.*

A derived class with multilevel base classes is declared as follows:

```
class A
{
    . . . .
    . . . .
}
class B : A // First level derivation
```

```
{
    . . . . .
    . . . . .
}
class C : B // Second level derivation
{
    . . . . .
    . . . . .
}
```

This process may be extended to any number of levels. The class **C** can inherit the members of both **A** and **B** as shown in Fig. 13.6.

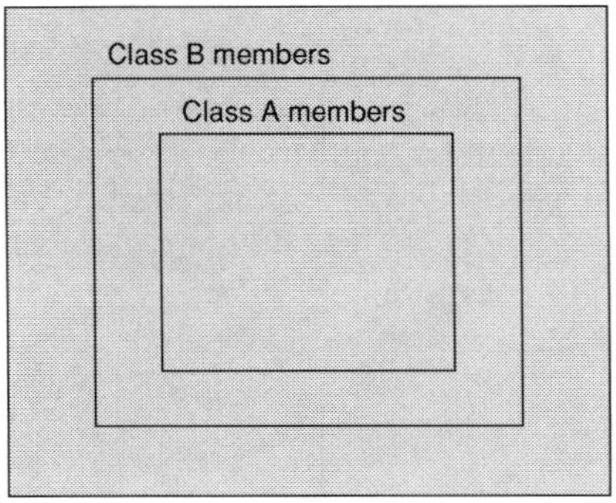

Figure 13.6 *C contains B which contains A*

As discussed earlier, the constructors are executed from the top downwards, with the bottom most class constructor being executed last. *Example:*

```
class A
{
    protected int a;
    public A (int x)
{
        a = x;
    }
}
class B : A
{
    protected int b;
    public B (int x, int y ) : base (x)
    {
        b = y;
    }
}
class C : B
{
    int c ;
    public C (int x, int y, int z) : base (x, y)
    {
        c = z;
    }
}
```

The constructors are implemented in the following order:

```
A ( ) // class A
B ( ) // class B
C ( ) // class C
```

Note that we have used **base** in both the derived classes without specifying the actual name of the **base** class. Since C# supports only 'single' inheritance, base means the 'immediate' base class constructor.

However, care should be exercised to place arguments in proper order so that correct values are passed to the data members.

Program 13.3 shows the multilevel inheritance of **Base Class** from three other inherited classes, calling attributes of the **Base Class** in them and manipulating them to display the desired output.

Program 13.3 Multilevel inheritance

```
// Multilevel Inheritance and Function Overriding

using System;
using System.Collections.Generic;
using System.Text;


namespace MultiLevel_Inheritance
{
  class BaseClass
   {
     public int num1, num2, num3;
     public virtual int function1()
       {
        Console.WriteLine("Base Class: function1() returning 1 called.");
          return 1;
       }
     public virtual int function2()
       {
        Console.WriteLine("Base Class: function2() returning 2 called.");
          return 2;
       }
     public virtual int function3()
       {
        Console.WriteLine("Base Class: function3() returning 3 called.");
          return 3;
       }
     public BaseClass()
       {
        Console.WriteLine("Base Class: Constructor called.");
         num1 = function1();
         num2 = function2();
         num3 = num1 + num2 + function3();
       }
     public static int Main()
       {
        InheritedClass1 ic = new InheritedClass1();
        InheritedClass2 ic2 = new InheritedClass2();
        InheritedClass3 ic3 = new InheritedClass3();
        Console.WriteLine("The value of num1 is " + ic.num1);
        Console.WriteLine("The value of num2 is " + ic2.num2);
```

```
        Console.WriteLine("The value of num3 is " + ic3.num3);
         return 0;
      }
   }

   class InheritedClass1 : BaseClass
   {
     public InheritedClass1()
       {
        Console.WriteLine("Inherited Class1: Constructor called.");
       }
     public override int function1()
       {
        Console.WriteLine("Inherited Class: function1() returning 10 called.");
         return 10;
       }


   }

   class InheritedClass2 : BaseClass
   {
     public InheritedClass2()
       {
        Console.WriteLine("Inherited Class2: Constructor called.");
       }
     public override int function2()
       {
        Console.WriteLine("Inherited Class: function2 returning 20 called.");
         return 20;
       }

   }

   class InheritedClass3 : BaseClass
   {
     public InheritedClass3()
       {
        Console.WriteLine("Inherited Class3: Constructor called.");
       }
     public override int function3()
       {
        Console.WriteLine("Inherited Class: function3 returning 30 called.");
         return 30;
       }

   }
}
```

```
C:\WINNT\system32\cmd.exe
Base Class: Constructor called.
Inherited Class: function1() returning 10 called.
Base Class: function2() returning 2 called.
Base Class: function3() returning 3 called.
Inherited Class1: Constructor called.
Base Class: Constructor called.
Base Class: function1() returning 1 called.
Inherited Class: function2 returning 20 called.
Base Class: function3() returning 3 called.
Inherited Class2: Constructor called.
Base Class: Constructor called.
Base Class: function1() returning 1 called.
Base Class: function2() returning 2 called.
Inherited Class: function3 returning 30 called.
Inherited Class3: Constructor called.
The value of num1 is 10
The value of num2 is 20
The value of num3 is 33
Press any key to continue . . .
```

13.8 HIERARCHICAL INHERITANCE

Another interesting application of inheritance is to use it as a support to the hierarchical design of a program. Many programming problems can be cast into a hierarchy where certain features of one level are shared by many others below the level. An an example, Fig. 13.7 shows a hierarchical classification of accounts in a commercial bank. This is possible because all the accounts possess certain common features.

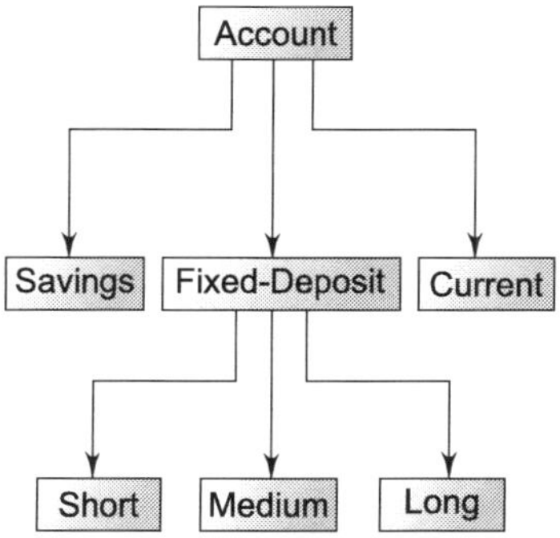

Figure 13.7 *Hierarchical classification of bank accounts*

13.9 OVERRIDING METHODS

We have seen that a method defined in a superclass is inherited by its subclass and is used by the objects created by the subclass. Method inheritance enables us to define and use methods repeatedly in subclasses.

However, there may be occasions when we want an object to respond to the same method but behave differently when that method is called. That means, we should override the method defined in the superclass. This is possible by defining a method in the subclass that has the same name, same arguments and same return type as a method in the superclass. Then, when that method is called, the method defined in the subclass is invoked and executed instead of the one in the superclass, provided that:

- We specify the method in base class as **virtual**
- Implement the method in subclass using the keyword **override**

This is known as *overriding*. Program 13.4 illustrates the concept of overriding.

Program 13.4 Illustration of method overriding

```
using System;
class Super                    //base class
{
    protected int x ;
    public Super (int x)
    {
          this.x = x ;


    }
    public virtual void Display () // method defined with virtual

    {
        Console.WriteLine ("Super x = " + x);
    }
}
class Sub : Super                  //derived class
{
    int y;
    public Sub (int x, int y) : base (x)
    {
        this.y = y;
    }
    public override void Display ( )   // method defined again
                                       //with override
    {
        Console.WriteLine("Super x = " + x) ;
        Console.WriteLine("Sub y = " + y);

    }
}
class OverrideTest
{
    public static void Main( )
    {
        Sub s1 = new Sub (100,200) ;
        s1.Display ( ) ;
    }
}
```

Output of Program 13.4 shows that the base class method has not been called.

```
Super x = 100
Sub y = 200
```

When overriding a method of a base class, we must be aware that we cannot change the accessibility level of the method. For some reason, if we want to call the base method we may do so using the **base** reference as shown below:

```
base.Display ( );
```

Note:

1. An override declaration may include the **abstract** modifier.
2. It is an error for an override declaration to include **new** or **static** or **virtual** modifier.
3. The overridden base method cannot be **static** or nonvirtual.
4. The overridden base method cannot be a **sealed** method.

13.10 HIDING METHODS

In the previous section, when we were overriding a base class method, we declared the base class method as **virtual** and the subclass method with the keyword **override**. This resulted in 'hiding' the base class method from the subclass.

Now, let us assume that we wish to derive from a class provided by someone else and we also want to redefine some methods contained in it. Here, we cannot declare the base class methods as virtual. Then, how do we override a method without declaring it virtual? This is possible in C#. We can use the modifier **new** to tell the compiler that the derived class method "hides" the base class method. Program 13.5 demonstrates how to use the **new** modifier to hide a base class method.

Program 13.5 Hiding a base class method

```
using System;
class Base
{
    public void Display( )
    {
        Console.WriteLine("Base Method");
    }
}

class Derived : Base
{
    public new void Display( )        // hides base method
    {
        Console.WriteLine("Derived Method");
    }
}
```

```
class HideTest
{
    public static void Main( )
    {
        Derived d = new Derived( );
        d.Display ( );
    }
}
```

Program 13.5 would display the following:

```
Derived Method
```

The **new** modifier indicates that the **Display ( )** in **Derived** is 'new', and it is intended to hide the inherited member.

Note however that hiding an inherited member using **new** does not remove the member; it only makes the member inaccessible in the derived class. If the **new** modifier is used in a declaration of a member that does not hide any inherited member, then the compiler will issue a warning. It is an error to use both the **new** and **override** modifiers in the same declaration although they independently achieve the same effect.

A declaration of a new method hides an inherited method only within the scope of the 'new' member. *Example:*

```
class A
{
    public static void F( )
    { }
}
class B : A
{
    new private static void F( )
    { }
}
class C : B
{
    static void F1( )
    {
        F( ); //Invokes F( )of A

    }
}
```

Since the **new** F( ) in **B** has private access, its scope does not extend to the class **C**. Thus, the call **F ()** in **C,** which is valid, invokes **F ( )** of A class.

13.11 ABSTRACT CLASSES

In a number of hierarchical applications, we would have one base class and a number of different derived classes. The top-most base class simply acts as a base for others and is not useful on its own. In such situations, we might not want any one to create its objects. We can do this by making the base class **abstract.**

The **abstract** is a modifier and when used to declare a class indicates that the class cannot be instantiated. Only its derived classes (that are not marked abstract) can be instantiated. *Example:*

```
abstract class Base
{
    . . . .
}

class Derived : Base
{
    . . . .
}
. . . .
. . . .
. . . .
Base b1;        //Error
Derived d1;     //OK
```

We cannot create objects of **Base** type but we can derive its subclasses which can be instantiated. Some characteristics of an abstract class are:

- It cannot be instantiated directly
- It can have abstract members
- We cannot apply a **sealed** modifier to it

13.12 ABSTRACT METHODS

Similar to abstract classes, we can also create abstract methods. When an instance method declaration includes the modifier **abstract**, the method is said to be an *abstract method.*

An abstract method is implicitly a virtual method and does not provide any implementation. Therefore, an abstract method does not have method body. *Example:*

```
public abstract void Draw(int x, int y);
```

Note that the method body simply consists of a semicolon. Some characteristics of an abstract method are:

- It cannot have implementation.
- Its implementation must be provided in non-abstract derived classes by overriding the method
- It can be declared only in abstract classes.
- It cannot take either **static** or **virtual** modifiers.
- An abstract declaration is permitted to override a virtual method.

13.13 SEALED CLASSES: PREVENTING INHERITANCE

Sometimes, we may like to prevent a class being further subclassed for security reasons. A class that cannot be subclassed is called a *sealed class*. This is achieved in C# using the modifier **sealed** as follows:

```
sealed class Aclass
{
    . . . .
    . . . .
}
sealed class Bclass: Someclass
{
    . . . .
    . . . .
}
```

Any attempt to inherit these classes will cause an error and the compiler will not allow it.

Declaring a class **sealed** prevents any unwanted extensions to the class. It also allows the compiler to perform some optimizations when a method of a sealed class is invoked. Usually standalone utility classes are created as sealed classes.

A **sealed** class cannot also be an **abstract** class.

13.14 SEALED METHODS

When an instance method declaration includes the **sealed** modifier, the method is said to be a *sealed method.* It means a derived class cannot override this method.

A sealed method is used to override an inherited virtual method with the same signature. That means, the **sealed** modifier is always used in combination with the **override** modifier. *Example:*

```
class A
{
    public virtual void Fun( )
    {
    . . . .
    }
}
class B : A
{
    public sealed override void Fun ( )
    {
    . . . .
    }
}
```

The **sealed** method **Fun( )** overrides the **virtual** method **Fun( )** defined in Class **A.** Any derived class of **B** cannot further override the method **Fun( ).**

13.15 POLYMORPHISM

As discussed earlier in Chapter 12, *polymorphism* means 'one name, many forms'. Essentially, polymorphism is the capability of one object to behave in multiple ways. Polymorphism can be achieved in two ways as shown in Fig. 13.8. C# supports both of them.

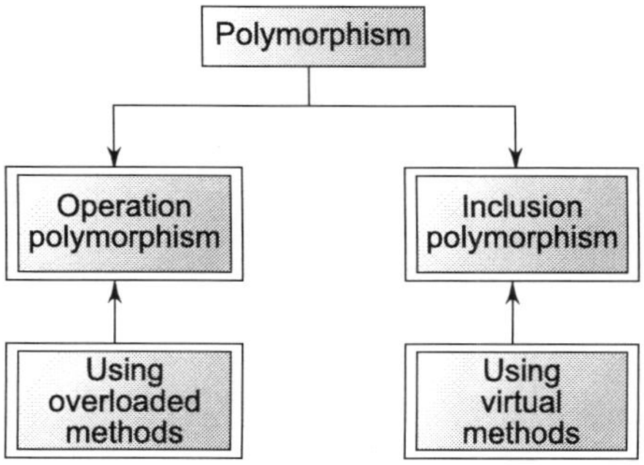

Figure 13.8 *Achieving polymorphism*

13.15.1 Operation Polymorphism

Operation polymorphism is implemented using overloaded methods and operators. We have already used the concept of overloading while discussing methods and constructors. The overloaded methods are 'selected' for invoking by matching arguments, in terms of number, type and order. This information is known to the compiler at the time of compilation and, therefore, the compiler is able to select and bind the appropriate method to the object for a particular call at *compile time* itself. This process is called *early binding,* or *static binding*, or *static linking*. It is also known as *compile time polymorphism.*

Early binding simply means that an object is bound to its method call at compile time. Program 13.6 shows how the compile time polymorphism is achieved using the concept of overloading. As we know, with overloading, multiple methods with the same name are created. However, the parameters on each method vary. When an object makes a method call using the method name, the method that matches the calling signature is invoked.

Program 13.6 Operation polymorphism

```
using System;
class Dog
{
}
class Cat
{
}
class Operation
{

    static void Call (Dog d)
    {
        Console.WriteLine ("Dog is called");
    }
    static void Call (Cat c)
    {
        Console.WriteLine (" Cat is called ");
    }
```

```
    public static void Main( )
    {
        Dog dog = new Dog( );
        Cat cat = new Cat ( );
        Call(dog); //invoking Call( )
        Call(cat); //again invoking Call( )
    }
}
```

Output of Program 13.6 would be:

```
Dog is called
Cat is called
```

Note that we have created two different **Call ( )** methods. One takes **Dog** type object as a parameter and the other takes **Cat** type object. The result is:

- Call (dog) invokes the method Call (Dog d)
- Call (cat) invokes the method Call (Cat d)

With the operation polymorphism, the multiple forms occur at the method level, rather than at class level. The same method name takes multiple implementation forms.

13.15.2 Casting Between Types

Let us look at one of the important aspects in the application of inheritance, namely, type casting between classes. There are a number of situations where we need to apply casting between the objects of base and derived classes. C# permits *upcasting* of an object of a derived class to an object of its base class. However, we cannot *downcast* implicitly an object of a base class to an object of its derived classes. *Examples:*

```
class Base { }
class Derived : Base { }
. . . .
. . . .
    Base b = new Derived ( ); //OK, upcasting
    . . . .
    . . . .
    Derived d = new Base ( ); //ERROR, downcasting
. . . .
. . . .
```

In case a downcast is required, it can be achieved using an explicit cast operation.

13.15.3 Inclusion Polymorphism

Inclusion polymorphism is achieved through the use of virtual functions. Assume that the class **A** implements a **virtual** method **M** and classes **B** and **C** that are derived from **A** override the virtual method **M.** When **B** is cast to **A**, a call to the method **M** from **A** is dispatched to **B**. Similarly, when **C** is cast to **A**, a call to **M** is dispatched to **C**. The decision on exactly which method to call is delayed until runtime and, therefore, it is also known as *runtime polymorphism.* Since the method is linked with a particular class much later after compilation, this process is termed *late binding*. It is also known as *dynamic binding* because the selection of the appropriate method is done dynamically at runtime. Program 13.7 illustrates the use of virtual methods to implement polymorphic behaviour of objects.

Program 13.7 Inclusion polymorphism

```
using System;
class Maruthi
{
    public virtual void Display ( ) //virtual method
    {
        Console.WritLine("Maruthi car");
    }
}
class Esteem : Maruthi
{
    public override void Display( )
    {
        Console.WriteLine("Maruthi Esteem");
    }
}
class Zen : Maruthi
{
    public override void Display ( )
    {
        Console.WriteLine("Maruthi Zen");
    }
}
class Inclusion
{
    public static void Main( )
    {
        Maruthi m = new Maruthi ( );
        m = new Esteem ( ); //upcasting
        m.Display ( );
        m = new Zen ( );    //upcasting
        m.Display ( )
    }
}
```

Program 13.7 outputs:

```
Maruthi : Esteem
Maruthi : Zen
```

In Program 13.7, a particular car object, whether it be **Esteem** or **Zen**, is cast to **m** which is of type **Maruthi**, the base class, using the casting statements.

```
m = new Esteem ( );
m = new Zen ( );
```

When an object of **Esteem** is cast to **m**, then, **m** behaves like an **Esteem** type object. Similarly, when a **Zen** object is cast to **m**, it behaves like a **Zen** type object. Therefore, the two identical calls produce two different outputs:

```
Maruthi Esteem
Maruthi Zen
```

Program 13.8 shows the concept of polymorphism while applying multilevel inheritance. A single method is called several times in the **Base** class and its nested classes and different output is shown for the same method based on the calling class.

Program 13.8 Polymorphism illustrated further

```
using System;
using System.Collections.Generic;
using System.Text;

namespace PolymorphismExample
{
   class PolymorphismDemo
    {
      static void Main(string[] args)
       {
        Automobile a = new Automobile();
        a.PowerSource = "100 H.P.";
        a.IncreaseVelocity(60);

         BMW car = new BMW();
        car.Color("Black");

        Console.WriteLine("\nPress Enter to exit...");
        Console.ReadLine();
       }
    }

   class Vehicle
    {

      string Vehiclepower = "";

      public string PowerSource
       {
```

```
        set
        {
          Vehiclepower = value;
         Console.WriteLine("Power of engine is set to \"{0}\"", value.ToString());
        }
        get
        {
          return Vehiclepower;
        }
    }//end property PowerSource

    public virtual void IncreaseVelocity(int i)
      {

       Console.WriteLine("The speed increases to " + i.ToString() + " m.p.h.");

    } //end method IncreaseVelocity



  } //end base class Vehicle

  class Automobile : Vehicle
   {

    public override void IncreaseVelocity(int i)
      {
        if (i > 120)
        {
         // an automobile shouldn't be going that fast, so reducing the speed to 120
m.p.h.
           i = 120;
        }
       Console.WriteLine("You press the accelerator of the automobile, increasing its
speed to {0}",
          i.ToString() + " m.p.h.");

      }

    public virtual void Color(string col)
      {
        col = "Blue";
       Console.Write("The color of BMW is ");
       Console.WriteLine(col);

    } //end method Color
```

```
    } //end derived class Automobile

    class BMW : Automobile
     {

       public override void Color(string col)
         {

              col = "Navy";
             Console.Write("The color of BMW is ");
             Console.WriteLine(col);
         }

    } //end derived class BMW

 }
```

```
C:\WINNT\system32\cmd.exe
Power of engine is set to "100 H.P."
You press the accelerator of the automobile, increasing its speed to 60 m.p.h.
The color of BMW is Navy

Press Enter to exit...
-
```

Case Study

Problem Statement

ArchBuild is a small organisation, which is involved in creating architectural designs for buildings such as homes, hospitals and educational institutes. While making the architectural designs, the architects working in ArchBuild need to calculate the area of different shapes such as rectangles, circles and squares. Till now, these architects have been calculating areas manually using calculator for complex calculations. Manual calculation of different shapes is time consuming. As a result, the efficiency of the architects working in ArchBuild has decreased considerably. How can ArchBuild solve this problem?

Solution

To automate the task of calculating the area of different shapes such as circles and rectangles, ArchBuild contacts ABC SoftDev Private Limited, which is a software development organisation. ArchBuild asks ABC SoftDev to create an application which will allow the architects working in ArchBuild to obtain the area of a shape. ABC SoftDev first performs a requirement analysis to understand the requirements of the architects. After the requirement analysis, it starts developing an application using the **inheritance** feature of the C# programming language. The **inheritance** feature allows a class in C# to inherit the properties and methods defined in another class besides having its own properties and methods. Inheritance allows programmers to reuse the code of one application in another application. ABC SoftDev creates the following application for ArchBuild to automate the task of calculating areas for different shapes.

```
using System;
using System.Collections.Generic;
using System.Text;

namespace CaseStudy13
{
   public abstract class Dimension_Shape
    {

      protected string colorValue;

      public Dimension_Shape(string color)
       {
          this.colorValue = color;
       }

      public string getColor()
       {
          return colorValue;
       }

      public abstract double getAreaof_Shape();


   }

   public class ShapeCircle : Dimension_Shape
   {

      private double radiusofCircle;
```

```
    public ShapeCircle(string colorValue, double radius)
        : base(colorValue)
    {
        this.radiusofCircle = radius;
    }

    public override double getAreaof_Shape()
    {
        return System.Math.PI * radiusofCircle * radiusofCircle;
    }
}

public class ShapeRectangle : Dimension_Shape
{

    private double lengthofRectangle;
    private double widthofRectangle;

    public ShapeRectangle(string colorValue, double length, double width)
        : base(colorValue)
    {
        this.lengthofRectangle = length;
        this.widthofRectangle = width;
    }

    public override double getAreaof_Shape()
    {
        return lengthofRectangle * widthofRectangle;
    }
}

public class ShapeSquare : Dimension_Shape
{

    private double lengthofSquare;

    public ShapeSquare(string colorValue, double length) : base(colorValue)
    {
        this.lengthofSquare = length;
    }

    public override double getAreaof_Shape()
    {
        return lengthofSquare * lengthofSquare;
    }
}
```

```
    class AreaandColor
    {
        static void Main(string[] args)
        {
            Console.WriteLine("This is an example where all the properties of different
classes\nlike Circle,Square can be defined in another class and can be inherited \nfrom it
when required without populating the classes with excess codes.");
            Console.WriteLine("————————");
            Dimension_Shape myCircle = new ShapeCircle("Maroon", 6);
            Dimension_Shape myRectangle = new ShapeRectangle("Blue", 12, 9);
            Dimension_Shape mySquare = new ShapeSquare("Gray", 15);
            System.Console.WriteLine("The Circle is of " + myCircle.getColor()
                + " color and its area is " + myCircle.getAreaof_Shape() + ".");
            Console.WriteLine("————————");
            System.Console.WriteLine("The Rectangle is of " + myRectangle.getColor()
              + " color and its area is " + myRectangle.getAreaof_Shape() + ".");
            Console.WriteLine("————————");
            System.Console.WriteLine("The square is of " + mySquare.getColor()
              + " color and its area is " + mySquare.getAreaof_Shape() + ".");
            Console.WriteLine("————————");

        }
    }
}
```

Remarks

The use of **inheritance** in the application for calculating the area helps programmers avoid defining excess code in one class. The properties and methods defined in one class can be easily used in another class because of inheritance.

Common Programming Errors

- *Attempting to create an object of an abstract class.*
- *Using **new** modifier in a non-nested class declarations.*
- *Using **new** modifier on an abstract class.*
- *Not declaring a non static constructor as public.*
- *Specifying a sealed class as the base class of another class.*
- *Declaring a sealed class as an abstract class as well.*
- *Using a **private** or **internal** class at the base class of a public class.*
- *Using **new** and **override** modifiers in the same declaration*
- *Trying to access private members of the base class in the derived class.*
- *Trying to override a sealed method.*
- *Attempting to change the accessibility level in the override method..*
- *Declaring abstract methods in nonabstract classes.*
- *Using **abstract** or **override** on virtual methods.*

- *Improper construction of a subclass constructor that has to pass values to its base class.*
- *Assigning an object of superclass to a subclass reference, without a cast.*
- *Attempting to use sealed classes such as* ***System.Array*** *as the base class of a class.*
- *Declaring a virtual method as static.*
- *Declaring static method as abstract.*
- *Declaring an override method as static.*
- *Trying to override a static or nonvirtual method.*

Review Questions

13.1 State whether the following statements are true or false.
 (a) A derived class inherits all the members of its base class.
 (b) A derived class can access all the data members of its base class.
 (c) A derived class can have only one class as its direct base class.
 (d) An abstract class is permitted to contain abstract members.
 (e) A sealed class can be declared as abstract.
 (f) A derived class can have a member with the same name or signature as an inherited member.
 (g) An instance of a derived class contains a copy of all instance fields of its base class.
 (h) A class member declared private is not inherited by a derive class.
 (i) A class member declared internal is visible only inside the containing class.
 (j) A member declared public inside a private class is visible everywhere in the containing program.
 (k) A class declared sealed cannot be used as a base class.
 (l) A class declared abstract cannot be used as a base class.
 (m) An abstract class cannot be instantiated directly.
 (n) The implementation of virtual method can be changed by derived classes.
 (o) It is legal to declare a virtual method as abstract.
 (p) An override method can be declared as abstract.
 (q) The overridden base method can be declared as static..
 (r) Abstract method declarations are only permitted in abstract classes.

13.2 What is reusability of code? How is it achieved in C#?

13.3 What is inheritance? How is it implemented in C#?

13.4 What are the two forms of inheritance? Give your own examples.

13.5 What is the difference between multilevel and hierarchical inheritance?

13.6 Describe the syntax of single inheritance in C#.

13.7 Describe important characteristics of inheritance.

13.8 What is class visibility? How is it implemented?

13.9 Describe the visibility of class members declared with the following modifiers.:
 (a) private
 (b) protected
 (c) protected internal

13.10 When do you use the modifier internal to a class member?

13.11 We do not normally use the public modifier to declare data members. Comment.

13.12 When do we declare a class private?

13.13 Describe the general form of a derived class constructor.

13.14 What is overriding a method? How is it achieved in C#?

13.15 What is hiding a method? How is it different from overriding?
13.16 What is the difference between overriding and overloading?
13.17 When do we declare a class abstract?
13.18 State the characteristics of abstract classes.
13.19 What are the characteristics of abstract methods?
13.20 When do we use sealed classes?
13.21 Compare and contract abstract and sealed classes.
13.22 What is a virtual method?
13.23 What is polymorphism?
13.24 What is early binding? How is it achieved?
13.25 What is late binding? How is it achieved?
13.26 Why is the late binding called dynamic binding?
13.27 What is wrong with the following code segments?

(a)
```
class A
{
    abstract void Display ( );
}
```
(b)
```
class A : B
{
    new override void display ( ) { }
}
```
(c)
```
sealed class A{ }
class B : A
{
    . . .
    . . .
}
```
(d)
```
class A { }
public class B : A
{
    . . .
    . . .
}
```
(e)
```
class A { }
class B { }
class C: A,B
{
    . . . .
    . . . .
}
```
(f)
```
abstract class XY
{
    abstract Sum (int x, int y) { }
}
```

(g)
```
abstract class Print
{
    abstract void Show ( );
}
class Display : Print
{
}
```
(h)
```
class A { }
public class B
{
    public A Method ( ) { }
}
```

13.28 How does a base class receive its arguments in order to construct its members properly?
13.29 What happens if a derived class fails to explicitly provide arguments to the base class constructor?
13.30 When do we need to use the new modifier on a class declaration?
13.31 Distinguish between the inclusion polymorphism and the operation polymorphism.
13.32 What is the use of inclusion polymorphism?

Debugging Exercises

13.1 The following program demonstrates single-level inheritance. Find errors in the program.

```
using System;
using System.Collections.Generic;
using System.Text;

namespace debugAppl_chap13
{
   class Program
    {
      static void Main(string[] args)
        {
         Derived d1 = new Derived();
         Console.WriteLine("{0},{1},{2}", d1.x, d1.y, d1.z); // displays 10,20,30
          d1.show();// displays 'Base Method'

        }
    }
   class Base
    {
       int x = 10;
       int y = 20;
      public void show()
        {
```

```
            Console.WriteLine("Base Method");
        }
    }
    class Derived : Base
    {
        public int z = 30;
    }

}
```

13.2 Debug the following program.

```
using System;
using System.Collections.Generic;
using System.Text;

namespace debugApp2_chap13
{
    class Program
    {
        static void Main(string[] args)
        {
            DerivedClass dc = new DerivedClass();

        }
    }
    class BaseClass
    {
        private BaseClass()
        {
            Console.WriteLine("This is Base class's default constructor.");
        }
    }
    public class DerivedClass : BaseClass
    {
    }

}
```

13.3 This program explains overriding in C#. Correct errors, if any.

```
using System;
using System.Collections.Generic;
using System.Text;
```

```
namespace debugApp3_chap13
{
   class BaseClass
   {
      public void Show()
      {
        Console.WriteLine("This is a BaseClass Method.");
      }
   }
   class DerivedClass :: BaseClass
   {
      public override void Show()
      {
        Console.WriteLine("This is a overriden method Show of Base Class in Derived
Class");
      }
   }

   class Program
   {
      static void Main(string[] args)
      {
        BaseClass bc = new DerivedClass();
        bc.Show(); // Displays 'Base Method'

      }
   }
}
```

Programming Exercises

13.1 Define a base class called Animal with following members:

- *A string type data member to store the name of the animal*
- *An integer member to store the age of the animal in years*
- *A method to display the name and age of the animal*

Derive two classes named Cat and Dog from Animal class. Then, write a driver program to create Cat and Dog objects with suitable values. Display the contents of the objects by calling the Display method on the derived objects.

13.2 Modify the program in Exercise 13.1 as follows:

- *Implement Display in Cat class*
- *Implement Display in Dog class with the new modifier*
- *Modify the driver program such that base version as well as derived versions of Display are called. Run the program. Did you face any problem? Discuss.*

13.3 Add the modifier new to the Display member of Cat in Exercise 13.2 and the run the program. Does it make any difference? Discuss.

13.4 Modify the program of Exercise 13.2 as follows:
- *Add the modifier override to Display in Cat*
- *Change the modifier new to override in the declaration of Display of Dog*
- *Make the Display method in Animal class virtual*

Execute the program and analyse the results.

13.5 Modify the program of Exercise 13.1 as follows:
- *Add another method named Sound in the base class and make it abstract*
- *Implement the Sound method in derived classes*
- *Make the base class Animal abstract*

Execute the program and analyse the results. What would be the result if the base class Animal is not made abstract?

13.6 Again consider the program of Exercise 13.1. Add the modifier sealed to the base class and run the program. Any changes! Comment.

13.7 Define a Person class with three data members: age, name and sex.
- *Derive a class called Employee from Person that adds a data member code to store employee code*
- *Derive another class called Specialist from Employee*
- *Add a method to each derived class to display the information about what it is.*

Write a driver program to generate an array of three ordinary employees and another array of three specialists and display information about them.

Also display the information of the specialists by calling the method inherited from Employee class.

13.8 Assume that a bank maintains two kinds of accounts for customers, one called savings account and the other current account. The savings account provides compound interest and withdrawal facilities but no cheque book facility. The current account provides cheque book facility but no interest. Current account holders should also maintain a minimum balance and if the balance falls below this level, a service charge is imposed.

Create a class Account that stores customer name, account number and type of account. From this derive the classes Cur-Acct and Sav-Acct to make them more specific to their requirements. Include the necessary member methods in order to accomplish the following tasks

(a) Accept deposit from a customer and update the balance.

(b) Display the balance

(c) Compute and deposit interest

(d) Permit withdrawal and update the balance

(e) Check for the minimum balance, impose penalty, if necessary, and update the balance.

Do not use any constructors. Use member methods to initialize the class members.

13.9 Modify the program of Question 13.8 to include constructors for all the three classes.

13.10 An educational institution wishes to maintain a database of its employees. The database is divided into a number of classes whose hierarchical relationships are shown in Fig 13.9. The figure also shows the minimum information required for each class. Specify all the classes and define methods to create the database and retrieve individual information as and when required.

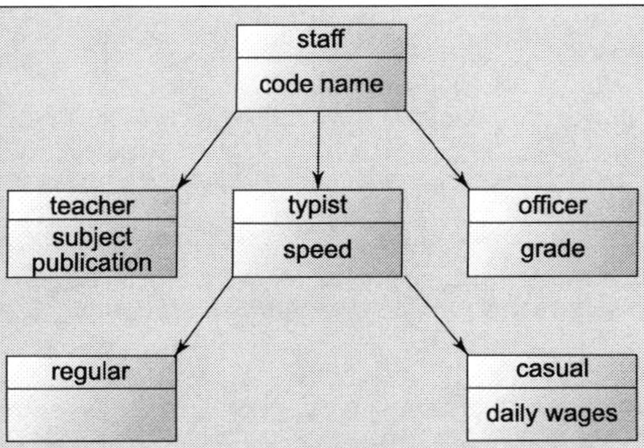

Figure 13.9 *Class relationships*

13.11 *Create a base class called Shape. Use this class to store two double type values that could be used to compute the area of figures*

- *Derive two specialized classes called Triangle and Rectangle from the base Shape.*
- *Add to the base class, a method Set to initialize base class data members and another method Area to compute and display the area of figures.*
- *Make Area as a virtual method and redefine this method suitable in the derived class.*

Using these classes, design a program that will accept dimensions of a triangle or rectangle interactively and display the area.

Note: *The two values given as input will be rated as lengths of two sides in the case of rectangles, and as base and height in the case of triangles, and used as follows in computing their areas:*

Area of rectangle = $x*y$
Area of triangle = $1/2*x*y$

13.12 *Run the above program by removing the definition of Area method in one of the derived classes. Comment on the output.*

Chapter

14

Interfaces: Multiple Inheritance

14.1 INTRODUCTION

In Chapters 12 and 13, we discussed the various features of classes and how they can be inherited by other classes. We also learned about various forms of inheritance and pointed out that C# does not support multiple inheritance. That is, classes in C# cannot have more than one superclass. For instance, a definition like

```
Class A : B, C
{
    . . . .
    . . . .
}
```

is not permitted in C#. However, the designers of C# could not overlook the importance of multiple inheritance. A large number of real-life applications require the use of multiple inheritance whereby we inherit methods and properties from several distinct classes. Since 'C++ - like' implementation of multiple inheritance proves difficult and adds complexity to the language, C# provides an alternate approach known as *interface* to support the concept of multiple inheritance. Although a C# class cannot be a subclass of more than one superclass, it can *implement* more than one interface, thereby enabling us to create classes that build upon other classes without the problems created by multiple inheritance.

An interface in C# is a reference type. It is basically a kind of class with some differences. Major differences include:

- All the members of an interface are implicitly **public** and **abstract.**
- An interface cannot contain constant fields, constructors and destructors.
- Its members cannot be declared **static.**
- Since the methods in an interface are abstract, they do not include implementation code.
- An interface can inherit multiple interfaces.

14.2 DEFINING AN INTERFACE

An interface can contain one or more methods, properties, indexers and events but none of them are implemented in the interface itself. It is the responsibility of the class that implements the interface to define the code for implementation of these members.

The syntax for defining an interface is very similar to that used for defining a class. The general form of an interface definition is:

```
interface InterfaceName
{
    Member declarations;
}
```

Here, **interface** is the keyword and *InterfaceName* is a valid C# identifier (just like class names). Member declarations will contain only a list of members without implementation code. Given below is a simple interface that defines a single method:

```
interface Show
{
    void Display ( ); // Note semicolon here
}
```

In addition to methods, interfaces can declare properties, indexers and events. *Example:*

```
interface Example
{
    int Aproperty
    {
        get ;
    }
    event someEvent Changed;
    void Display ( );
}
```

The accessibility of an interface can be controlled by using the modifiers **public, protected, internal** and **private.** The use of a particular modifier depends on the contact in which the interface declaration occurs.

We may also apply the modifier **new** on nested interfaces. It specifies that the interface hides an inherited member by the same name.

14.3 EXTENDING AN INTERFACE

Like classes, interfaces can also be extended. That is, an interface can be subinterfaced from other interfaces. The new subinterface will inherit all the members of the superinterface in the manner similar to subclasses. This is achieved as follows:

```
interface name2 : name1
{
```

```
    Members of name2
}
```

For example, we can put all members of particular behaviour category in one interface and the members of another category in the other.
Consider the code below:

```
interace Addition
{
    int Add (int x, int y) ;
}

interface Compute : Addition
{
    int Sub (int x, int y);
}
```

The interface **Compute** will have both the methods and any class implementing the interface **Compute** should implement both of them; otherwise, it is an error.

We can also combine several interfaces together into a single interface. Following declarations are valid:

```
interface I1
{
    . . . .
}
interface I2
{
    . . . .
}
interface I3 : I1, I2 //multiple inheritance
{
    . . . .
}
```

While interfaces are allowed to extend other interfaces, subinterfaces cannot define the methods declared in the superinterfaces. After all, subinterfaces are still interfaces, not classes. It is the responsibility of the class that implements the derived interface to define all the methods. Note that when an interface extends two or more interfaces, they are separated by commas.

It is important to remember that an interface cannot extend classes. This would violate the rule that an interface can have only abstract members.

14.4 IMPLEMENTING INTERFACES

Interfaces are used as 'superclasses' whose properties are inherited by classes. It is therefore necessary to create a class that inherits the given interface. This is done as follows:

```
class classname : interfacename
{
```

```
    body of classname
}
```

Here the class **classname** 'implements' the interface **interfacename.** A more general form of implementation may look like this:

```
class classname : superclass, interface1, interface2. . . .
{
    body of classname
}
```

This shows that a class can extend another class while implementing interfaces.

In C#, we can derive from a single class and, in addition, implement as many interfaces as the class needs. When a class inherits from a superclass, the name of each interface to be implemented must appear after the superclass name. *Example:*

```
class A : B, I1, I2, . . . . .
{
    . . . . .
}
```

where **B** is a base class and I1, I2, ... are interfaces. The base class and interfaces are separated by commas.

Program 14.1 illustrates an application implementing two interfaces. The class **Computation** implements two interfaces, **Addition** and **Multiplication.** It declares two data members and defines the code for the methods **Add ( )** and **Mul ( ).** Note the syntax and how the Computation object is cast to interface types to reference their members.

```
Addition add = (Addition ) com;

. . . .

Multiplication mul = (Multiplication ) com;
```

Program 14.1 Implementation of multiple interfaces

```
using System;

interface Addition
{
    int Add ( );
}
interface Multiplication
{
    int Mul ( );
}
class Computation  : Addition, Multiplication
{

    int x, y;
```

```
    public Computation (int x, int y )        //Constructor
    {
            this.x = x;
            this.y = y;
    }
    public int Add ( )                        //Implement Add ( )
    {
            return ( x + y );
    }
    public int Mul ( )                        //Implement Mul ( )
    {
            return ( x * y );
    }
}

class InterfaceTest1
{
    public static void Main( )
    {
            Computation com = new Computation (10,20);

            Addition add = (Addition ) com;                 // casting
            Console.WriteLine ("Sum = " + add.Add ( ));

            Multiplication mul = (Multiplication) com;      // casting
            Console.WriteLine("Product = " + mul.Mul ( ) );
    }
}
```

Output of Program 14.1 will be:

```
Sum = 30
Product = 200
```

Note that we cannot instantiate an interface directly. That is, we cannot write:

```
Addition add = new Addition ( );
```

We should always create an instance of the implementing class and then cast the object to the interface type.

Let us consider another example where an interface is implemented by multiple classes. Program 14.2 shows how two classes **Circle** and **Square** implement an interface **Area.**

Program 14.2 Multiple implementation of an interface

```
using System;
interface Area
{
```

```
    double Compute ( double x ):
}
class Square  : Area
{
    public double Compute (double x)
    {
        return (x * x)
    }
}
class Circle : Area
{
    public double Compute (double x)
    {
        return (Math.PI * x * x):
    }
}
class InterfaceTest2
{

    public static void Main( )
    {
            Square sqr = new Square ( );
            Circle cir = new Circle ( );

            Area area;
            area = sqr as Area; //casting
            Console.WriteLine ("Area of Square ="
                        + area.Compute(10.0));

            area = cir as Area; //casting

            Console.WriteLine ("Area of Circle="
                        + area.Compute(10.0));
    }

}
```

Program 14.2 will output the following two lines:

```
Area of Square = 100
Area of circle = 314.159265
```

In Program 14.2, we create an instance of each class using the **new** operator. Then, we declare an object of type **Area** interface. Now, we cast the **Square** object to the interface type using **as** operator as shown below:

```
area = sqr as Area;
```

Then, the call

```
area.Compute(10.0)
```

will invoke the **Compute** method of **Square** class. We repeat a similar procedure with the **Circle** object.

Note that any number of classes can implement an interface. However, to implement the methods, we need to refer to the class objects as types of the interface rather than their respective classes.

14.5 INTERFACES AND INHERITANCE

Most often we have situations where the base class of a derived class implements an interface. In such situations, when an object of the derived class is converted to the interface type, the inheritance hierarchy is searched until it finds a class that directly implements the interface. Consider the code in Program 14.3.

Program 14.3 Inheriting a class that implements an interface

```
using System;
interface Display
{
    void Print ( );
}
class B : Display    // implements Display
{
    public void Print ( )
    {
        Console.WriteLine("Base Display");
    }
}
class D : B // inherits B class
{
    public new void Print ( )
    {
        Console.WriteLine("Derived Display");
    }
}


class InterfaceTest3
{
    public static void Main( )
    {
        D d = new D ( );
        d.Print ( );

        Display dis = (Display) d;
        dis.Print ( );
    }
}
```

Program 14.3 would produce the following output:

```
Derived Display
Base Display
```

Note that the statement

```
dis.Print ( );
```

calls the method **Print ( )** in base class B but not the one available in the derived class itself. This is because the derived class does not implement the interface. That is, the use of modifier **new** in the derived class "hides" the **Print** method implemented in the base class.

14.6 EXPLICIT INTERFACE IMPLEMENTATION

One of the reasons that C# does not support multiple inheritance is the problem of name collision. However, this problem still persists in C# when it implements more than one interface. *Example:*

```
interface I1 { void Display ( ); }
interface I2 { void Display ( ); }
class C1 : I1, I2
{
    public void Display { }
}
```

Does **C1.Display( )** implement **I1.Display( ),** or **I2.Display( )**? It is ambiguous and so the compiler reports an error. Such problems of name collision may occur when we implement interfaces from different sources.

C# supports a technique known as *explicit interface implementation*, which allows a method to specify explicitly the name of the interface it is implementing. Program 14.4 illustrates the use of explicit interface implementation.

Program 14.4 Explicit interface implementation

```
using System;
interface I1
{
    void display ( );
}
interface I2
{
    void Display ( );
}
class C1 : I1, I2

{
    void I1.Display ( )   //no access modifier
```

```
    {
            Console.WriteLine("I1 Display");
    }
    void I2.Display ( )  //no access modifier
    {
            Console.WriteLine("I2 Display");
    }
}
class InterfaceTest4
{
    public static void Main( )
    {
        C1 c = new C1( );

        I1 i1 = (I1) c;
        i1.display;

        I2 i2 = (I2)c;
        i2.Display ( );
    }

}
```

Output of Program 14.4 would be:

```
I1 Display
I2 Display
```

Note that while implementing the interface members, we have qualified explicitly with the interface name.

```
void I1. Display ( );
void I2. Display ( );
```

Also note that access modifers are prohibited on explicit interface implementations.

Program 14.5 demonstrates the usage of interfaces and mainly explicit inheritance. Two Interfaces are created in the program, each containing two methods **read, write**, and **view, read** respectively. The methods are being explicitly overridden in an inner class and are called in a **Main** class method **CreateDoc()** and the results are displayed.

Program 14.5 Another example of explicit interface

```
using System;

using System.Collections.Generic;
using System.Text;

namespace ExplicitInterface
```

```
{
    class ExplicitInterface_Implementation
    {
        interface readFile
        {
            void Read();
            void Write();
        }

        interface writeFile
        {
            void View();
            void Read();
        }


        public class Document : readFile, writeFile
        {
            // the document constructor
            public Document(string doc)
            {
            Console.WriteLine("Creating a document with: {0}", doc);
            }

            // Implicit implementation
            public virtual void Read()
            {
                Console.WriteLine("Reading from  Read method in interface readFile.");
            }

            public void Write()
            {
                Console.WriteLine("Writing in Write Method in interface readFile.");
            }
            // Explicit implementation
            void writeFile.Read()
            {
                Console.WriteLine("Reading from  Read method in interface writeFile.");
            }

            public void View()
            {
                Console.WriteLine("Viewing from View method in interface writeFile.");
            }
        }

        public void CreateDoc()
        {
```

```
            // Creating a Document object
            Document doc1 = new Document("This Document is getting written \n in the
document created.");
            readFile iFile = doc1 as readFile;
            if (iFile != null)
            {
                iFile.Read();
            }

            // Cast to an writeFile interface
            writeFile wFile = doc1 as writeFile;
            if (wFile != null)
            {
                wFile.Read();
            }

            doc1.Read();
            doc1.View();
        }

        public static void Main()
        {
            ExplicitInterface_Implementation ei = new ExplicitInterface_Implementation();
            ei.CreateDoc();
        }

    }
}
```

```
C:\WINNT\system32\cmd.exe
Creating a document with: This Document is getting written
 in the document created.
Reading from  Read method in interface readFile.
Reading from  Read method in interface writeFile.
Reading from  Read method in interface readFile.
Viewing from View method in interface writeFile.
Press any key to continue . . .
```

14.7 ABSTRACT CLASS AND INTERFACES

Like any other class, an abstract class can use an interface in the base class list. However, the interface methods are implemented as abstract methods. *Example:*

```
interface A
{
    void Method ( );
}
abstract class B : A
{
    . . . .
    . . . .
    public abstract void Method ( );
}
```

Note that the class B does not implement the interface method; it simply redeclares as a public abstract method. It is the duty of the class that derives from B to override and implement the method.

Note that interfaces are similar to abstract classes. In fact, we can convert an interface into an abstract class. Consider the following interface:

```
interface A
{
    void Print ( );
}
```

This is identical to the following abstract class:

```
abstract class B
{
    abstract public void Print ( );
}
```

Now, a class can inherit from **B** instead of implementing the interface **A**. However, the class that inherits B cannot inherit any other class directly. If it is an interface, then the class can not only implement the interface but also use another class as base class thus implementing in effect multiple inheritance.

Program 14.6 uses the concepts of abstract classes, multilevel inheritance, calling the methods in inheriting classes **Class1**, **Class2** and manipulating the methods and getting the output displayed.

Program 14.6 Inheritance using abstract class

```
using System;
using System.Collections.Generic;
using System.Text;

namespace InheritanceusingAbstractClass
```

```
{
    class Program
    {
        static void Main()
        {
            ABS[] aArray = new ABS[3];
            aArray[0] = new Class1(1, 2, "This is the first value in array.");
            aArray[1] = new Class1(3, 4, "This is the next value in array.");
            aArray[2] = new Class2(5, 6);

            for (int i = 0; i < 3; i++)
            {
                aArray[i].show();
            }
        }
    }           // end class Program


    public abstract class ABS
    {
        // constructor takes two integers to fix location on the console
        public ABS(int top, int left)
        {
            this.top = top;
            this.left = left;
        }


        public abstract void show();

        protected int top;
        protected int left;
    }       // end class Window

    // Class1 derives from ABS
    public class Class1 : ABS
    {
        // constructor adds a parameter
        public Class1(
        int top,
        int left,
        string contents)
            : base(top, left) // call base constructor
        {

            class1values = contents;
```

```
        }

        // an overridden version implementing the abstract method
      public override void show()
        {
            Console.WriteLine("This is overriding the method show declared in \n abstract
class ABS : {0}.\n", class1values);
        }
        private string class1values; // new member variable
    }    // end class Class1

    public class Class2 : ABS
    {
            public Class2(
        int top,
        int left)
            : base(top, left) { }

        // implement the abstract method
        public override void show()
        {
            Console.WriteLine("Overriding again the show method in Class2. \n",
            top, left);
        }
    }       // end class Class2

}
```

```
C:\WINNT\system32\cmd.exe
This is overriding the method show declared in
 abstract class ABS : This is the first value in array..

This is overriding the method show declared in
 abstract class ABS : This is the next value in array..

Overriding again the show method in Class2.

Press any key to continue . . .
```

Case Study

Problem Statement

HPStore is a computer store, which mainly sells different varieties of Hewlet Packard (HP) computers. While selling HP computers, it has recently become a tedious task for the marketing staff of HPStore to remember the features, price and names of different varieties of HP computers. As a result, the marketing staff in HPStore is facing problems in providing correct information to customers who come to enquire about HP computers. Because of this, the HPStore is losing its customers and running in a loss. How can HPStore solve the problem of the marketing staff and increase its customers?

Solution

In order to solve the problem of its marketing staff, the HPStore decides that it needs an application that will allow the staff to view the details such as features and prices related to different varieties of HP computers. It approaches ABC Software Consultants to ask them to create the application which will display the details about different HP computers. ABC Software Consultants, after analysing the requirements of HPStore, decide to use the **Interface** feature of C# programming language for developing the application. **Interface** is a reference type in C# that contains declaration of methods. **Interface** in C# is implemented in different classes where the methods declared in the interface are defined and used. ABC Software Consultants create the following application to make the task of the marketing staff working in HPStore easy:

```
/* How in areal life situation all the different properties(color,computername,price and
processor of a single product like

  HP Computer can be stored in different methods and called separately for their use when
required*/
using System;
using System.Collections.Generic;
using System.Text;

namespace CaseStudy14
{
    interface HPSystem
    {
        void ListComputer();
        void ShowComputerColor(string str);
        void ShowComputerFeature(string feature);
        void ShowComputerPrice(double price);

    }

    class ComputerList : HPSystem
    {
        string comptercolor = "";
        string computernames = "";
        public void ListComputer()
```

```
{
    computernames = "1.      HP Pavilion Media Center m8000n";
    computernames = computernames + "\n2.      HP Compaq DC5750 PC";
    computernames = computernames + "\n3.      HP Compaq DX2300 pc";

Console.WriteLine("The HP Computers are :\n\n" + computernames);
}

public void ShowComputerColor(string color)
{
    this.comptercolor = color;
    Console.WriteLine("\nThe color of the systems are : " + comptercolor + "\n");
    }
    public void ShowComputerFeature(string feature)
    { }
    public void ShowComputerPrice(double price) { }

}

class ShowHPCompterDetails : ComputerList
{
    string Computerfeature = "";
    double ComputerPrice = 0;
    public void ShowComputerFeature(string feature)
    {
        this.Computerfeature = feature;
        Console.WriteLine("The features of the systems are " +Computerfeature);
    }

    public void ShowComputerPrice(double price)
    {
        this.ComputerPrice = price;
        Console.WriteLine("\nThe Prices of the systems are:" +ComputerPrice);
    }
}

class ShowComputerFeatures
{
    static void Main(string[] args)
    {
        ShowHPCompterDetails hpm = new ShowHPCompterDetails();
        hpm.ListComputer();
        hpm.ShowComputerColor("Steel Gray");
        hpm.ShowComputerColor("Black");
        hpm.ShowComputerColor("Gray");
        hpm.ShowComputerFeature("Pentium PV PC Dual Core2");
```

```
            hpm.ShowComputerFeature("Pentium PIV 2.5GHz processor");
            hpm.ShowComputerFeature("Pentium PV PC Dual Core");
            hpm.ShowComputerPrice(63000);
            hpm.ShowComputerPrice(30000);
            hpm.ShowComputerPrice(45000);

        }
    }
}
```

Remarks

The **Interface** feature allows us to implement the concept of multiple inheritance in a C# application. It also provides the benefit of reusing the code, which is defined in one application, for another application. In an **Interface**, the methods are declared as abstract without any implementation code. **Interface** can be implemented in multiple classes in C#.

Common Programming Errors

- *Declaring data fields as members.*
- *Declaring a member as static.*
- *Using a visibility modifiers to declare a member.*
- *Forgetting to place a semicolon at the end of a method declaration.*
- *Forgetting to implement a member in the implementing class.*
- *Forgetting to cast the class object to the interface type.*
- *Using an access modifier on an explicit implementation method.*
- *Attempting to create an instance of an interface.*
- *Forgetting to declare an implementation method as public.*

Review Questions

14.1 State whether the following statements are true or false.
  (a) Like classes, an interface can inherit only one interface.
  (b) All the members of an interface are implicitly abstract.
  (c) A member of an interface may be declared private.
  (d) An interface is normally used to declare constant data fields.
  (e) An interface cannot extend classes.
  (f) A method declaredin a super interface must be implemented in the subinterface.
  (g) A class can implement any number of interfaces.
  (h) Any number of classes can implement an interface.
  (i) An interface may be considered as an alternative to an abstract class.
  (j) An abstract class can implement an interface.
  (k) We can create an instance of an interface using the new operator.

14.2 What is an interface?

14.3 What are the major differences between a class and an interface?

14.4 What are the similarities between a class and an interface?

14.5 All the members of an interface are made implicitly public. Why?

14.6 All the members of an interface are implicitly abstract. What is the implication of this?

14.7 When an interface inherits another interface, we call it as extending an interface. But when a class inherits an interface, we call it as implementing an interface. Why?

14.8 What is explicit interface implementation? When is it used?

14.9 Does C# support multiple inheritance? If yes, how can it be implemented?

14.10 Using interfaces, we can implement multiple inheritance characteristics in a class. How?

14.11 How do we indicate that class **Class1** derives from base class **BaseClass** and implements the interfaces **FileRead** and **FileWrite**?

14.12 We can combine two or more interfaces together. Discuss.

14.13 Describe various forms of implementing interfaces. Give examples of C# code for each case.

14.14 Give an example where interfaces can be used to support multiple inheritance.

14.15 Can an abstract class implement interfaces? If yes, how are the interface methods implemented?

14.16 Distinguish between an abstract class and an interface?

14.17 Find errors in the following interface definitions and implementations:

(a)
```
new interface A
{
    void M1 ( );
} ;
```
(b)
```
private interface A { }
public interface B : A { }
```
(c)
```
interface A { }
interface B { }
interface C extends A, B
{
}
```
(d)
```
interface A { }
class B { }
class C : A, B
{
}
```
(e)
```
interface A { }
class B implements A
{
}
```
(f)
```
interface A
{
    void M1( );
    int n;
}
```
(g)
```
interface X
{
    static void Fun ( );
}
```
(h)
```
interface Y
{
```

```
    public void Display ( );
    int count {get;}
}
```

(i)
```
interface P
{
    abstract void Paint ( );
}
```

(j)
```
interface A
{
    void Mul ( );
}
interface B : A
{
    void Mul ( );
}
```

Debugging Exercises

14.1. Locate and correct errors in the program shown below.

```
using System;
using System.Collections.Generic;
using System.Text;

namespace debugAppl_chap14
{
    public abstract class Car
    {

        abstract void RemoveAllFuses();

        abstract void StartEngine();

    }
    class Seat : Car
    {

        public override void RemoveAllFuses()
        {

        }

        public override void StartEngine()
        {
```

```
            }

        }

        class Program
        {
            static void Main(string[] args)
            {
                Seat c = new Seat();
                    c.RemoveAllFuses();

            }
        }
    }
```

14.2. Find the errors in the following program.

```
using System;
using System.Collections.Generic;
using System.Text;

namespace debugApp2_chap14
{
   interface Dimensions
    {
      float Length();
      float Width();
    }

   interface MetricDimensions
    {
      int Length();
      int Width();
    }


   class Box : Dimensions, MetricDimensions
    {
      float lengthInches;
      float widthInches;

      public Box(float length, float width)
       {
         lengthInches = length;
         widthInches = width;
       }
```

```
        // Explicitly implement the members of IEnglishDimensions:
        float Dimensions.Length()
        {
          return lengthInches;
        }

        float Dimensions.Width()
        {
          return widthInches;
        }


        float MetricDimensions.Length()
        {
          return lengthInches * 2.54f;
        }

        float MetricDimensions.Width()
        {
          return widthInches * 2.54f;
        }

        static void Main()
        {

          Box box1 = new Box(30.0f, 20.0f);

          // Declare an instance of the English units interface:
          Dimensions eDimensions = (Dimensions)box1;

          // Declare an instance of the metric units interface:
          MetricDimensions mDim = (MetricDimensions)box1;

          // Print dimensions in English units:
          System.Console.WriteLine("Length(in): {0}", eDimensions.Length());
          System.Console.WriteLine("Width (in): {0}", eDimensions.Width());

          // Print dimensions in metric units:
          System.Console.WriteLine("Length(cm): {0}", mDim.Length());
          System.Console.WriteLine("Width (cm): {0}", mDim.Width());
        }
    }

}
```

14.3. Check for syntax errors in the program and correct them.

```
using System;
using System.Collections.Generic;
using System.Text;

namespace debugApp3_chap14
{
    class Program
    {
        public class ShowValues
        {
            void Show();
        }

        public class CharContainer :: ShowValues
        {
            public void Show()
            {
                // Do character type sorting here.

                Console.WriteLine("From CharContainer.");
            }
        }

        public class NumberContainer :: ShowValues
        {
            public void Show()
            {
                // Do numeric type sorting here.
                Console.WriteLine("From NumberContainer.");
            }
        }

        public void ShowAll(ShowValues[] sortobj)
        {
            foreach (ShowValues sv in sortobj)
            {
                sv.Show();
                }
            }


        static void Main(string[] args) {
            Program Obj = new Program();
            obj.ShowAll(new ShowValues[2] { new CharContainer(), new
NumberContainer() }
```

```
                );

            }
    }
}
```

14.4. Debug the following program for the use of **private** keyword.

```
using System;
using System.Collections.Generic;
using System.Text;

namespace debugApp4_chap14
{
    interface Inter1
    {
        void Display();
    }

    private class Class1
    {
        public void Display()
        {
            Console.WriteLine("This message is from Class1.");
        }
    }

    private class Class2 : Inter1
    {
        public void Display()
        {
            Console.WriteLine("This message is from Class2.");
        }
    }

    private class Class3
    {
        void ShowMessage(object obj)
        {
            Inter1 show;
            show = obj as Inter1;
            if (show != null)
                show.Display();
        }

        void Display()
```

```
        {
            class1 cl1 = new class1();
            class2 cl2 = new class2();
            ShowMessage(cl1);
            ShowMessage(cl2);
        }
    }

    class Program
    {

        static void Main(string[] args)
        {
            class3 cl3 = new class3();
            cl3.ShowMessage();
        }
    }
}
```

Chapter 15

Operator Overloading

15.1 INTRODUCTION

Operator overloading is one of the many exciting features of object-oriented programming. C# supports the idea of operator overloading. It means that C# operators can be defined to work with the user-defined data types such as structs and classes in much the same way as the built-in types. For instance, C# permits us to add two class objects with the same syntax that is applied to the basic types. *Example:*

```
Vector u1, u2, u3;
. . . .     //initialize u1 and u2 here
. . . .
u3 = u1 + u2; //adding two objects, u1 and u2
```

where **Vector** is a class or a struct. Two vectors **u1** and **u2** are added (like a simple type) to give a third vector **(u3).** This means, C# has the ability to provide the operators with a special meaning for a data type. This mechanism of giving such special meaning to an operator is known as *operator overloading.*

Operator overloading provides a flexible option for the creation of new definitions for most of the C# operators. We can almost create a new language of our own by the creative use of the method and operator overloading techniques.

15.2 OVERLOADABLE OPERATORS

There are quite a number of operators in C# that can be overloaded. There are also many others that cannot be overloaded. They are listed in Tables 15.1 and 15.2.

Table 15.1 *Overloadable operators*

<table>
<tr><th>Category</th><th>Operators</th></tr>
<tr><td>Binary arithmetic</td><td>+ , * , / , - , %</td></tr>
<tr><td>Unary arithmetic</td><td>+ , - , + + , - -</td></tr>
<tr><td>Binary bitwise</td><td>& , | , ^ , < < , > ></td></tr>
<tr><td>Unary bitwise</td><td>! , ~ , true , false</td></tr>
<tr><td>Logical operators</td><td>= = , ! = , > = , < , < = , ></td></tr>
</table>

Table 15.2 *Operators that cannot be overloaded*

<table>
<tr><th>Category</th><th>Operators</th></tr>
<tr><td>Conditional operators</td><td>& & , | |</td></tr>
<tr><td>Compound assignment</td><td>+ = , - = , * = , / = , % = ,</td></tr>
<tr><td>Other operators</td><td>[ ], ( ), = , ? : , - >, new, sizeof, typeof, is, as</td></tr>
</table>

Note:

- When we overload a binary operator, its compound assignment equivalent is implicitly overloaded.
- Logical operators must be overloaded in pairs, meaning that = = and ! = must be done together.
- Operators that are currently not defined in C# cannot be overloaded.

Although the semantics of an operator can be extended, we cannot change its syntax, the grammatical rules that govern its use such as the number of operands, precedence and associativity. For example, the multiplication operator will enjoy higher precedence than the addition operator. Remember, when an operator is overloaded, its original meaning is not lost. For instance, the operator +, which has been overloaded to add two vectors, can still be used to add two integers.

15.3 NEED FOR OPERATOR OVERLOADING

Although operator overloading gives us syntactical convenience, it also help us greatly to generate more readable and intuitive code in a number of situations. These include:

- Mathematical or physical modeling where we use classes to represent objects such as co-ordinates, vectors, matrices, tensors, complex numbers and so on.
- Graphical programs where co-ordinate-related objects are used to represent positions on the screen.
- Financial programs where a class represents an amount of money.
- Text manipulations where classes are used to represent strings and sentences.

15.4 DEFINING OPERATOR OVERLOADING

To define an additional task to an operator, we must specify what it means in relation to the class (or struct) to which the operator is applied. This is done with the help of a special method called *operator method,* which describes the task. The general form of an operator method is:

```
public static retval operator op (arglist)
{
    Method body //task defined
}
```

The operator is defined in much the same way as a method, except that we tell the compiler it is actually an operator we are defining by the **operator** keyword, followed by the operator symbol *op*. The key features of operator methods are:

- They must be defined as **public** and **static.**
- The *retval* (return value) type is the type that we get when we use this operator. But, technically, it can be of any type.
- The *arglist* is the list of arguments passed. The number of arguments will be one for the unary operators and two for the binary operators.
- In the case of unary operators, the argument must be the same type as that of the enclosing class or struct.
- In the case of binary operators, the first argument must be of the same type as that of the enclosing class or struct and the second may be of any type.

Examples of overloaded operators are:

```
// vector addition
public static Vector operator + (Vector a, Vector b)
// unary minus
public static Vector operator -(Vector a)
// comparison
public static bool operator ==(Vector a, Vector b)
```

Vector is a data type of **class** and may represent both magnitude and direction (as in physics and engineering) or a series of points called elements (as in mathematics). The process of overloading involves the following steps:

- Create a class (or struct) that defines the data type that is to be used in the overloading operation.
- Declare the operator method **operator** op( ) using **public** and **static** modifiers.
- Define the body of the operator method to implement the required operation.

15.5 OVERLOADING UNARY OPERATORS

Let us consider the unary minus operator. A minus operator, when used as unary, takes just one argument. We know that this operator changes the sign of an operand when applied to a basic data item. We shall see here how to overload this operator so that it can be applied to an object in much the same way as is applied to an **int** or **float** variable. The unary minus when applied to an object should change the sign of each of its data items. Program 15.1 shows how the unary minus operator is overloaded.

The method **operator –( )** takes one argument of type **Space** and changes the sign of data members of the object **s.** Since it is a member method of the same class, it can directly access the members of the object which activated it. Remember, a statement like

```
s2 = -s1;
```

will not work, because the operator method does not return any value. It can work if the method is modified to return an object.

Program 15.1 Overloading unary minus

```
using System;
class Space
{
    int x, y, z;
    public Space ( int a, int b, int c )
    {
        x = a;
        y = b;
        z = c;
    }
    public void Display ( )
    {
        Console.Write(" " + x);
        Console.Write(" " + y);
        Console.Write(" " + z);
        Console.WriteLine( );

    }
    public static Space operator - (Space s)
    {
        s.x = -s.x;
        s.y = -s.y;
        s.z = -s.z;
    }
}

class SpaceTest
{
    public static void Main( )
    {
        Space s = new Space ( 10, -20, 30 );

        Console.Write(" s : ");
        s.Display( );

        -s;        //activates opeator -( ) method
```

```
        Console.Write(" s : ");
        s.Display ( );
    }
}
```

Program 15.1 will produce the following output:

```
S : 10 -20 30
S : -10 20 -30
```

15.6 OVERLOADING BINARY OPERATORS

We have just seen how to overload a unary operator. The same mechanism can be used to overload a binary operator. We can always use methods to add two objects. A statement like

```
C = sum ( A, B); //functional notation
```

is possible. The functional notation can be replaced by a natural looking expression

```
C = A + B; //arithmetic notation.
```

by overloading the + operator using an **operator + ( )** method. Here is the syntax used to define the **operator + ( )** method.

```
public static Vector operator + (Vector u1, Vector u2)
    {
        //Create a new Vector object
        //Add the contents of u1 and u2
        //to the new Vector object
        //Return the new vector object
    }
```

Program 15.2 illustrates how this is accomplished. This program overloads the binary plus operator to add two complex numbers of type:

```
x = a + jb
```

Program 15.2 Overloading + Operator

```
using System;
class Complex
{

    double x;       //real part
    double y;       //imaginary part
    public Complex ( )
    {
```

```
}
public Complex(double real, double imag)
{

    x = real;
    y = imag;
}
public static Complex operator + (Complex c1, Complex c2)
{
        Complex c3 = new Complex ();
        c3.x = c1.x + c2.x;
        c3.y = c1.y + c2.y;
        return (c3);
    }
    public void Display( )
    {

        Console.Write(x);
        Console.Write(" + j" + y);
        Console.WriteLine( );

    }
}
class ComplexTest
{

    public static void Main( )
    {

        Complex a, b, c;
        a = new Complex (2.5 , 3.5);
        b = new Complex (1.6, 2.7);
        c = a + b;

        Console.Write(" a = ");
        a.Display( );
        Console.Write("b = ");
        b.Display( );
        Console.Write("c =");
        c.Display( );
    }

}
```

The output of Program 15.2 would be:

```
a = 2.5 + j3.5
b = 1.6 + j2.7
c = 4.1 + j6.2
```

15.7 OVERLOADING COMPARISON OPERATORS

C# supports six comparison operators that can be considered in three pairs:

- = = and ! =
- > and > =
- < and < =

The significance of pairing is two-fold.

1. Within each pair, the second operator should always give exactly the opposite result to the first. That is, whenever the first returns **true**, the second returns **false** and vice versa.
2. C# always requires us to overload the comparison operators in pairs. That is, if we overload = =, then we must overload ! = also, otherwise it is an error.

There is one fundamental difference between overloading comparison operators and overloading arithmetic operators. Comparison operators must return a **bool** type value. Apart from these differences, overloading comparison operators follows the same principles as overloading the arithmetic operators.

Program 15.3 illustrates how the comparison operators == and != are overloaded and used for comparing two vectors. The program also overloads + and – operators so that two vector quantities may be added or subtracted as illustrated. The method **ToString( )** is overridden to display the contents of vectors.

Program 15.3 Overloading comparison operators

```
using system;
class Vector
{

    int x, y, z;
    public Vector () {}
    public Vector ( int a, int b, int c)
    {

        x = a;
        y = b;
        z = c;
    }
    public override string ToString( )
    {
        return("("+x+","+y+","+z+")");
```

```
    }
    public static Vector operator + (Vetor u1, Vector u2)
    {
        Vector u3 = new Vector ();
        u3.x = u1.x + u2.x;
        u3.y = u1.y + u2.y;
        u3.z = u1.z + u2.z;
        return (u3);
    }
    public static Vector operator-(Vector u1, Vector u2)
    {
        Vector u3 = new Vector ();
        u3.x = u1.x - u2.x;
        u3.y = u1.y - u2.y;
        u3.z = u1.z - u2.y;
        return (u3 );
    }


    public static bool operator == (Vector u1, Vector u2)
    {
        if (u1.x == u2.x && u1.y == u2.y && u1.z == u2.z)
            return (true);
        else
            return (false);

    }
    public static bool operator != (Vector u1, Vector u2)
    {
        return ( !(u1 == u2));
    }
}


class CompareTest
{
    public static void Main( )
    {
        Vector u1 = new Vector (1,2,3);
        Vector u2 = new Vector (4,5,6);
        Vector u3 = u1 + u2;
        Vector u4 = u2 - u1;

        Console.WriteLine("u1 = " + u1);
```

```
        Console.WriteLine("u2 = " + u2);
        Console.WriteLine("u3 = " + u3);
        Console.WriteLine("u4 = " + u4);

        If (ul == u2)
        {
            Console.WriteLine("ul is equal to u2");
        }
        else
        {
            Console.WriteLine("ul is not equal to u2");
        }
    }
}
```

Note that the code segment

```
Vector u3 = new Vector ();
u3.x = ul.x + u2.x;
u3.y = ul.y + v2.y;
u3.z = ul.z + u2.z;
return (u3);
```

may be replaced by a single statement as follows:

```
return new Vector (ul.x+u2.x, ul.y+u2.y, ul.z+u2.z);
```

The overloading of == operator can also be done by using the method **Equal ( )** as shown below:

```
Public static bool operator == (Vector ul, Vector u2)
{
return (ul.Equals(u2));
}
```

In this case, we must override the method **Equals ( )** defined in **System** namespace as given below:

```
public override bool Equals (object value)

{
    Vector u = (Vector) value;
    return ((this.x == u.x) &&
          (this.y == u.y) &&
          (this.z == u.z));
}
```

Program 15.3 would produce the following output:

```
ul = (1, 2, 3)
u2 = (4, 5, 6)
```

```
u3 = (5, 7, 9)
u4 = (3, 3, 3)
u1 is not equal to u2
```

Program 15.4 overloads the == Boolean operator and displays the value on the basis of the division done and the value of the output is compared with another value to check whether they are equal. Another method **Equal()** uses the same functionality of the == operator to display the data.

Program 15.4 Equal operator overloaded

```
using System;

public class Fraction
{
    private int numerator;
    private int denominator;

    // create a fraction by passing in the numerator
    // and denominator
    public Fraction( int numerator, int denominator )
    {
        this.numerator = numerator;
        this.denominator = denominator;
    }

    // overloaded operator+ takes two fractions
    // and returns their sum
    public static Fraction operator +( Fraction lhs, Fraction rhs )
    {
        // like fractions (shared denominator) can be added
        // by adding thier numerators
        if ( lhs.denominator == rhs.denominator )
        {
            return new Fraction( lhs.numerator + rhs.numerator,
            lhs.denominator );
        }

        // simplistic solution for unlike fractions
        // 1/2 + 3/4 == (1*4) + (3*2) / (2*4) == 10/8
        // this method does not reduce.
        int firstProduct = lhs.numerator * rhs.denominator;
        int secondProduct = rhs.numerator * lhs.denominator;
        return new Fraction(
        firstProduct + secondProduct,
        lhs.denominator * rhs.denominator
        );
    }
```

```
    // test whether two Fractions are equal
    public static bool operator ==( Fraction lhs, Fraction rhs )
    {
        if ( lhs.denominator == rhs.denominator &&
        lhs.numerator == rhs.numerator )
        {
            return true;
        }
        // code here to handle unlike fractions
        return false;
    }
    // delegates to operator ==
    public static bool operator !=( Fraction lhs, Fraction rhs )
    {
        return !( lhs == rhs );
    }
    // tests for same types, then delegates
    public override bool Equals( object o )
    {
        if ( !( o is Fraction ) )
        {
            return false;
        }
        return this == (Fraction)o;
    }
    // return a string representation of the fraction
    public override string ToString( )
    {
        String s = numerator.ToString( ) + "/" +
        denominator.ToString( );
        return s;
    }
}


public class Tester
{
    public void Run( )
    {
        Fraction f1 = new Fraction( 3, 4 );
        Console.WriteLine( "f1: {0}", f1.ToString( ) );

        Fraction f2 = new Fraction( 2, 4 );
        Console.WriteLine( "f2: {0}", f2.ToString( ) );

        Fraction f3 = f1 + f2;
        Console.WriteLine( "f1 + f2 = f3: {0}", f3.ToString( ) );
```

```
        Fraction f4 = new Fraction( 5, 4 );
        if ( f4 == f3 )
        {
            Console.WriteLine( "f4: {0} == F3: {1}",
            f4.ToString( ),
            f3.ToString( ) );
        }

        if ( f4 != f2 )
        {
            Console.WriteLine( "f4: {0} != F2: {1}",
            f4.ToString( ),
            f2.ToString( ) );
        }


        if ( f4.Equals( f3 ) )
        {
            Console.WriteLine( "{0}.Equals({1})",
            f4.ToString( ),
            f3.ToString( ) );
        }


    }
    static void Main( )
    {
        Tester t = new Tester( );
        t.Run( );
    }
}
```

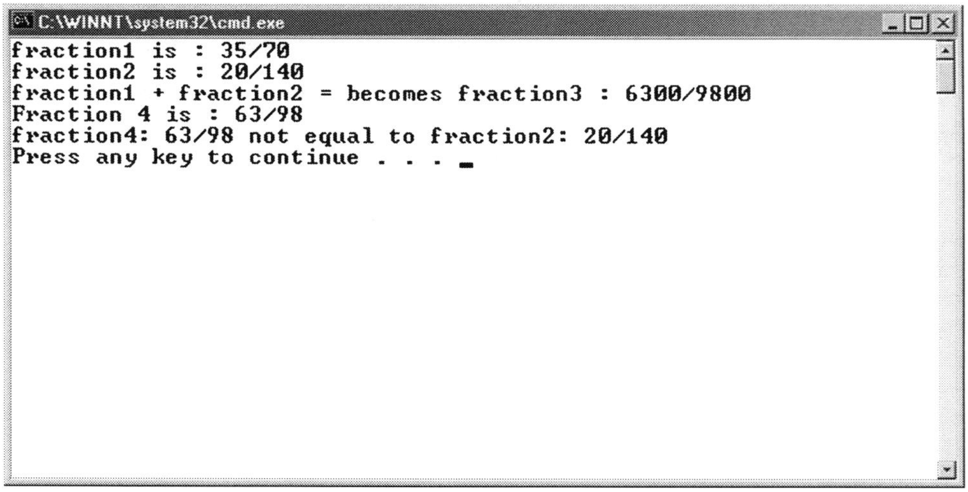

Program 15.5 overloads the + operator. A static method **Fraction** takes two integer parameters and returns a fraction output and displays the output as string type.

Program 15.5 Overloading plus operator

```
using System;
using System.Collections.Generic;
using System.Text;

namespace OverloadingPlusOperator
{
    class Calculate
    {
        public void calc()
        {
          Fraction Fraction1 = new Fraction(388, 6677);
          Console.WriteLine("The fraction 1 is : {0}", Fraction1.ToString());

          Fraction Fraction2 = new Fraction(205, 47);
          Console.WriteLine("The fraction 2 is : {0}", Fraction2.ToString());

          Fraction sum = Fraction1 + Fraction2;
          Console.WriteLine("The output of fraction 1 + fraction 2 is : {0}",
sum.ToString());

        }
        static void Main()
        {
            Calculate ca = new Calculate();
            ca.calc();
        }
    }

    public class Fraction
    {
        private int numerator;
        private int denominator;

        // create a fraction by passing in the numerator and denominator
        public Fraction(int numerator, int denominator)
        {
            this.numerator = numerator;
            this.denominator = denominator;
        }
```

```
        // overloaded operator + takes two fractions
        // and returns their sum
        public static Fraction operator +(Fraction frac1, Fraction frac2)
        {
            // like fractions (shared denominator) can be added by adding their numerators
            if (frac1.denominator == frac2.denominator)
            {
                return new Fraction(frac1.numerator + frac2.numerator, frac1.denominator);
            }

            int product1 = frac1.numerator * frac2.denominator;
            int product2 = frac2.numerator * frac1.denominator;
            return new Fraction(product1 + product2, frac1.denominator *
frac2.denominator);
        }

        // return a string representation of the fraction
        public override string ToString()
        {
            String show = numerator.ToString() + "/" +
            denominator.ToString();
            return show;
        }
    }
}
```

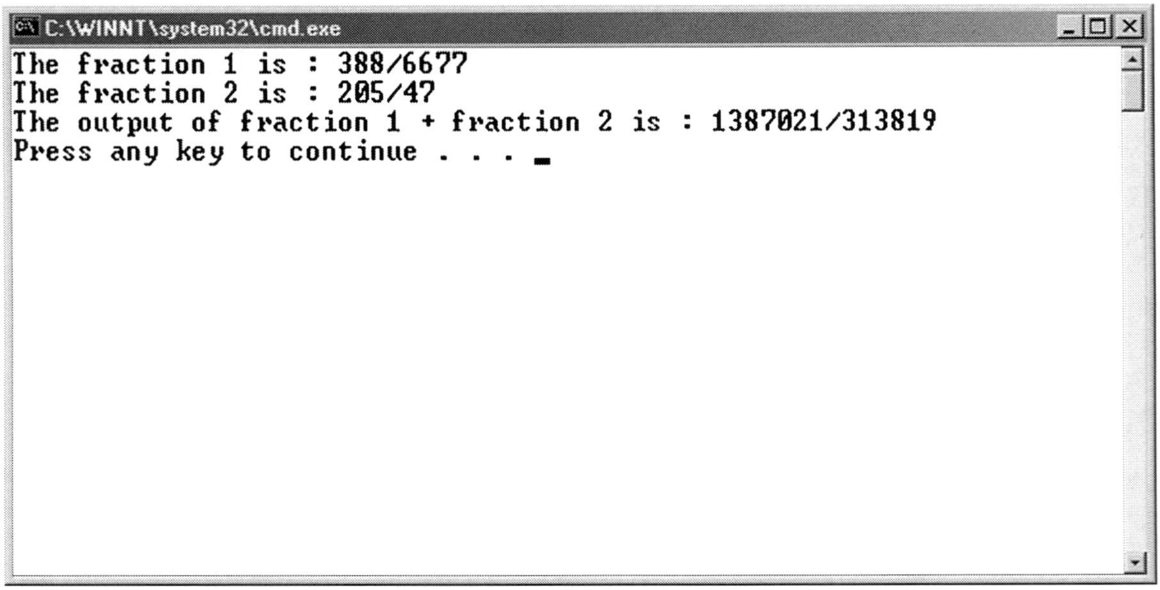

Case Study

Problem Statement

UInfo Private Limited is involved in the development of new aircrafts for companies like AeroFly and FlyHigh. When developing the designs for the new aircrafts, the designing staff of UInfo has to do a number of complex calculations. Manually performing these complex calculations has become a tedious task for the designing staff of UInfo. How can UInfo automate the task of performing complex calculations?

Solution

The management of UInfo, after much discussion, came to the conclusion that an application must be developed to automate the task of performing complex calculations. For this purpose, it contacts a software development company called ABC Software Solutions Private Limited. This software development company first performs a requirement analysis to understand the requirements of UInfo. It then decides that the **operator overloading** feature of C# should be used in a C# application to overload operators such as + and _ so that different functions can be performed using the same operator. After 10 days, ABC Software Solutions Private Limited comes up with the following C# application, which helps perform mathematical operations such as adding and multiplying two numbers.

```
//Operator Overloading using inheritance

using System;
using System.Collections.Generic;
using System.Text;

namespace OperatorOverloading
{

    class StoreValue
    {
        private int num1;
        private int num2;
        public StoreValue()
        {
        }
        public StoreValue(int m, int n)
        {
            num1 = m+m;
            num2 = n+n;
        }
        public void DisplayWindow()
        {
```

```
            Console.WriteLine("The sum of numbers is {0} {1}", num1, num2);
        }

    }

    class InheritValue : StoreValue
    {
        private double num1 ;
        private double num2 ;
        public InheritValue(double a, double b)
        {
            num1 = a*a;
            num2 = b*b;
        }
        public InheritValue()
        {
        }
        public new void DisplayWindow()
        {
        Console.WriteLine("Now the square of the numbers 355 and 455 are {0} and
{1}",num1,num2);
        }
    }

    class InheritMoreValue : InheritValue
    {
        private double num1;
        private double num2;
        public InheritMoreValue(double a, double b)
        {
            num1 = (a * b) / (a+b);
            num2 = (a * b) / (b-a);
        }
        public InheritMoreValue()
        {
        }
        public new void DisplayWindow()
        {
            Console.WriteLine("The value of the numbers 48 and 62 are {0} and {1}", num1,
num2);
        }
    }

    class Program
```

```
    {
        static void Main(string[] args)
        {
            Console.WriteLine("Operator Overloading with Inheritance.");
            StoreValue Obj1 = new StoreValue(119,280);
            Obj1.DisplayWindow();
            InheritValue Obj2 = new InheritValue(355,455);
            Obj2.DisplayWindow();
            InheritMoreValue Obj3 = new InheritMoreValue(150,250);
            Obj3.DisplayWindow();

        }
    }
}
```

Remarks

By the use of operator overloading in a C# application, a new definition for operators such as +, _ and ++ can be provided. Operator overloading allows you to use operators supported by C# to work with classes and structs in the same way as these operators function with built-in data types.

Common Programming Errors

- *Forgetting to declare an operator method as public.*
- *Forgetting to declare an operator method as static.*
- *While overloading a relational operator, forgetting to overload its partner operator.*
- *Forgetting to return a bool type value while overloading a relational operator.*
- *Attempting to compare two float values while overloading comparison operators.*
- *Improper use of relational operators. For example, writing "greater than or equal to" symbol as => instead of >=.*

Review Questions

15.1 State whether the following statements are true or false.
  (a) All conditional operators can be overloaded.
  (b) Logical operators must be overloaded in pairs.
  (c) We can change the precedence of operators through overloading.
  (d) An operator method used for defining operator overloading must be public.
  (e) Operator methods never return a value.
  (f) Comparison operator methods must return a bool type value.
15.2 What is the difference between unary operators and binary operators?
15.3 What should we also do if we overload the == operator?
15.4 What is operator overloading?
15.5 Why is it necessary to overload an operator?
15.6 What is an operator method? Describe its syntax.
15.7 How many arguments are used in the definition of an unary operator method?

15.8 Discuss the differences between overloading comparison operators and overloading arithmetic operators.
15.9 List the operators that can be overloaded.
15.10 List the operators that are not overloadable.
15.11 Give some examples where application of operator overloading might be useful.
15.12 Find errors, if any, in the following header lines of operator methods:

(a) `public Matrix operator + (Matrix x, Matrix y)`
`{ }`

(b) `static Vector operator -(Vector a, Vector b)`
`{ }`

(c) `public static Vector *(Vector a)`
`{ }`

(d) `public static Vector *=(Vector a, Vector b)`
`{ }`

(e) `public static Vector *(double x, Vector a)`
`{ }`

Debugging Exercises

15.1. What is wrong with the program below?

```
using System;
using System.Collections.Generic;
using System.Text;

namespace debugAppl_chap15
{
class OvrLoadingBinary
{
    int x;
    int y;
    OvrLoadingBinary() { }

    OvrLoadingBinary(int i, int j)
    {
        x = i;
        y = j;
    }
    public void DisplayWindow()
    {
        Console.WriteLine("{0} {1}",+x,+y);
    }
    public static OvrLoadingBinary operator +(OvrLoadingBinary olb1,
OvrLoadingBinary olb2)
    {
        OvrLoadingBinary pos = new OvrLoadingBinary();
```

```
            pos.x = x + olb2.x;
            pos.y = y + olb2.y;
            return pos;
        }
    }

    class Program
    {
        public static void Main()
        {
            OvrLoadingBinary olb1 = new OvrLoadingBinary(46, 200);
            Console.Write("First Window Position at : ");
            olb1.DisplayWindow(); // displays 10 & 20
            Console.Write("Then the Window is moved to position : ");
            OvrLoadingBinary olb2 = new OvrLoadingBinary(125, 310);
            olb2.DisplayWindow(); // displays 125 & 310
            OvrLoadingBinary olb3 = new OvrLoadingBinary();
            olb3 = olb1 + olb2;
            Console.Write("Displaying Window at : ");
            olb3.DisplayWindow();
        }
    }
}
```

15.2. Find the errors in the following program for **unary** operator overloading.

```
using System;
using System.Collections.Generic;
using System.Text;

namespace debugApp2_chap15
{
    class UnaryOperatorOvrload
    {
        private int val1;
        private int val2;
        public UnaryOperatorOvrload()
        {
        }
        public UnaryOperatorOvrload(int i, int j)
        {
            i = val1;
            j = val2;
        }
```

```
            public void ShowCommandWindowPosition()
            {
                Console.WriteLine("The values keep changing due to operator
overloading");
                Console.WriteLine("The values are {0} {1}", val1, val2);
            }
            public static UnaryOperatorOvrload operator -(UnaryOperatorOvrload uoo)
            {
                UnaryOperatorOvrload pos = new UnaryOperatorOvrload();
                pos.val1 = -uoo.val1;
                pos.val2 = -uoo.val2;
                return pos;
            }
        }
        class MyClient
        {
            public static void Main()
            {
                UnaryOperatorOvrload uo = new UnaryOperatorOvrload(450, 300);
                uo.ShowCommandWindowPosition(); // displays the command window at 450,
300 pixels
                UnaryOperatorOvrload uo1 = new UnaryOperatorOvrload();
                uo1.ShowCommandWindowPosition(); // displays command window
at 0 & 0
                -uo = -uo1;
                Console.WriteLine("Again values are changed.");
                uo.ShowCommandWindowPosition();
            }
        }

    }
```

15.3. Debug the following program for comparison of different sets of coordinates.

```
    using System;
    using System.Collections.Generic;
    using System.Text;

    namespace debugApp3_chap15
    {
        class Program
        {
            class Coordinates
```

```
        {
            int X;
            int Y;
            public static bool operator ==(Coordinates CX, Coordinates CY)
            {
              if (CX != CY)
                return false;
              if (CX != CY)
                return false;
             return true;
            }
            public override bool Equals(object Obj)
            {
               return true;
            }

               public override int GetHashCode()
            {
               return true;
            }

               publicstaticbool operator !=(Coordinates CX, Coordinates CY)
            {
               if (CX != CY)
               return true;
               if (CY != CY)
               return true;
               return false;
            }
            static void Main(string[] args)
            {
               Coordinates coord1 = new Coordinates();
               Coordinates coord2 = new Coordinates();
               Coordinates coord3 = new Coordinates();
               coord1.X = 225;
               coord1.Y = 280;
               coord2.X = 225;
               coord2.Y = 280;
               coord3.X = 400;
               coord3.Y = 500;
               if (coord1 == coord2)
                  System.Console.WriteLine("The first coordinate and the second
are at the same coordinates " + coord1.X + ","+ coord1.Y);
              else
```

```
                        System.Console.WriteLine("The first coordinate and
the second coordinate are not at the same coordinates.");
                if (coord1 == coord3)
                        System.Console.WriteLine("The first coordinate and the third
coordinate are at the same coordinates " + coord3.X + "," + coord3.Y);
                else
                        System.Console.WriteLine("The first coordinate and the third
coordinate are not at the same coordinates.");

                if (coord2 == coord3)
                    System.Console.WriteLine("The second coordinate and the third
coordinate are at the same coordinates " + coord2.X + "," + coord2.Y);
                else
                    System.Console.WriteLine("The second coordinate and the third
coordinate are not at the same coordinates.");

                        Console.WriteLine("");
                        Console.WriteLine("Now changing the second coordinates of X to
400 and Y to 500.");
                    X = 400;
                    Y = 500;

                    if (coord2 == coord3)
                        System.Console.WriteLine("The second and the third coordinate
are now at the same coordinates " + coord2.X + "," + coord2.Y);

            }
        }
    }
}
```

Programming Exercises

15.1 *Create a class Float that contains one float data member. Overload all the four basic arithmetic operators so that they can be applied to the objects of Float.*
Use the overload ToString method to provide output in the required format.

15.2 *Define a class Vector to hold three integer values. Include operator methods such that we shall be able to perform the following operations on the vector objects, V1, V2 and V3*

```
V3 = V1 + V2;
V3 = 5 * V1;
V3 = V2 * 5;
int m = V1 * V2;
```

15.3 *Create a class Matrix of size m x n. Define all possible matrix operations for Matrix type objects.*

15.4 *Modify the program of Exercise 15.2 to including overloading operator methods for = = and != operators.*

15.5 *Modify the program of Exercise 15.4 so that we may use the vectors to handle double type elements.*

Note that comparing two double type values may lead to incorrect results, because these values may differ by very small amounts. How would you overcome this problem? Incorporate your solution into the program.

15.6 *Define a class called MyString with the following features:*

- *An overloaded binary + operator that concatenates two MyString objects and create a third object.*
- *An overloaded comparison operator == to compare two strings.*
- *An overloaded unary – (minus) operator that returns true if the string object is empty; otherwise false.*

15.7 *Write class called Date that includes data members day, month and year and methods that could implement the following tasks:*

- *Read a date from keyboard*
- *Display a date*
- *Increment a date by one day*
- *Compare two dates to see which one is greater*

Use operator overloading where applicable.

15.8 *Design a class Polar which describes a point in the plane using polar coordinates, radius and angle. A point in polar coordinates is shown below. Include an overloaded operator method to add two objects of Polar.*

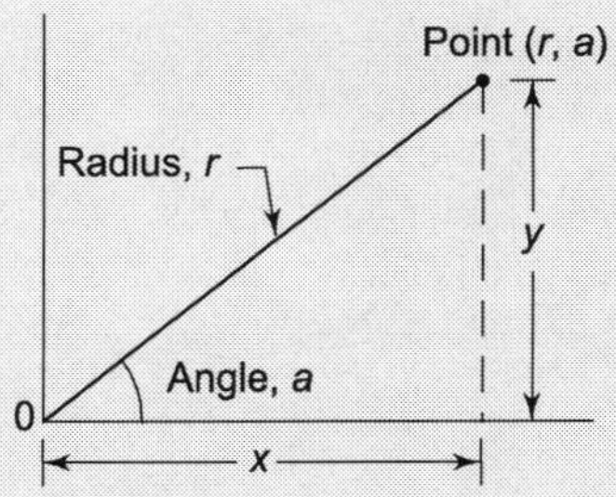

Note:

We cannot add the values of two polar points directly. This requires first conversion of points into rectangular coordinates, then adding the corresponding rectangular coordinates, and finally converting the result back to polar coordinates. We need to use the following trigonometric formulae:

```
x = r * cos (a)
y = r * sin (a)
a = atan (y/x) //arc tangent
r = sqrt (x *x + y * y)
```

Chapter

16

Delegates and Events

16.1 INTRODUCTION

In object-oriented programming, it is the usual practice for one object to send messages to other objects. However, in real-life applications, it is quite common for an object to report back to the object that was responsible for sending a message. This, in effect, results in a two-way conversation between objects. The methods used to call back messages are known as *callback methods.*

Languages like C and C++ implement callback techniques using what are known as *function pointers.* Function pointers simply represent memory addresses and they do not include 'type-safe' information such as:

- number of parameters
- types of parameters
- return type and
- calling convention

Further, in object-oriented programming, methods rarely exist in isolation. They are usually associated with a class instance before they can be called. Because of these problems, C# implements the callback technique in a much safer and more object-oriented manner, using a kind of object called **delegate** object.

A delegate object is a special type of object that contains the details of a method rather than data. Delegates in C# are used for two purposes:

- Callback
- Event handling

We shall discuss briefly how delegates are created and used in these applications.

16.2 DELEGATES

The dictionary meaning of **delegate** is "a person acting for another person". In C#, it really means a method acting for another method. As pointed out earlier, a delegate in C# is a class type object and is used to invoke a method that has been encapsulated into it at the time of its creation. Creating and using delegates involve four steps. They include:

- Delegate declaration
- Delegate methods definition
- Delegate instantiation
- Delegate invocation

A delegate declaration defines a class using the class **System.Delegate** as a base class. Delegate methods are any functions (defined in a class) whose signature matches the delegate signature exactly. The delegate instance holds the reference to delegate methods. The instance is used to invoke the methods indirectly.

An important feature of a delegate is that it can be used to hold reference to a method of any class. The only requirement is that its signature must match the signature of the method.

16.3 DELEGATE DECLARATION

A delegate declaration is a type declaration and takes the following general form:

modifier **delegate** *return-type delegate-name (parameters*);

delegate is the keyword that signifies that the declaration represents a class type derived from **System.Delegate**. The *return-type* indicates the return type of the delegate. *Parameters* identifies the signature of the delegate. The *delegate-name* is any valid C# identifier and is the name of the delegate that will be used to instantiate delegate objects.

The *modifier* controls the accessibility of the delegate. It is optional. Depending upon the context in which they are declared, delegates may take any of the following modifiers:

- **new**
- **public**
- **protected**
- **internal**
- **private**

The **new** modifier is only permitted on delegates declared within another type. It signifies that the delegate hides an inherited member by the same name.

Some examples of delegates are:

```
delegate void SimpleDelegate( );
delegate int MathOperation(int x, int y);
public delegate int CompareItems(object o1, object o2);
private delegate string GetAString( );
delegate double DoubleOperation(double x);
```

Although the syntax is similar to that of a method definition (without method body), the use of keyword **delegate** tells the compiler that it is the definition of a new class using the **System.Delegate** as the base class. Since it is a class type, it can be defined in any place where a class definition is permitted. That is, a delegate may be defined in the following places:

- Inside a class
- Outside all classes
- As the top level object in a namespace

Depending on how visible we want the delegate to be, we can apply any of the visibility modifiers to the delegate definition.

Delegate types are implicitly **sealed** and therefore it is not possible to derive any type from a delegate type. It is also not permissible to derive a non-delegate class type from **System.Delegate.**

16.4 DELEGATE METHODS

The methods whose references are encapsulated into a delegate instance are known as *delegate methods* or *callable entities.* The signature and return type of delegate methods must exactly match the signature and return type of the delegate.

One feature of delegates, as pointed out earlier, is that they are type-safe to the extent that they ensure the matching of signatures of the delegate methods. However, they do not care

- what type of object the method is being called against, and
- whether the method is a static or an instance method.

For instance, the delegate

```
delegate string GetAString( )
```

can be made to refer to the method **ToString( )** using an **int** object **N** as follows:

```
. . . .
. . . .
int N = 100
GetAString s1 = new GetAString(N.ToString);
```

The delegate

```
delegate void Delegate1();
```

can encapsulate references to the following methods:

```
public void F1( )          //instance method
{
   Console.WriteLine("F1");
}
static public void F2( )      //static method
{
   Console.Writeline("F2");
}
```

Similarly, the delegate

```
delegate double MathOp(double x, double y);
```

can refer to any of the following methods:

```
public static double Multiply(double a, double b)
{
   return (a*b);
}
```

```
public double Divide(double a, double b)
{
    return (a/b);
}
```

Note that in both the cases, the signature and return type of methods match the signature and type of the delegate.

16.5 DELEGATE INSTANTIATION

Although delegates are of class types and behave like classes, C# provides a special syntax for instantiating their instances. A *delegate-creation-expression* is used to create a new instance of a delegate.

new *delegate-type (expression)*

The *delegate-type* is the name of the delegate declared earlier whose object is to be created . The *expression* must be a method name or a value of a *delegate-type* . If it is a method name its signature and return type must be the same as those of the delegate. If no matching method exists, or more than one matching method exists, an error occurs. The matching method may be either an instance method or a static method. If it is an instance method, we need to specify the instance as well as the name of the method. If it is a static one, then it is enough to specify the class name and the method name.

The method and the object to which a delegate refers are determined when the delegate is instantiated and then remain constant for the entire lifetime of the delegate. It is, therefore, not possible to change them, once the delegate is created.

It is also not possible to create a delegate that would refer to a constructor, indexer, or user-defined operator.

Consider the following code:

```
//delegate declaration
delegate int ProductDelegate (int x, int y);
class Delegate
{
    static float Product (float a, float b)    //signature does
                                               //not match
    {
        return ( a * b);
    }

    static int Product (int a, int b)   //signature matches
    {
        return (a * b);
    }

    //delegate instantiation
    ProductDelegate p = new ProductDelegate(Product);
}
```

Here, we have two methods with the same name but with different signatures. The delegate **p** is initialized with the reference to the second **Product** method because that method exactly matches the signature and return type of **ProductDelegate.** If this method is not present, an error will occur. Note that since the method and the instantiation statement are within the same class, we simply use the method name for creating the instance.

Consider another example code:

```
//delegate declaration

delegate void DisplayDelegate( );

class A
{
    //delegate method
    public void DisplayA ( )                //instance method
    {
        Console.WriteLine("Display A");
    }
}

class B
{
    //delegate method
    static public void DisplayB( )          //static method
    {
        Console.WriteLine("Display B");
    }
}
. . . .
. . . .
// delegate instance
A a = new A( );                             //create object a
DisplayDelegate d1 = new DisplayDelegate(a.DisplayA);
. . . .
. . . .
//Another delegate instance
DisplayDelegate d2 = new DisplayDelegate(B.DisplayB);
. . . .
. . . .
```

The code defines two delegate methods in two different classes. Since class **A** defines an instance method, an **A** type object is created and used with the method name to initialize the delegate object **d1**. The delegate method defined in class **B** is static and therefore the class name is used directly with the method name in creating the delegate object **d2.**

16.6 DELEGATE INVOCATION

C# uses a special syntax for invoking a delegate. When a delegate is invoked, it in turn invokes the method whose reference has been encapsulated into the delegate, (only if their signatures match). Invocation takes the following form:

delegate_object (parameters list)

The optional *parameters list* provides values for the parameters of the method to be used.

- If the invocation invokes a method that returns **void,** the result is nothing and therefore it cannot be used as an operand of any operator. It can be simply a statement_expression. *Example:*

```
delegate1(x, y); //void delegate
```

  This delegate invokes a method that does not return any value.

- If the method returns a value, then it can be used as an operand of any operator. Usually, we assign the return value to an appropriate variable for further processing. *Example:*

```
double result = delegate2(2.56, 45.73);
```

  This statement invokes a method (that takes two **double** values as parameters and returns **double** type value) and then assigns the returned value to the variable **result**.

Note: Sometimes the term 'delegate' is used to denote both the delegate type and delegate object. We need to be aware of the context to know which meaning we are using when we talk about delegates.

16.7 USING DELEGATES

Program 16.1 illustrates how a delegate is created and used in a program. The program implements two delegate methods and therefore we need to create two delegate objects to invoke these methods independently.

Program 16.1 Creating and implementing a delegate

```
using System;

//delegate declaration
delegate int ArithOp(int x, int y);

class MathOperation
{
    //delegate methods definition
    public static int Add(int a, int b)
    {
        return (a + b);
    }
    public static int Sub(int a, int b)
    {
        return (a - b);
    }
}
```

```
class DelegateTest
{
    public static void Main( )
    {
        //delegate instances

        ArithOp operation1 = new ArithOp (MathOperation.Add);
        ArithOp operation2 = new ArithOp(MathOperation.Sub);

        //invoking delegates

        int result1 = operation1(200, 100);
        int result2 = operation2(200,100);

        Console.WriteLine("Result1 = " + result1);
        Console.WriteLine("Result2 = " + result2);
    }
}
```

Output of Program 16.1:

```
Result 1 = 300
Result 2 = 100
```

In Program 16.1, we created two delegates of the same type. If we need to implement more delegate methods, we have to create more delegate objects. In such cases, we may create an array of delegate objects and then use them in a **for** loop to invoke the methods. *Example:*

```
ArithOp [ ] operation =
    {new ArithOp(MathOperation.Add),
    new ArithOp(MathOperation.Sub)
    };
```

This creates two delegates, **operation[0]** to invoke **Add** method and **operation [1]** to invoke **Sub** method.

It is also allowed to use delegate objects as method parameters. For instance:

```
ProcessMethod(operation[0], 200, 100);
```

is valid. This statement invokes the method **ProcessMethod** using three parameters including the delegate object **operation1.** The **ProcessMethod** would be defined as under:

```
static void ProcessMethod (ArithOp operation, int x, int y)

{
    int result1 = operation(x,y);
    Console.WriteLine("Result1 = " + result);
}
```

When **operation [0]** is passed as a parameter, the invocation

```
operation (x,y)
```

invokes **Add** method. Similarly, we can pass **operation [1]** to invoke **Sub** method.

16.8 MULTICAST DELEGATES

We have seen so far that a delegate can invoke only one method (whose reference has been encapsulated into the delegate). However, it is possible for certain delegates to hold and invoke multiple methods. Such delegates are called *multicast delegates.* Multicast delegates, also known as *combinable delegates*, must satisfy the following conditions:

- The return type of the delegate must be **void.**
- None of the parameters of the delegate type can be declared as output parameters, using **out** keyword.

If **D** is a delegate that satisfies the above conditions and **d1, d2, d3** and **d4** are the instances of **D**, then the statements

```
d3 = d1 + d2; //d3 refers to two methods
d4 = d3 - d2; //d4 refers to only d1 method
```

are valid provided that the delegate instances **d1** and **d2** have already been initialized with method references and **d3** and **d4** contain **null** reference.

For a multicast delegate instance that was created by combining two delegates, the invocation list is formed by concatenating the invocation list of the two operands of the addition operation. Delegates are invoked in the order they are added.

Program 16.2 illustrates the application of multicast delegates. The program demonstrates the use of both addition and removal of delegates. Note the order in which the delegates **m3** and **m4** invoke the methods.

Program 16.2 Implementing multicast delegates

```
using System;

delegate void MDelegate( );

class DM
{

    static public void Display( )
    {
        Console.WriteLine("NEW DELHI");
    }
    static public void Print( )
    {
```

```
            Console.WriteLine("NEW  YORK");
        }
    }
    class MTest
    {
        public static void Main( )
        {
            MDelegate m1 = new MDelegate(DM.Display);
            MDelegate m2 = new MDelegate (DM.Print);
            MDelegate m3 = m1 + m2;
            MDelegate m4 = m2 + m1;
            MDelegate m5 = m3 - m2;

            //invoking delegates

            m3( );
            m4( );
            M5( );
        }
    }
```

The output of Program 16.2 would be:

```
NEW  DELHI
NEW YORK
NEW YORK
NEW DELHI
NEW DELHI
```

16.9 EVENTS

An *event* is a delegate type class member that is used by the object or class to provide a notification to other objects that an event has occurred. The client object can act on an event by adding an *event handler* to the event.

Events are declared using the simple *event declaration* format as follows:

modifier **event** *type event-name;*

The *modifier* may be **new**, a valid combination of the four access modifiers, and a valid combination of **static, virtual, override, abstract** and **sealed**. The *type* of an event declaration must be a delegate type and the delegate must be as accessible as the event itself. The *event-name* is any valid C# variable name. **event** is a keyword that signifies that the *event-name* is an event.

Examples of event declaration are:

```
public event EventHandler Click;
public event RateChange Rate;
```

EventHandler and **RateChange** are delegates and **Click** and **Rate** are events.

Since events are based on delegates, we must first declare a delegate and then declare an instance of the delegate using the keyword **event**. Program 16.3 illustrates a simple implementation of an event handler.

Program 16.3 Implementing an event handler

```
using System;

//delegate declaration first
public delegate void Edelegate(string str);

class EventClass
{
        //declaration of event
        public event Edelegate Status;

        public void TriggerEvent( )
    {
        if(Status != null)
      Status (" Event Triggered");
    }
}

class EventTest
{
    public static void Main( )
    {
        EventClass ec = new EventClass( );
        EventTest  et = new EventTest( );

        ec.Status += new EDelegate(et.EventCatch);
        ec.TriggerEvent( );
    }


    public void EventCatch(string str)
    {
        Console.WriteLine(str);
    }

}
```

Program 16.3 would produce the following output:

```
Event Triggered
```

Program 16.3 declares first a delegate which is used to declare the event **Status** in the class **EventClass**. This class has a method to trigger the event when it occurs. Note that we have to check the event **Status** against **null** because no event might have occurred earlier.

The class **EventTest** defines, in addition to the **Main**, a method **EventCatch** whose signature matches that of the event. The **EventClass** is instantiated, and the method is subscribed to the **Status** event:

```
ec.Status += new Edelegate(et.EventCatch);
```

From now on, this method is called when the event is triggered.

Program 16.4 shows another application of delegates which store values in an array, sorts and then, reverses the array.

Program 16.4 An application of delegates

```
using System;
using System.Collections.Generic;
using System.Text;

namespace Delegates
{
    public enum Comparison
    {
        Student1 = 1,
        Student2 = 2
    }

    // a simple array to hold 2 items
    public class Pair<array1>
    {

        // private array to hold the two objects
        private array1[] pairObj = new array1[2];

        // the delegate declaration
        public delegate Comparison
            WhichIsFirst(array1 obj1, array1 obj2);

        // passed in constructor take two objects, added in the order as
            received in the array
        public Pair(
            array1 arr1,
            array1 arr2)
        {
```

```
        pairObj[0] = arr1;
        pairObj[1] = arr2;
    }

    // public method which reverse orders the two objects
    public void ReverseSort(
        WhichIsFirst theDelegatedFunc)
    {
        if (theDelegatedFunc(pairObj[0], pairObj[1])
            == Comparison.Student1)
        {
            array1 temp = pairObj[0];
            pairObj[0] = pairObj[1];
            pairObj[1] = temp;
        }
    }
    // public method which orders the two objects
    public void Sort(
        WhichIsFirst theDelegatedFunc)
    {
        if (theDelegatedFunc(pairObj[0], pairObj[1]) == Comparison.Student2)
        {
            array1 temp = pairObj[0];
            pairObj[0] = pairObj[1];
            pairObj[1] = temp;
        }
    }

    // ask the two objects to give their string value
    public override string ToString()
    {
        return pairObj[0].ToString() + ", "
            + pairObj[1].ToString();
    }
}       // end class Pair

public class Student
{
    private string name;

    public Student(string name)
    {
```

```
            this.name = name;
        }
        // students are ordered alphabetically
        public static Comparison  sortStudentComesFirst(Student s1, Student s2)
        {
            return String.Compare(s1.name, s2.name) < 0 ? Comparison.Student1 :
Comparison.Student2;
        }

        public override string ToString()
        {
            return name;
        }
    }       // end class Student

    public class ShowDelegate
    {
        public static void Main()
        {
            // create two students and add them to Pair objects
            Student st1 = new Student("Jim");
            Student st2 = new Student("Mark");
            Pair<Student> studentPair = new Pair<Student>(st2, st1);

            Console.WriteLine("Pair of students are \t\t\t: {0}",
                studentPair.ToString());

            // Instantiate  the delegates
            Pair<Student>.WhichIsFirst theStudentDelegate =
                new Pair<Student>.WhichIsFirst(Student.sortStudentComesFirst);

            // sort using the delegates
            studentPair.Sort(theStudentDelegate);
            Console.WriteLine("After Sorting  Pair of students are\t\t: {0}",
                studentPair.ToString());
            studentPair.ReverseSort(theStudentDelegate);
            Console.WriteLine("After ReverseSorting Pair of students are\t: {0}",
                studentPair.ToString());

        }
    }
}
```

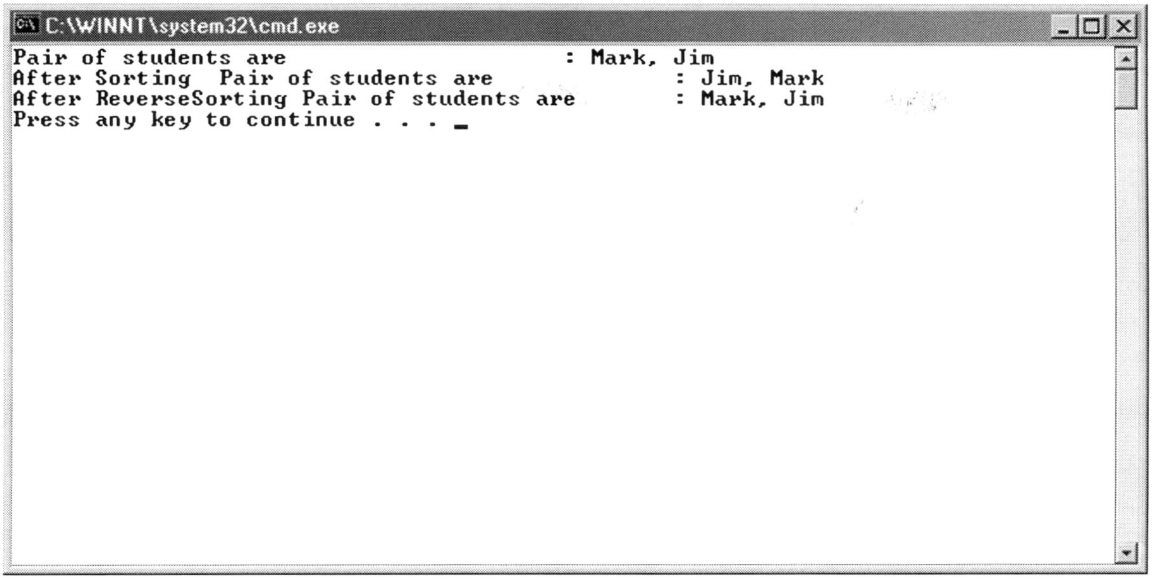

Program 16.5 demonstrates the application of **event** with delegates, which handles the access to the delegate. The **event handler** stores the clock time and displays time after every second.

Program 16.5 Application of event with delegates

```
using System;
using System.Collections.Generic;
using System.Text;

namespace EventwithDelegate
{
    public class ShowTime : EventArgs
    {
        public readonly int hour;
        public readonly int minute;
        public readonly int second;
        public ShowTime(int hour, int minute, int second)
        {
            this.hour = hour;
            this.minute = minute;
            this.second = second;
        }
    }

    public class ClockTime
```

```
{
    private int hour;
    private int minute;
    private int second;

    // the delegate the subscribers must implement
    public delegate void Handler2(object clock, ShowTime timeinfo);

    // the keyword event controls access to the delegate
    public event Handler2 handl2;

    // set the clock running
    // it will raise an event for each new second
    public void Display()
    {
        int count=5;
        for (;;)
        {
            count = 0;
            // get the current time
            System.DateTime dt = System.DateTime.Now;

            // if the second has changed notify the subscribers
            if (dt.Second != second)
            {
                // create the ShowTime object to pass to the subscriber
                ShowTime time = new ShowTime(dt.Hour, dt.Minute, dt.Second);

                // if anyone has subscribed, notify them
                if (handl2 != null)
                {
                    handl2(
                        this, time);
                }
            }

            // update the state
            this.second = dt.Second;
            this.minute = dt.Minute;
            this.hour = dt.Hour;
```

```
            }
        }
    }

    // DisplayClock subscribes to the clock's events. The job of DisplayClock is to display
the current time
    public class DisplayClock
    {
        // given a clock, list to its Handler2 event
        public void ListClock(ClockTime clock)
        {
            clock.handl2 += new ClockTime.Handler2(ChangeTime);
        }
        // the method that implements the delegated functionality
        public void ChangeTime(
            object clock, ShowTime time)
        {
            Console.WriteLine("Current Time is : {0}:{1}:{2}",
                time.hour.ToString(),
                time.minute.ToString(),
                time.second.ToString());
        }
    }

    public class UsingEvent
    {
        public static void Main()
        {
            // creates a new clock
            ClockTime clock = new ClockTime();

            // create the display
            DisplayClock dc = new DisplayClock();
            dc.ListClock(clock);

            // Get the clock started
            clock.Display();
        }
    }
}
```

```
Current Time is : 17:17:16
Current Time is : 17:17:17
Current Time is : 17:17:18
Current Time is : 17:17:19
Current Time is : 17:17:20
Current Time is : 17:17:21
Current Time is : 17:17:22
Current Time is : 17:17:23
Current Time is : 17:17:24
Current Time is : 17:17:25
Current Time is : 17:17:26
Current Time is : 17:17:27
Current Time is : 17:17:28
Current Time is : 17:17:29
Current Time is : 17:17:30
Current Time is : 17:17:31
Current Time is : 17:17:32
Current Time is : 17:17:33
Current Time is : 17:17:34
Current Time is : 17:17:35
Current Time is : 17:17:36
Current Time is : 17:17:37
Current Time is : 17:17:38
Current Time is : 17:17:39
```

Program 16.6 uses the **inheritance** concept with delegates. The delegate is called inside a base class and subsequently called in the inherited class and displays the output.

Program 16.6 Inheritance of delegates

```
using System;
using System.Collections.Generic;
using System.Text;

namespace UsingDelegate_Inheritance
{
    public delegate void delegl();

    //The base class having object of delegl delegate
    class baseclass
        {
            public delegl del;
            public void show()
            {
                del();
            }
        }
        class derivedclass : baseclass
        {
            //Constructor of derived class
        public derivedclass()
            {
```

```
                del = new delegl (funcl);
                show();

            del = new delegl (func2);
                show();
            }
            public void funcl()
            {
                Console.WriteLine ("This is in funcl() in the derived class.") ;
            }
            public void func2()
            {
                Console.WriteLine ("This is in func2() in the derived class.") ;
            }
        }

    class Program
    {
        static void Main(string[] args)
        {
            derivedclass dc = new derivedclass();
        }
    }
}
```

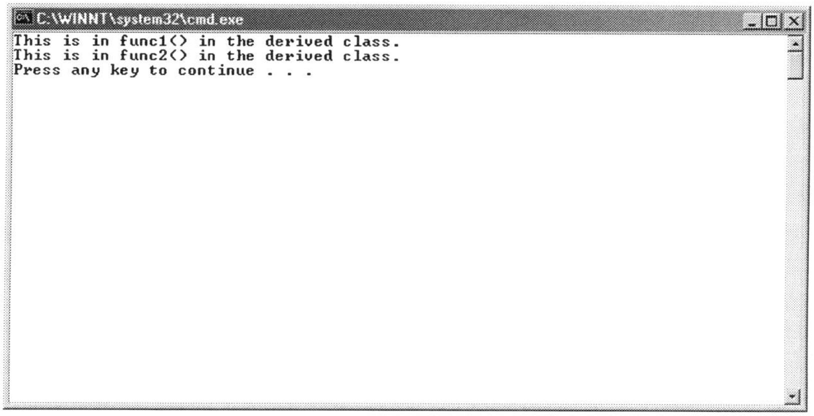

Case Study

Problem Statement

CapFinance Corporation is a finance organisation that provides loans to individuals for a time period ranging from 1 year to 5 years. The staff working in CapFinance manually calculates interest and the total amount for a loan taken by an individual. A large amount of errors were occurring during the calculation of interest and total amount. As a result, CapFinance Corporation were facing financial loss. In addition, the task of calculating interest and total amount manually was time consuming. How can CapFinance solve these problems and prevent financial loss?

Solution

In order to avoid financial loss, CapFinance has to prevent the errors from occurring during the calculation of interest and total amount. For this, the organisation has to automate the task of calculating interest and total amount through an application. CapFinance contacts S B Private Limited, which is a software development company for developing the application, which allow computerised calculation of interest and total amount related to a specific loan. SB Private Limited first conducts a requirement analysis and then decides to use the **delegate** feature of C# programming language in the application for calculating interest and total amount for a loan. A **delegate** in C# is a type of object that contains method details such as arguments passed to a method. In C#, **delegate** is mainly used for callback and event handling. After a period of 15 days, SB Private Limited comes up with the following C# application.

```
using System;
using System.Collections.Generic;
using System.Text;

namespace CalculateInterest
{
    delegate double ShowInterest(double val1, double val2);
    class Interest
    {
        public double Principal()
        {
            Console.Write("Enter the Principal (Rs.):   ");
            double principal = double.Parse(Console.ReadLine());
            return principal;
        }

        public double InterestRate()
        {
            Console.Write("Enter the Interest Rate (in %): ");
            double rate = double.Parse(Console.ReadLine());

            return rate;
        }

        public double Time(double time)
        {
            Console.Write("Enter the time in years: ");
            time = double.Parse(Console.ReadLine());
```

```
    return time;

    }

    // Interest = Principal * rate * time in years
    public double CalculateInterest(double principal, double rate, double time)
    {
        return principal * (rate / 100) * time;
    }
}


class CalculateLoan
{
        static int Main(string[] args)
        {
        double principal, rate, time, amount;
        double duration = 0;
        string durationname = null;
        Interest interest = new Interest();

        ShowInterest si = new ShowInterest(AddupAmount);

        Console.WriteLine("This is a program to Calculate Loan Amount in Simple
Interest.");

        Console.WriteLine("");
        principal = interest.Principal();
        rate   = interest.InterestRate();
        time    = interest.Time(duration);
        amount = interest.CalculateInterest(principal, rate, time);
        double Amount = si(principal, amount);


        durationname = "years";

        Console.WriteLine("");
        Console.WriteLine("==================================");
        Console.WriteLine("Calculated Loan");
        Console.WriteLine("------------------");
        Console.WriteLine("Principal: {0}", principal);
        Console.WriteLine("Interest:  {0}", rate / 100);
        Console.WriteLine("Period:    {0} {1}", time, durationname);
        Console.WriteLine("------------------");
        Console.WriteLine("Interest to be paid on Loan (Rs.): {0}", amount);
```

```
            Console.WriteLine("Total Amount to pay after "+time+" years
(Rs.) : {0}", Amount);

            Console.WriteLine("===================================\n");

            return 0;
        }

        static double AddupAmount(double val1, double val2)
        {
            return val1 + val2;
        }

    }
}
```

Remarks

Delegates allow you to implement the **callback** technique in a C# application in which an object sends a callback message to the object, which has sent a message. A **delegate** in C# acts as a method for another method. The main function of a **delegate** is to call the method encapsulated in it at the time of creation.

Review Questions

16.1 State whether the following statements are true or false.

(a) A delegate represents a data member of a class.
(b) A delegate can hold reference to a method of any class.
(c) A delegate can be defined only inside a class.
(d) A delegate type may be derived from another delegate type.
(e) It is optional to use a accessibility modifier to a delegate type.
(f) The new modifier may be used to declare any delegate.
(g) The signature of delegate methods must match the signature of the delegate.
(h) The return type of delegate methods must match the return type of the delegate.
(i) A delegate can hold reference to only instance methods.
(j) A delegate method must always return a value.
(k) It is always allowed to perform addition operation on the instances of a delegate.
(l) A event is a delegate type member.

16.2 What is a delegate? What is it used for?
16.3 Are delegates value types or reference types?
16.4 Enumerate the steps involved in creating and using a delegate.
16.5 Discuss the syntax of a delegate declaration.
16.6 What are the modifiers that can be applied to a delegate?
16.7 When do we need to use the new modifier?
16.8 Give atleast two examples of delegate declaration.
16.9 What are the places a delegate can be defined in a program?

16.10 What is a delegate method?
16.11 What are the requirements of a delegate method?
16.12 What are the special features of a delegate?
16.13 How do we create an instance of a delegate?
16.14 How do we invoke a delegate? What happens when a delegate is invoked?
16.15 What is a multicast delegate?
16.16 What are the characteristics of a multicast delegate?
16.17 What is an event?
16.18 Describe the syntax of an event declaration.
16.19 Give two typical examples where events are used.
16.20 What is an event handler? How is it designed?
16.21 What does the event keyword do?
16.22 Find errors, if any, in the following delegate declarations:
   (a) delegate D1( );
   (b) int delegate D2 ( );
   (c) delegate void D3 (x,y);
   (d) delegate int D4 (int a, int b) { }
   (e) private double D5 (double x);
   (f) public delegate int D6 (int x, int y)
16.23 Given the delegate
      delegate double D1 (int x, int y);
   which of the following methods can be encapsulated by it?
   (a) int M1 (int x, int y) { }
   (b) double M2 (double a, double b) { }
   (c) static double M3 (int x, int y) { }
   (d) public double M4 (int x, y) { }
   (e) private double M5 (int a, int b) { }
16.24 Given the following code segment

```
delegate void D1 ( );
class A
{
   public void M ( ) { }
}
```

   which of the following delegate instantiation statements are correct?
   (a) D1 d1 = new D1 (M);
   (b) D1 d1 = new (A.M);
   (c) D1 d1 = new D1 (A,M);
   (d) D1 D1 = new ((new A( )).M);

Debugging Exercises

16.1. Debug the following program for demonstrating return types of a method in C#.

```
using System;
using System.Collections.Generic;
```

```
using System.Text;

namespace debugApp1_chap16
{
   int ResultDisplay(int i, int j);

   class Show
   {
       public static MathsFunc(int x, int y)
       {
           return ((x * x + 2*x*y + y * y) / (x+y));
       }
   }
   class Program
   {
       static void Main(string[] args)
       {
           ResultDisplay rd = new ResultDisplay(Show.Mathsfunc());
           int getResult = rd(1, 2);
           Console.WriteLine("The value of the function is " + getResult);
       }
   }
}
```

16.2. Find errors in the following example of delegates in C#.

```
using System;
using System.Collections.Generic;
using System.Text;

namespace debugApp2_chap16
{
   public delegate void MsgHandler();
   class Class1
   {
       public MsgHandler msg;
       public void Show()
       {
               msg();
       }
   }
   class Class2 : Class1
   {
```

```
        public Class2()
        {
            msg += MsgHandler(Display());
            Show() ;
        }
        public void Display()
        {
            Console.WriteLine ("This is an example of working of a delegate.");
        }
    }

        class Program
        {

            static void Main(string[] args)
            {
                Class2 c1 = new Class2();
                c1.Display();
            }
        }
}
```

16.3. Debug the given C# program.

```
using System;
using System.Collections.Generic;
using System.Text;

namespace debugApp3_chap16
{

    delegate void Delegate1(int i);

    class DelegateExample
    {
        string name;
        static void Main(string[] args)
        {
            Delegate1 d11 = new
Delegate1(DelegateExample.Show);

            DelegateExample de = new DelegateExample();
            de.name = "Display method returns ";
            Delegate1 d12 = new Delegate1(de.Display);
```

```
            Console.WriteLine (dl1(455));
            Console.WriteLine (dl2(77));

        }
        string Show (int i)
        {
            return string.Format("Show method returns : {0}", i);
        }

        void Display(int i)
        {
            return string.Format("{0}: {1}", name, i);
        }

    }
}
```

16.4. Correct the following program which employs static keyword.

```
using System;
using System.Collections.Generic;
using System.Text;

namespace debugApp4_chap16
{
    class Program
    {
        delegate void MulticastDelegate(int a);
        public class DelegateClass
        {
            public void Method1(int a)
            {
                Console.WriteLine(a +" is passed in the Method1.");
            }
            static public void staticMethod1(int b)
            {
                Console.WriteLine(b +" is passed in the staticMethod1.");
            }

        }
        static void Main(string[] args)
        {
            DelegateClass dc = new DelegateClass();
```

```
            MulticastDelegate md = new MulticastDelegate(dc.Method1());
            md("447");
            Console.WriteLine("");
            md += new MulticastDelegate(DelegateClass.staticMethod1());
            md("689");

        }
    }
}
```

Chapter

17

Managing Console I/O Operations

17.1 INTRODUCTION

Programs are written to manipulate a given set of data and to generate processed data as output following the familiar *input-process-output* cycle. It is therefore, essential that we understand how to provide input data and how to present the results in a desired format.

We have already seen the use of the **Console.WrieLine( )** method to produce simple outputs and **Console.ReadLine( )** method for reading input from the keyboard. In this chapter, we shall consider in more detail the console I/O operations supported by C#.

17.2 THE CONSOLE CLASS

The methods for reading from and writing to the console are provided by the **System.Console** class. This class gives us access to the standard input, standard output and standard error streams as shown in Table 17.1.

Table 17.1 *Input/Output streams*

Stream Object	Represents
Console.In	Standard input
Console.Out	Standard output
Console.Error	Standard error

The standard input system **Console.In** gets input by default from the keyboard. We can also redirect it to receive input from a file. The standard output stream **Console.Out** sends output to the screen by default. We can also redirect the output to a file.

The standard error stream **Console.Error** usually sends error messages to the screen. Even when the standard output is sent to a file, error messages, if any, will be displayed on the screen.

17.3 CONSOLE INPUT

The console input stream object supports two methods for obtaining input from the keyboard:

- **Read ( )** Returns a single character as **int**. Returns -1 if no more characters are available.
- **ReadLine ( )**Returns a string containing a line of text. Returns **null** if no more lines are available.

These methods can be invoked using either **Console.In** object or **Console** class itself. The following code snippet reads a character from the keyboard and displays it on the screen.

```
int x = Console.Read ( );
Console.WriteLine ( (char) x );
```

We have used a casting operator (char) to **x** to convert it to a character type. Remember, **x** was read as an **int**.

The following code reads entire line of text as a single string and displays it on the screen.

```
string str = Console.ReadLine( );
Console.WriteLine(str);
```

Program 17.1 illustrates the use of **ReadLine( )** method to read interactively strings from the keyboard and to echo them to the screen.

Program 17.1 Using ReadLine ( ) to read strings

```
using System;
class ReadString
{

    public static void Main ( )
    {

        Console.WriteLine("What is your name?");
        string s;
        s = Console.ReadLine ( );
        Console.WriteLine("How are you Mr." + s);
    }
}
```

17.4 CONSOLE OUTPUT

The console output stream supports two methods for writing to the console:

- **Write ( )** Outputs one or more values to the screen without a newline character
- **WriteLine ( )** Outputs one or more values to the screen (same as Write ( ) method) but adds a newline character at the end of the output

Both the methods have various overloaded forms for handling all the predefined types and therefore we can easily output many different types of data. **WriteLine( )** also allows us to output data in many different formats (in a way, comparable to the **printf** method of C).

The **Write( )** method sends information into a buffer. This buffer is not flushed until a newline (or end-of-line) character is sent. As a result, this method prints output on one line until a newline character is encountered. For example, the statements,

```
Console.Write("Hello ");
Console.Write("C Sharp!");
```

will display the words Hello C Sharp! on one line and wait for displaying of further information on the same line. We may force the display to be brought to the next line by printing a newline character as shown below:

```
Console.Write("\n");
```

For example, the statements

```
Console.Write("Hello");
Console.Write("\n");
Console.Write("C Sharp!");
```

will display the output in two lines as follows:

```
Hello
C Sharp!
```

The **WriteLine( )** method, by contrast, takes the information provided and displays it on a single line followed by a line feed (carriage return). This means that the statements

```
Console.WriteLine("Hello");
Console.WriteLine("C Sharp!");
```

will produce the following output:

```
Hello
C Sharp!
```

The statement

```
Console.WriteLine( );
```

will print a blank line. Program 17.2 illustrates the application of **Write( )** and **WriteLine( )** methods.

Program 17.2 Behaviour of output methods

```
using System;
class Displaying
{
    public static void Main ( )
    {
```

```
        Console.WriteLine("Screen Display");
        for (int i = 1; i <= 9; i++)
        {
            for(int j = 1; j <= i; j++)
            {
                Console.Write(" ");
                Console.Write(i);
            }
            Console.Write("\n");
        }
        Console.WriteLine("Screen Display Done");
    }
}
```

Program 17.2 displays the following on the screen:

```
Screen Display
    1
    2  2
    3  3  3
    4  4  4   4
    5  5  5   5   5
    6  6  6   6   6   6
    7  7  7   7   7   7   7
    8  8  8   8   8   8   8   8
    9  9  9   9   9   9   9   9   9

Screen Display Done
```

17.5 FORMATTED OUTPUT

We can produce formatted output using the overloaded **WriteLine( )** method. This method takes a string containing a format and a list of variables whose values have to be printed out. The general form of formatted **WriteLine( )** method is:

```
Console.WriteLine(format-string, v1, v2 ....);
```

Format string contains both static text and *markers* which indicate:

- Where the values are to be printed
- How the values are to be formatted

In its simplest form, a marker is an index number in curly brackets, that indicates which variable of the argument list is to be substituted. *Examples:*

```
Console.WriteLine("Total is {0}" , total);
Console.WriteLine("Sum of {0} and {1} is {2}", a,b,c);
```

In the examples, {0} represents the first argument in the list, {1} represents the second, and so on. For instance, if a = 100, b = 200 and c = a + b, then the second statement will display the following output:

```
Sum of 100 and 200is 300
       ↑       ↑      ↑
      {0}     {1}    {2}
       ↑       ↑      ↑
       a       b      c
```

We can also specify a width for the value using the format:

```
{ n, w }
```

Where **n** is the index number and **w** is the width for the value. If **w** is positive, the value will be printed right-justified in the width, and, if it is negative, the value will be left-justified. *Example:*

```
int a = 45 ;
int b = 976 ;
int c = a + b ;
Console.WriteLine ("{0,5}\n+{1,5}\n- - - - - -\n{2,5}" , a,b,c) ;
```

This will display the following output:

```
    4 5
  + 9 7 6
 - - - - - - -
  1 0 2 1
```

Note how one **WriteLine( )** statement prints four lines of output. This is due to use of line character '\n' in the format string.

We can also print an array of objects using markers in a format string.

Example:

```
object [ ] array = { "abc", 12.3, 100, "123"};
Console.WriteLine("Array:{0}, {1}, {2}, {3}", array);
```

Program 17.3 illustrates how a format string is implemented for displaying a simple formatted output.

Program 17.3 Using format strings

```
using System;
class FormatString
{
    public static void Main ( )
    {
        string s = "Items";
        int a = 100;
        float x = 99.9F;
        Console.WriteLine("{0}\nIntValue: {1}\nFloatValue: {2}",
                                                              s,a,x);
    }
}
```

The output of Program 17.3 would be:

```
Items
IntValue: 100
FloatValue : 99.9
```

17.6 NUMERIC FORMATTING

We have seen how a simple format string can be used to output different types of values with the help of markers "{ }". Each marker (also known as *placeholder)* can optionally contain various *format characters* to specify the nature of numeric format. C# supports two methods of numeric format:

- Standard format
- Custom format

The most commonly used one is the *standard format* that converts a numeric type to a specific string representation. We may use *custom format* if further refinement of the output format is required.

17.7 STANDARD NUMERIC FORMAT

The standard numeric format consists of a numeric format character and optionally a precision specifier. The general format of a marker would appear like this:

```
{n:fc[p]}
```

where **n** is the index number of the argument to be substituted, **fc** is the format character and **p** is the precision specifier. Table 17.2 shows the list of format characters that can be used for formatting the output of numbers.

Table 17.2 *C# format characters*

Format Character	Description
C or c	Currency formatting. By default, the dollar sign $ will be fixed to the value. A precision specifier may be used to specify the number of decimal places.
D or d	Formats integer numbers, and converts integers to base 10. A precision specifier may be used to pad with leading zeros, if necessary.
E or e	Exponential (or Scientific) formatting. A precision specifier may be used to set the number of decimal places, which is 6 by default. There is always one digit before the decimal point.
F or f	Fixed-point formatting. The precision specifier decides the number of decimal places. Zero is acceptable.
G or g	General formatting, used to format a number to a fixed or exponential format depending on which is the most compact.
N or n	Number formatting; produces values with embedded commas.
X or x	Hexadecimal formatting. A precision specifier may be used to pad with leading zero. If we use uppercase X, hex format will also contain uppercase characters.

17.7.1 Currency Formatting

The currency format converts the numerical value to a string containing a locale-specific currency symbol. By default, the dollar symbol is added. However, it can be changed using a **NumberFormatInfo** object. *Examples:*

```
("{ 0 : C }" , 4567.899) ⟶ $4,567.90
("{ 0 : C}" , -4567.899) ⟶ ($4,567.90)
```

When the number is negative, the value is printed as positive inside simple brackets. Note that the decimal part, by default, is rounded to two places.

17.7.2 Integer Formatting

Integer format converts a given numerical value to an integer of base 10. The minimum number of digits is determined by the precision specifier. If the precision specifier is less than the actual digits, the specifier is ignored and all the digits are printed. If it is greater, then the result is left-padded with zeros to obtain the specified number of digits. *Examples:*

```
("{ 0 : D }" , 45678) ⟶ 45678
("{ 0 : D8 }" , 45678) ⟶ 00045678
```

17.7.3 Exponential Formatting

Exponential formatting character (E or e) converts a given value to a string in the form of:

```
m.dddd E+xxx
```

The output will contain only one digit m before the decimal point. The number of decimal places (dddd) is decided by the precison specifier. By default, six places are used. The format character E (or e) will appear in the output. *Examples:*

```
("{ 0 : E }" , 34567.899)    ⟶    3.456790E+004
("{ 0 : E9 }" , 34567.899)   ⟶    3.456789900E+004
("{ 0 : e5 }" , 34567.899)   ⟶    3.45679e+004
```

Note that the decimal part is rounded (not truncated) to the specified places. If the specified places are more, then the decimal part is padded at the right with zeros.

17.7.4 Fixed-Point Format

The fixed-point format converts the given value to a string containing decimal places as decided by the precision specifier. By default, two decimal places are assumed. Zero as a precision specifier is allowed. *Examples:*

```
("{ 0 : F  }" , 3456.7899)   ⟶    3456.79
("{ 0 : F6 }" , 3456.7899)   ⟶    3456.789900
("{ 0 : F0 }" , 3456.7899)   ⟶    3457
```

Note that the decimal part is always rounded (or padded with zeros) to the specified precision places. A zero precision converts the value to the nearest integer.

17.7.5 Number Format

This format produces the output with the embedded commas, like:

```
45,678.45
```

Examples:

```
("{ 0 : N }" , 34567.899)   ⟶  34,567.90
("{ 0 : N }" , 34567)       ⟶  34,567.00
("{ 0 : N4 }" , 34567.899)  ⟶  34,567.8990
```

The precision specifier decides the number of decimal places. It is two by default. Note that the decimal part is rounded to the specified precision places. If the specified places are more, the decimal part is padded with zeros.

17.8 CUSTOM NUMERIC FORMAT

Although the standard numeric format is good enough for most situations, C# also supports what is known as *custom format strings* to provide more control over the output format, if desired. Here, special characters are used to form a 'template' that decides the nature of output. . The commonly used special characters include:

- Zero (0)
- Pound Character (#)
- Period (.)
- Comma (,)
- Percent Character (%)

The zero and pound characters are called *placeholders.*
We consider in this section how these placeholders and special characters are used to format a given number.

17.8.1 Zero Placeholder

The zero placeholders are used to format integer outputs. It takes the following form :

```
"{ 0 : 0000 }"
```

If the number has a digit in the position at which the "0" appears in the format string, the digit will appear in the output; otherwise, zero will appear. If the number of zeros is less than the number of digits, then all the digits will appear. *Examples:*

```
("{ 0   : 0000 }" , 123) ⟶   0123
("{ 0   : 000 }" , 1234) ⟶   1234
```

17.8.2 Space Placeholder

We can also use the pound character (#) to format integer output. It takes the following form:

```
"( 0 : #### }"
```

It works exactly the same way as the zero placeholder, except that a blank (instead of zero) appears if there is no digit in that position. *Examples:*

```
("{ 0 : #### }" , 123) ⟶ 123
("{ 0 : ### }" , 1234) ⟶ 1234
```

17.8.3 Decimal Separator

The location of the decimal point in the output may be decided by placing a period (.) character in the format string, as shown below:

```
"{ 0 : ##.000 }"
```

Examples:

```
("{ 0 : ##.000 }" , 12.34567) ⟶ 12.346
("{ 0 : ##.000 }" , 1234.5)   ⟶ 1234.500
```

Note that the number of decimal places is decided by the number of zeros appearing after the period in the format string. If zeros are less, the decimal part is rounded to the required places and if it is more, the decimal part is padded with zeros as illustrated.

17.8.4 Comma Separator

The character comma (,) can be used along with the placeholders (zero and pound characters) to separate groups of digits in the integer part of the output. *Examples:*

```
("{ 0 : ##, ###.00 }", 1234567.899)  ⟶  1,234,567.90
("{ 0 : #, #.000 }" , 1234567.899)   ⟶  1,234,567.899
("{ 0 : #, ## }" , 1234567)          ⟶  1,234,567
```

17.8.5 Percent Notation

We can display a number in percentage form by using the percent character (%) at the end of the format string. *Examples:*

```
("{ 0 : 00% }" , 0.01234)        01 %
("{ 0 : ##.000% }", 0.12345) 12.345 %
```

Note that the number is multiplied by 100 before it is formatted using the format string. If the format string does not have a decimal point, the decimal part is truncated.

Program 17.4 displays a number in a number format with a $(dollar sign and in a formatted decimal way of ###.###.

Program 17.4 Custom number format example 1

```
using System;
using System.Collections.Generic;
using System.Text;
```

```
namespace customNumberFormatExample
{
    class Program
    {
        static void Main(string[] args)
        {
                    StringBuilder str = new StringBuilder();
                    Console.WriteLine("This is an example of CustomNumberFormat!\n");
                    string oldString = "{0:$#,###0.00;($#,###0.00);Zero}";
                    Console.Write("This is the Original Number Format string : ");
                    Console.WriteLine(oldString);
                    double val = 335.72;
                    Console.WriteLine("Number entered is : " + val);
                    str.Append("\nThis is the formatted output : ");
                    str.AppendFormat(oldString, val);
                    Console.WriteLine(str.ToString());
        }
    }
}
```

```
C:\WINNT\system32\cmd.exe
This is an example of CustomNumberFormat!

This is the Original Number Format string : {0:$#,###0.00;($#,###0.00);Zero}
Number entered is : 335.72

This is the formatted output : $335.72
Press any key to continue . . .
```

Program 17.5 shows a number 1973458783 to be formatted in ## + ## * #### – ##, breaks the first number in 1 9 + 73 * 4587 – 83. The second number 2662.63 with a decimal will be rounded off to the nearest decimal 2662.6, if #.# number format is passed. Similarly, if only # is passed, the second number will be rounded off with no decimal displayed as 2663.

Program 17.5 Custom number format example 2

```
using System;
using System.Collections.Generic;
using System.Text;

namespace CustomNumberFormatExample2
{
   class Program
   {
      static void Main(string[] args)
      {
             double val1 = 1973458783;
             Console.WriteLine(“First number is : ” +val1);
             Console.WriteLine(“To be formatted in ## + ## * #### - ##”);
             String val2 = val1.ToString(“## + ## * #### - ##”);
             Console.WriteLine(“\nFormatted first number : ” + val2);


             double val3 = 2662.63;
             Console.WriteLine(“\nSecond number is :” + val3);
             Console.WriteLine(“\nTo be formatted in #.#”);
             val2 = val3.ToString(“Formatted second number = #.#”);
             Console.WriteLine(val2);


      }
   }
}
```

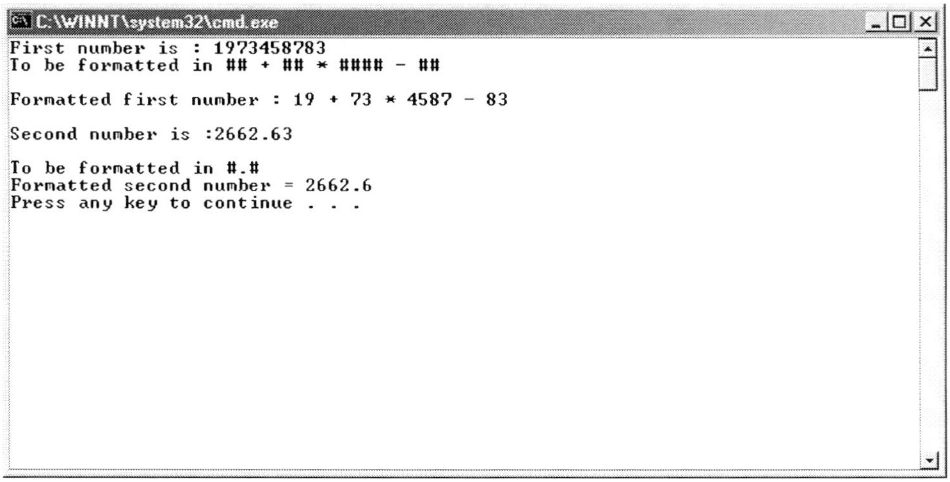

Case Study

Problem Statement

Modern Engineering College provides courses for streams such as mechanical, electrical, electronics and civil engineering. When a student takes admission in the Modern Engineering College and comes to the administrative staff to pay the fees, the staff has to manually fill a form specifying details such as name, address, course and fees. A large amount of time is consumed in the filling of the form. In addition, the efficiency of the administrative staff also decreases as a lot time is spent in filling the form. The students who are taking admission in the Modern Engineering College have to wait for a long time in a queue for paying their fees. How can Modern Engineering College solve this problem of the students and the administrative staff ?

Solution

The principal of Modern Engineering College along with some of the important management staff conducts a meeting and tries to find the solution to the problem, which is being faced by the students taking admission in the college and the administrative staff. By the end of the meeting, the principal and the management staff come to the conclusion that an application must be developed for automating the task of entering student information such as name and address. The principal calls the Head of the Electronics department after a few days and asks the Head to develop the application. The Head of the Electronics department first conducts a requirement analysis through interviews with the administrative staff of the college. After the analysis, the Head of the Electronics Department comes to the conclusion that methods such as **ReadLine( )** and **WriteLine( )** available in C# must be used in a C# application for obtaining details such as name and address and displaying a message to confirm the admission of the student. The methods such as **ReadLine( )** and **WriteLine( )** help in receiving inputs from users and displaying output. The following C# application is developed by the Head of the Electronics department.

```
using System;
using System.Collections.Generic;
using System.Text;

namespace StudentAdmissionSystem
{
    public enum Degree
    {
        Mechanical = 1,
        Electrical,
        Electronics,
        Aeronautics,
        Civil
    }
```

```
    class Select
    {

        public Degree degree;

        public Select()
        {
          degree = Degree.Civil;
        }
      public Select(Degree degree)
        {

          degree = Degree.Civil;
        }

    }

    class GetDetails
    {
        Select degree;
        string name,address,telephone,course,fees;
      public GetDetails()
        {
          degree = new Select();
        }


        public void GetAdmission()
        {
           int choice = 0;

          Console.WriteLine("Enter the Degree course to take admission in Modern
Engineering College : ");
          Console.WriteLine("----------------");
          Console.WriteLine("1. Mechanical");
          Console.WriteLine("2. Electrical");
          Console.WriteLine("3. Electronics");
          Console.WriteLine("4. Aeronautics");
          Console.WriteLine("5. Civil");

          Console.Write("Please enter your Choice: ");
          choice = int.Parse(Console.ReadLine());

           switch (choice)
```

```
        {
        case 1:
          degree.degree = Degree.Mechanical;
            break;
        case 2:
          degree.degree = Degree.Electrical;
            break;
        case 3:
          degree.degree = Degree.Electronics;
            break;
        case 4:
          degree.degree = Degree.Aeronautics;
            break;
        default:
          degree.degree = Degree.Civil;
            break;

      }
  }

public void GetStudentDetails()
  {
    Console.Write("Enter name : ");
    name = Console.ReadLine();
    Console.WriteLine("");
    Console.Write("Enter address : ");
    address = Console.ReadLine();
    Console.WriteLine("");
    Console.Write("Enter telephone : ");
    telephone = Console.ReadLine();
    Console.WriteLine("");
    Console.WriteLine("");
    Console.Write("Course : " + degree.degree);
    course = degree.degree.ToString();
    Console.WriteLine("");
    Console.Write("Enter Fees : ");
    fees = Console.ReadLine();
  }



class StudentMenu
```

```
        {
            static void Main(string[] args)
            {
                GetDetails gd = new GetDetails();
                gd.GetAdmission();

                Console.WriteLine();

                gd.GetStudentDetails();
                Console.WriteLine();
                Console.WriteLine("Thank You " + gd.name + " for getting
Admission in " + gd.course+ " course.");
            }
        }
    }
}
```

Remarks

In C#, methods such as **Read( )** and **Write( )** allow programmers to perform input and output operations. This means that these methods can be used to receive inputs such as text and numbers from a user in a specific format and display the output in a particular format.

Review Questions

17.1 Distinguish between Write and WriteLine methods.
17.2 Distinguish between Read and ReadLine methods.
17.3 What is a format string? How does it help produce outputs?
17.4 What is numeric formatting?
17.5 Explain the general format of the standard numerical format.
17.6 Why do we need exponential formatting?
17.7 What is a number formatting? When do we use it?
17.8 What is a custom numeric format? Why do we need custom format strings?
17.9 Distinguish between the zero space holder and the pound character (#) space holder in producing integer outputs.
17.10 When do we use the percent character in a format string? How does it work?

Debugging Exercises

17.1 Correct the syntax errors in the given program.

```
using System;
using System.Collections.Generic;
using System.Text;
namespace debugAppl_chap17
```

```
{
    class Program
    {
        static void Main(string[] args)
        {
            Console.WriteLine("This is a console application.");
            Console.Write("Please enter your lucky number: ");
            string val1 = Console.Readline();
            int val2 = System.Convert.ToInt32(val1, 10);
            val2 = val2 * val2;
            Console.WriteLine("The square of the number you entered is " + val2);

        }
    }
}
```

17.2 Find errors in the following program for reading and writing to the console.

```
using System;
using System.Collections.Generic;
using System.Text;

namespace debugApp2_chap17
{
   class Program
    {
      static void Main(string[] args)
       {
        Console.WriteLine("This is a Console Application in C#!");
        Console.Write("Please enter your Name: ");
        string name = Console.ReadLine();
        Console.WriteLine("Hi! " , name , ". This program takes input from the
users.");
        Console.Write();
       }
    }
}
```

17.3 Correct the program for calculating the sum of two numbers.

```
using System;
using System.Collections.Generic;
using System.Text;

namespace debugApp3_chap17
```

```
{
    class Program
    {
        int num1 = 0;
        int num2 = 0;
        int sum = 0;

        static void Main(string[] args)
        {
            Console.WriteLine("Please enter the first number : ");
            num1 = Int32.Parse(Console.ReadLine());
            Console.WriteLine("Please enter the second number : ");
            num2 = Int32.Parse(Console.ReadLine());

            sum = addValues(num1 + num2);

            Console.WriteLine("Sum of " + num1 + " + " + num2 + " = " + sum);
            Console.ReadLine();
        }
        int addValues(int m, int n)
        {
            return m + n;
        }
    }
}
```

Programming Exercises

17.1 Write a program to print a multiplication table from 1 × 1 to 9 × 9 as shown below using only one Write statement and one WriteLine statement:

1 2 3 4 5 6 7 8 9
2 4 6 18
3 6 9 27
. . . .
. . . .
. . . .
9 18 27. 81

17.2 Modify the above program so that the display includes the title
MULTIPLICATION TABLE
at the center.

17.3 Write a program that stores the integers 111, 222 and 333 in an array and displays them as follows without using any looping statements.
(a) 111 222 333

(b) 111
222
333

17.4 Write a program that would produce the following output:
Product of 1 and 1 is 1
Product of 3 and 3 is 9
Product of 5 and 5 is 25
Product of 7 and 7 is 49
Product of 9 and 9 is 81
Use nested for loops and only one WriteLine statement.

17.5 Write a program that prints the value 1234 in a column width of 10:
(a) Right-justified
(b) Left-justified
(c) Right-justified with padded zeros on left

17.6 Write a program that prints the value 10.45678 in exponential format with the following specifications:
(a) correct to two decimal places
(b) correct to four decimal places
(c) correct to eight decimal places

17.7 Write a program that would print the value 345.6789 in fixed-point format with the following specifications.
(a) correct to two decimal places
(b) correct to five decimal places
(c) correct to zero decimal place

17.8 Write a program to print the integer 3456789 using the following format specifications:
(a) using the format character D
(b) using the format character N
(c) using the format character E
(d) using zero placeholder
(e) using the pound character
(f) using the comma separator

17.9 Write a program to illustrate the use of the following format characters:
(a) Decimal separator (.)
(b) Percent character (%)

17.10 Write a program to read and display the following table of data:

Item-name	Item-code	Price
Fan	67831	1234.50
Motor	935642	5786.70

The item-name and item-code must be left-justified and the price should be right-justified.

Chapter

18

Managing Errors and Exceptions

18.1 INTRODUCTION

Rarely does a program run successfully in the very first attempt. It is common to make mistakes while developing as well as typing a program. A mistake might lead to an error causing the program to produce unexpected results. *Errors* are mistakes that can make a program go wrong.

An error may produce an incorrect output or may terminate the execution of the program abruptly or even may cause the system to crash. It is therefore important to detect and manage properly all the possible errors and error conditions in the program so that they do not terminate or cause the system to crash during execution.

18.2 TYPES OF ERRORS

Errors may be broadly classified into two categories:

- Compile-time errors
- Run-time errors

18.2.1 Compile-Time Errors

All syntax errors will be detected and displayed by the C# compiler and therefore these errors are known as *compile-time errors*. Whenever the compiler displays an error, it will not create the **.cs** file. It is therefore necessary that we fix all the errors before we can successfully compile and run the program.

Program 18.1 Illustration of compile-time errors

```
/* This program contains an error */
using Sytem;
class Error1
```

```
{
    public static void Main()
    {
        Console.WriteLine("Hello C#!");
    }
}
```

The C# compiler does a nice job of telling us where the errors are in the program. For example, if we have misspelled the **System** namespace in Program 18.1, the following message will be displayed on the screen:

```
Error1.cs(2.7) : error CS0234 : The type or namespace name 'Sytem' does not
exist in the class or namespace.
```

We can now go to the appropriate line, correct the error, and recompile the program. Sometimes, a single error may be the source of multiple errors later in the compilation. For example, use of an undeclared variable in a number of places will cause a series of errors of type 'undefined variable'. We should generally consider the earliest errors as the major source of our problem. After we fix such an error, we should recompile the program and look for other errors.

Most of the compile-time errors are due to typing mistakes. Typographical errors are hard to find. We may have to check the code word by word, or even character by character. The most common problems are:

- Missing semicolons
- Missing (or mismatch of) brackets in classes and methods
- Misspelling of identifiers and keywords
- Missing double quotes in strings
- Use of undeclared variables
- Incompatible types in assignments / initialization
- Bad references to objects
- Use of = in place of == operator, etc.

18.2.2 Run-time Errors

Sometimes, a program may compile successfully creating the **.exe** file but may not run properly. Such programs may produce wrong results due to wrong logic or may terminate due to errors such as stack overflow. Most common run-time errors are:

- Dividing an integer by zero.
- Accessing an element that is out of the bounds of an array.
- Trying to store a value into an array of an incompatible class or type.
- Passing a parameter that is not in a valid range or value for a method.
- Attempting to use a negative size for an array.
- Using a null object reference as a legitimate object reference to access a method or a variable.
- Converting an invalid string to a number or vice versa.
- Accessing a character that is out of bounds of a string, and so on.

When such errors are encountered, C# typically generates an error message and aborts the program. Program 18.2 illustrates how a run-time error causes termination of execution of the program.

Program 18.2 Illustration of run-time errors

```
using System;
class Error2
{
    public static void Main( )
    {
        int a = 10;
        int b = 5;
        int c = 5;

        int x = a/(b-c);
        Console.WriteLine("x = " + x);

        int y = a/(b+c);
        Console.WriteLine("y = " + y);
    }
}
```

Program 18.2 is syntactically correct and therefore does not cause any problem during compilation. When C# run-time tries to execute a division by zero, it generates an error condition which causes the program to stop after displaying an appropriate message. The following statement is never executed:

```
int y = a/(b + c);
```

18.3 EXCEPTIONS

An *exception* is a condition that is caused by a run-time error in the program. When the C# compiler encounters an error such as dividing an integer by zero, it creates an exception object and throws it (i.e., informs us that an error has occurred).

If the exception object is not caught and handled properly, the compiler will display an error message and will terminate the program. If we want the program to continue with the execution of the remaining code, then we should try to catch the exception object thrown by the error condition and then display an appropriate message for taking corrective actions. This task is known as *exception handling*.

The purpose of the exception handling mechanism is to provide a means to detect and report an 'exceptional circumstance' so that appropriate action can be taken. The mechanism suggests incorporation of a separate error handling code that performs the following tasks:

- Find the problem (***Hit*** the exception)
- Inform that an error has occurred (***Throw*** the exception)
- Receive the error information (***Catch*** the exception)
- Take corrective actions (***Handle*** the exception)

The error handling code basically consists of two segments, one to detect errors and to throw exceptions and the other to catch exceptions and to take appropriate actions.

When writing programs, we must always be on the lookout for places in the program where an exception could be generated. Some common exceptions that we must catch are listed in Table 18.1.

Table 18.1 *Common C# exceptions*

Exception Class	Cause of Exception
SystemException	A failed run-time check; used as a base class for other exceptions
AccessException	Failure to access a type member, such as a method or field
ArgumentException	An argument to a method was invalid
ArgumentNullException	A null argument was passed to a method that does not accept it
ArgumentOutofRangeException	Argument value is out of range
ArithmeticException	Arithmetic over-or underflow has occurred
ArrayTypeMismatchException	Attempt to store the wrong type of object in an array
BadImageFormatException	Image is in the wrong format
CoreException	Base class for exceptions thrown by the runtime
DivideByZeroException	An attempt was made to divide by zero
FormatException	The format of an argument is wrong
IndexOutofRangeException	An array index is out of bounds
InvalidCastException	An attempt was made to cast to an invalid class
InvalidOperationException	A method was called at an invalid time
MissingMemberException	An invalid version of a DLL was accessed
NotFiniteNumberException	A number is not valid
NotSupportedException	Indicates that a method is not implemented by a class
NullReferenceException	Attempt to use an unassigned reference
OutofMemoryException	Not enough memory to continue execution
StackOverflowException	A stack has overflowed

18.4 SYNTAX OF EXCEPTION HANDLING CODE

The basic concepts of exception handling are *throwing* an exception and *catching* it. This is illustrated in Fig. 18.1.

C# uses a keyword **try** to preface a block of code that is likely to cause an error condition and 'throw' an exception. A catch block defined by the keyword **catch** 'catches' the exception 'thrown' by the try block and handles it appropriately. The catch block is added immediately after the try block. The following example illustrates the use of simple **try** and **catch** statements:

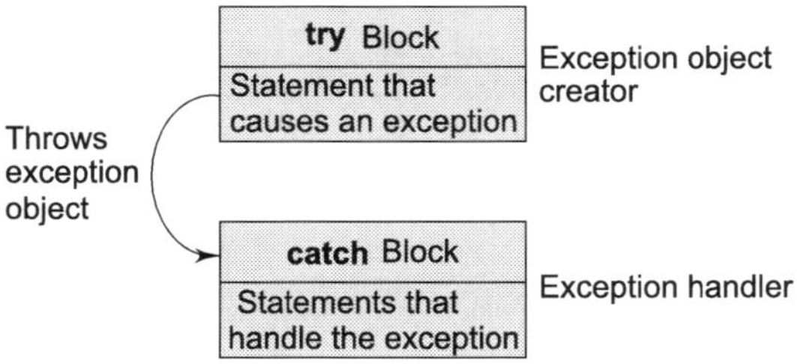

Figure 18.1 *Exception handling mechanism*

```
..........
..........
try
{
    statement;      // generates an exception
}
catch (Exception e)
{
    statement;      // processes the exception
}
..........
..........
```

The try block can have one or more statements that could generate an exception. If any one statement generates an exception, the remaining statements in the block are skipped and execution jumps to the catch block that is placed next to the try block.

The catch block too can have one or more statements that are necessary to process the exception. Remember that every **try** statement should be followed by at least one **catch** statement; otherwise compilation error will occur.

Note that the **catch** statement works like a method definition. The **catch** statement is passed a single parameter, which is the reference to the exception object thrown (by the try block). If the catch parameter matches with the type of exception object, then the exception is caught and statements in the catch block will be executed. Otherwise, the exception is not caught and the default exception handler will cause the execution to terminate.

Program 18.3 illustrates the use of try and catch blocks to handle an arithmetic exception. Note that it is a modified version of Program 18.2.

Program 18.3 Using try and catch for exception handling

```
using System;
class Error3
{
    public static void Main( )
    {
        int a = 10;
        int b = 5;
        int c = 5;
        int x, y ;

        try
        {
            x = a / (b-c);  / / Exception here
        }

        catch (Exception e)
```

```
            {
                Console.WriteLine("Division by zero");
            }

            y = a / (b+c);
            Console.WriteLine("y = " + y);
        }
    }
```

Program 18.3 displays the following output:

```
Division by zero
y = 1
```

Note that the program did not stop at the point of exceptional condition. It catches the error condition, prints the error message and then continues the execution as if nothing has happened. Compare this with the execution of Program 18.2 which did not give the value of **y**.

Most often, exceptions are thrown by methods that are invoked from within the **try** blocks. The point at which an exception is thrown is called the *throw point*. Once an exception is thrown to the **catch** block, control cannot return to the throw point. This kind of relationship is shown in Fig. 18.2.

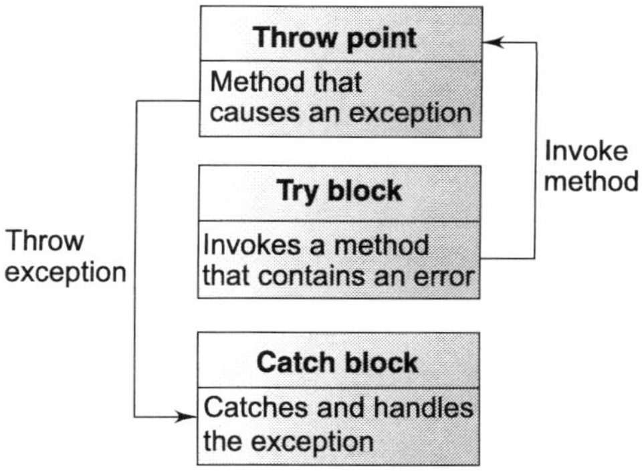

Figure 18.2 *Invoking a method that contain exceptions*

18.5 MULTIPLE CATCH STATEMENTS

It is possible to have more than one catch statement in the catch block as illustrated below:

```
..........
..........
try
{
        statement ;                 // generates an exception
}

catch (Exception-Type-1 e)
{
        statement;                  // processes exception type 1
}

catch (Exception-Type-2 e)
{
```

```
        statement;              / / processes exception type 2
}
.
.
.
catch (Exception-type-N e)
{
        statement ;             / / processes exception type N
}
..........
..........
```

When an exception in a **try** block is generated, the C# treats the multiple **catch** statements like cases in a **switch** statement. The first statement whose parameter matches with the exception object will be executed, and the remaining statements will be skipped.

Note that C# does not require any processing of the exception at all. We can simply have a catch statement with an empty block to avoid program abortion.

Example:

```
catch (Exception e){ }
```

This statement will catch an exception and then ignore it.

Program 18.4 Using multiple catch blocks

```
using System;
class Error4
{
    public static void Main( )
    {
        int [ ] a = {5,10};
        int b = 5;

        try
        {
            int x = a[2] / b - a[1];
        }

        catch(ArithmeticException e)
        {
            Console.WriteLine("Division by zero");
        }

        catch( IndexOutOfRangeException e)
        {
            Console.WriteLine("Array index error");
```

```
        }

        catch(ArrayTypeMismatchException e)
        {
            Console.WriteLine("Wrong data type");
        }
        int y = a[1] / a[0];
        Console.WriteLine("y = " + y);
    }
}
```

Program 18.4 uses a chain of catch blocks and, when run, produces the following output:

```
Array index error
y = 2
```

Note that the array element a[2] does not exist because array **a** is defined to have only two elements, a[0] and a[1]. Therefore, the index 2 is outside the array boundary thus causing the block

```
catch(IndexOutOfRangeException e)
```

to catch and handle the error. The remaining catch blocks are skipped.

18.6 THE EXCEPTION HIERARCHY

All C# exceptions are derived from the class **Exception**. When an exception occurs, the proper **catch** handler is determined by matching the type of exception to the name of the exception mentioned. If we are going to catch exceptions at different levels in the hierarchy, we need to put them in the right order. The rule is that we must always put the handlers for the most derived exception class first. Consider the following code snippet:

```
. . . .
. . . .
try
{
. . .//throw Divide by Zero Exception
{
catch(Exception e)
{
. . . .
}
catch (DivideByZeroException e)
{
. . . .
}
```

This code will generate a compiler error, because the exception is caught by the first catch (which is a more general one) and the second catch is therefore unreachable. In C#, having unreachable code is always an error. The code must be rewritten as follows:

```
try
{
. . . . //throw Divide By Zero Exception
}
catch(DivideByZeroException e)
{
. . . .
}
catch(Exception e)
{
. . . .
}
```

The order of catch blocks is important. We must start with the **catch** blocks that are designed to trap very specific exceptions and finish with more general blocks that will cover any other exceptions for which we have not provided handlers.

18.7 GENERAL CATCH HANDLER

A **catch** block which will catch any exception is called a general catch handler. A general catch handler does not specify any parameter and can be written as:

```
try
{
        . . . . //causes an exception

}
catch       //no parameters
{
    . . . . //handles error
    . . . .
```

Note that **catch (Exception e)** can handle all the exceptions thrown by the C# code and therefore can be used as a general catch handler. However, if the program uses libraries written in other languages, then there may be an exception that is not derived from the class **Exception.** Such exceptions can be handled by the parameter-less **catch** statement. This handler is always placed at the end. Since there is no parameter, it does not catch any information about the exception and therefore we do not know what went wrong.

18.8 USING FINALLY STATEMENT

C# supports another statement known as a **finally** statement that can be used to handle an exception that is not caught by any of the previous catch statements. A **finally** block can be used to handle any exception generated within a try block. It may be added immediately after the try block or after the last catch block shown as follows:

```
try                     try
{                       {
   ..........              ..........
   ..........              ..........
}                       }
finally                 catch (....)
{                       {
   ..........              ..........
   ..........              ..........
}                       }

                        catch (....)
                        {
                           ..........
                           ..........
                        }
                        .
                        .
                        .
                        .
                        finally
                        {
                           ..........
                           ..........
                        }
```

When a **finally** block is defined, the program is guaranteed to execute, regardless of how control leaves the try, whether it is due to normal termination, due to an exception occurring or due to a jump statement. Figure 18.3 illustrates this. As a result, we can use it to perform certain house-keeping operations such as closing files and releasing system resources.

In Program 18.4, we may include the last two statements inside a finally block as shown below:

```
finally
{
   int y = a[1]/a[0];
   Console.WriteLine("y = " +y);
}
```

This will produce the same output.

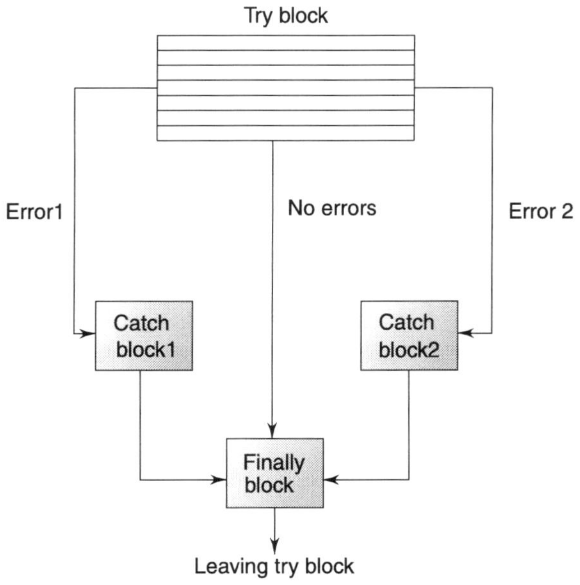

Figure 18.3 *Execution paths of try-catch-finally blocks*

18.9 NESTED TRY BLOCKS

C# permits us to nest **try** blocks inside each other. *Example:*

```
try
{
        . . . . (Point P1)
        . . . .
        try
        {
                . . . . (Point P2)
                . . . .
        }
        catch
        {                                   Inner
                . . . . (Point P3)          try block
                . . . .
        }
        finally
        {
                . . . .
        3       . . . .
        }
        . . . . (Point P4)
        . . . .
}
catch
{
        . . . .
}
finally
{
        . . . .
}
```

For simplicity, we have shown only one **catch** handler in each **try** block. However, we can string several **catch** handlers together in each place.

When nested try blocks are executed, the exceptions that are thrown at various points are handled as follows:

- The points P1 and P4 are outside the inner try block and therefore any exceptions thrown at these points will be handled by the **catch** in the outer block. The inner block is simply ignored.
- Any exception thrown at point P2 will be handled by the inner **catch** handler and the inner **finally** will be executed. The execution will continue at point P4 in the program.
- If there is no suitable **catch** handler to catch an exception thrown at P2, the control will leave the inner block (after executing the inner **finally**) and look for a suitable **catch** handler in the outer

block. If a suitable one is found, then that handler is executed followed by the outer **finally** code. Remember, the code at point P4 will be skipped.

- If an exception is thrown at point P3, it is treated as if it had been thrown by the outer try block and, therefore, the control will immediately leave the inner block (of course, after executing the inner **finally**) and search for a suitable **catch** handler in the outer block.
- In case, a suitable **catch** handler is not found, then the system will terminate program execution with an appropriate message.

Program 18.5 shows a simple implementation of nested **try** blocks. The program uses nested methods, each having its own **try …. catch** blocks.

The **Main** method invokes the **Division( )** method which throws an exception. Although the **Division ( )** method has a **catch** block, it does not match the type of exception. The control is therefore transferred to the **try** block in **Main** and the program looks for a matching **catch** handler. It finds one there and executes it.

Program 18.5 Implementing nested try blocks

```
using System;
class NestedTry
{
        static int m = 10;
        static int n = 0;
        static void Division ( )
       {
                try
                {

                    int k = m/n;
                }
                catch (ArgumentException e)
                {
                    Console.WriteLine("Caught an exception");
                }
                finally
                {
                Console.WriteLine("Inside Division Method");
                }
        }

        public static void Main( )
        {

                try
                {
        Divison ( );
                }
```

```
            catch (DivideByZeroException e)
            {
    Console.WriteLine("Caught an exception");
            }
            finally
            {
    Console.Write("Inside Main Method");
            }
        }
}
```

18.10 THROWING OUR OWN EXCEPTIONS

There may be times when we would like to throw our own exceptions. We can do this by using the keyword **throw** as follows:

```
throw new Throwable_subclass;
```

Examples:

```
throw new ArithmeticException( );
throw new FormatException( );
```

Program 18.6 demonstrates the use of a user-defined subclass of Exception class. An object of a class that extends **Throwable** can be thrown and caught.

Program 18.6 Throwing our own exception

```
using System;
class MyException:Exception
{
    public MyException (string Message) : base (Message)
    {
    }

    public MyException ( )
    {
    }
    public MyException (string Message, Exception inner)
        : base (Message, inner)
    {
    }
}
```

```
class TestMyException
{
    public static void Main( )
    {
        int x = 5, y = 1000;

        try
        {
          float z = (float) x / (float) y ;
          if(z < 0.01)
          {
              throw new MyException("Number is too small");
          }
        }

        catch (MyException  e)
        {
           Console.WriteLine("Caught my exception");
           Console.WriteLine(e.Message);
        }

        finally
        {
           Console.WriteLine("I am always here");
        }
    }
}
```

A run of Program 18.6 produces:

```
Caught my exception
Number is too small
I am always here
```

The object **e** which contains the error message "Number is too small" is caught by the **catch** block which then displays the message using the **Message** property.

Note that Program 18.7 also illustrates the use of **finally** block. The last line of the output is produced by the **finally** block.

A rule of thumb when creating our own exceptional classes is that we must implement all the three **System.Exception** constructors as illustrated in the example program.

Program 18.7 creates a new type of user-defined exception **MyException** and it is being used to catch an exception with its own customized message.

Program 18.7 Custom exception

```
using System;
using System.Collections.Generic;
using System.Text;

namespace CustomException
{
  public class MyCustomException :
  System.Exception
    {
      public MyCustomException(string txt) :  base(txt) // pass the message up to the
base class
        {

        }
    }
    class CustomeExceptionExample
    {
       public void DisplayAll()
        {
            try
            {
                Console.WriteLine("Application started!!");
                double a = 718;
                double b = 0;
                Console.WriteLine("{0} / {1} = {2}",  a, b, DivideValues(a, b));
                Console.WriteLine("The error is handled in a try catch block and customised
message shown above.");
            }

          // most derived exception type first
          catch (System.DivideByZeroException e)
            {
                Console.WriteLine("\nDivideByZeroException! Msg: {0}", e.Message);

            }

          // catch custom exception
          catch (MyCustomException e)
            {
                Console.WriteLine(
                "\nMyCustomException! Msg: {0}",
                e.Message);

            }

           finally
```

```
            {
                Console.WriteLine("Application executed!!");
            }
        }

      // do the division if legal
      public double DivideValues(double a, double b)
        {
           if (b == 0)
            {
                MyCustomException e =
                new MyCustomException(
                "Can't be divided by zero, result is Infinity.");
                Console.WriteLine(e);
            }
           if (a == 0)
            {
                // create a custom exception instance
                MyCustomException e =
                new MyCustomException(
                "Can't have a zero divisor");
                Console.WriteLine(e);
            }
          return a / b;
        }
      static void Main(string[] args)
        {
         Console.WriteLine("This is an example of a customised exception developed in
C#.");
         CustomeExceptionExample cee = new CustomeExceptionExample();
         cee.DisplayAll();

        }
    }
}
```

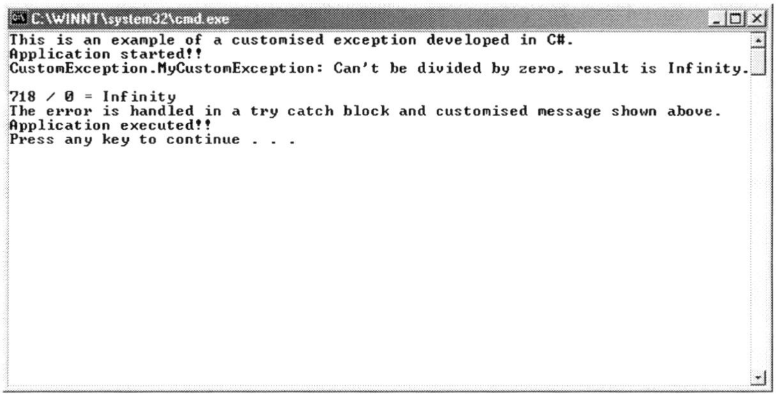

Program 18.8 uses different types of system-defined exception classes in multiple catch blocks. These blocks catch the exceptions thrown and display the exception messages.

Program 18.8 Multiple catch

```
using System;
using System.Collections.Generic;
using System.Text;

namespace MultipleCatch
{
   class MultipleCatchBlockExample
    {
      public void execute()
       {
          try
          {
                 double val1 = 0;
                 double val2 = 121;
                 Console.WriteLine("Dividing {0} by {1}...", val1, val2);
                 Console.WriteLine("{0} / {1} = {2}",
                 val1, val2, DivideValues(val1, val2));
          }

         // most specific exception type first
         catch (DivideByZeroException ex)
          {
                 Console.WriteLine("DivideByZeroException caught!", ex);
          }

         catch (ArithmeticException e)
          {
                 Console.WriteLine("ArithmeticException caught!", e);
          }

         // generic exception type last
          catch
          {
                 Console.WriteLine("Unknown exception caught");
          }

       }

      // do the division if legal
      public double DivideValues(double val1, double val2)
```

```
        {
            if (val2 == 0)
             {
                      DivideByZeroException dx = new DivideByZeroException();
                      Console.WriteLine(dx.Message);
             }

            if (val1 == 0)
             {
                      ArithmeticException ax = new ArithmeticException();
                      Console.WriteLine(ax.Message);
             }


            return val1 / val2;
        }
      static void Main(string[] args)
        {
                      Console.WriteLine("This is an example of multiple catch blocks and each
catch block \nhandles errors/exceptions of a certain type thrown in the application.\n");
                      MultipleCatchBlockExample mcbObj = new MultipleCatchBlockExample();
                      mcbObj.execute();
        }
    }
}
```

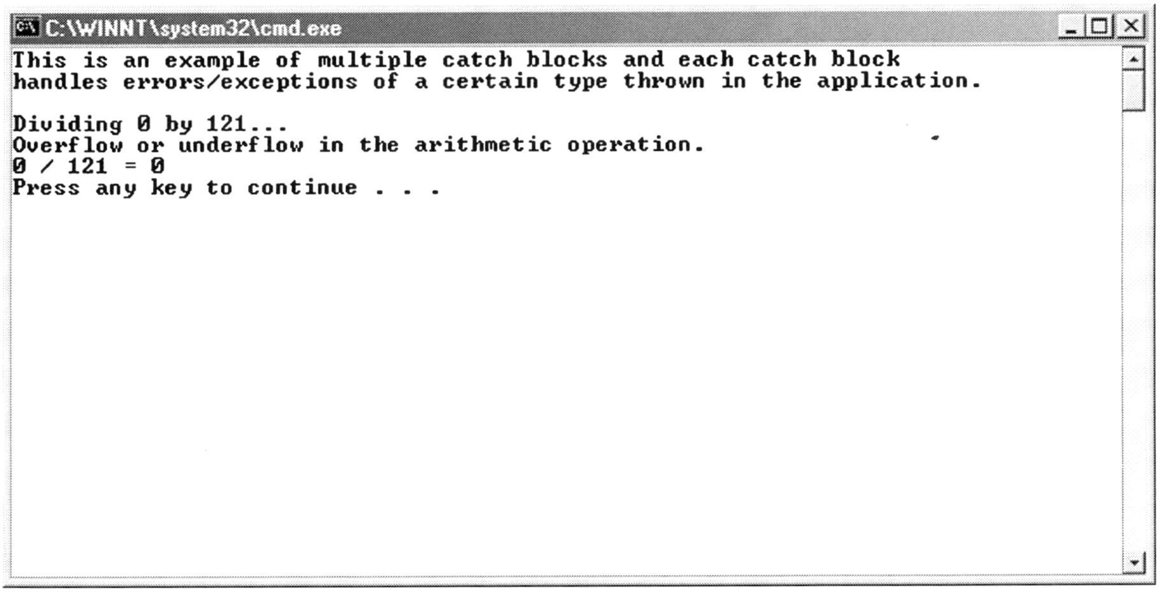

18.11 CHECKED AND UNCHECKED OPERATORS

Stack overflows are usual problems during arithmetic operations and conversion of integer types. C# supports two operators, **checked** and **unchecked,** which can be used for checking (or unchecking) stack overflows during program execution. If an operation is checked, then an exception will be thrown if overflow occurs. If it is not checked, no exception will be raised but we will lose data. For example, consider the code:

```
int a = 200000
int b = 300000
try
{
    int m = checked ( a * b);
}
catch (OverflowException e)
{
    Console.WriteLine (e);
}
```

Since **a*b** produces a value that will easily exceed the maximum value for an **int**, an overflow occurs. As the operation is checked with operator **checked,** an overflow exception will be thrown. In this case, we will get output like this:

```
System.OverflowException : An exception
Of type System.OverflowException was
thrown at…….
```

If we want to suppress the overflow checking, we can mark the code as **unchecked**

```
int n = unchecked (a * b);
```

In this case, no exception will be raised, but we will lose data.

18.12 USING EXCEPTIONS FOR DEBUGGING

As we have seen, the exception-handling mechanism can be used to hide errors from rest of the program. It is possible that the programmers may misuse this technique for hiding errors rather than debugging the code. Exception handling mechanism may be effectively used to locate the type and place of errors. Once we identify the errors, we must try to find out why these errors occur before we cover them up with exception handlers.

Case Study

Problem Statement

Hemant is working as a senior programmer in BlueTech Software Solutions Private Limited, which is involved in developing software and Web sites for different multinational organisations. Currently, Hemant is involved in developing an application for BlueMoon courier company. The application should allow the staff of BlueMoon to input data such as numbers and text for storing it in a database. Hemant wants that a user of the application must be allowed to enter only numeric values for the number field and character values for the text field such as **Name** and **Address**. He also wants that a proper message should be displayed if a user enters character values in the number field. How can Hemant apply this functionality in the application?

Solution

Hemant can use the exception-handling feature of C# programming language to allow users to enter a numeric value in the number field and a character value in a text field such as **Name** and **Address**. The exception handling feature of C# also allows Hemant to display appropriate messages to handle the exception that may occur because of character value being entered in a number field. Hemant creates a C# program to understand the exception-handling feature of C# so that he can implement it as per the requirement in the application for BlueMoon courier company. The following C# program is created by Hemant.

```
using System;
using System.Collections.Generic;
using System.Text;

namespace ExceptionHandling
{
   class ExceptionHandlers
    {
      int val1=0;
      public void DisplayErrors_type1()
        {
          Console.WriteLine("This is an example of how to handle input from users by
catching exceptions.");
          Console.WriteLine("Enter any string value instead of a number and see the
results.");
         Console.WriteLine("-----------------------------------------------------------
---------");
            try
             {
          Console.Write("Please enter a number: ");
          val1 = Int32.Parse(Console.ReadLine());
             }
           catch (Exception e)
             {
              Console.WriteLine(e.Message);
             }
         Console.WriteLine("-----------------------------------------------------------
---------");
        }
```

```
    public void DisplayErrors_type2()
      {
    Console.WriteLine("This section will show how to catching exceptions and \ndisplay
them in a proper way.");

         try
          {
        Console.Write("Please enter a number: ");
        val1 = Int32.Parse(Console.ReadLine());
          }
        catch (FormatException fe)
          {
        Console.WriteLine("The error is : " + fe.Message);
          }

        Console.WriteLine("--------------------------");

      }

    public void DisplayErrors_type3()
      {
    Console.WriteLine("This section will show customised messages and \ndisplay them in
a proper way.");

         try
          {
        Console.Write("Please enter a number: ");
        val1 = Int32.Parse(Console.ReadLine());
          }
         catch
          {
           Console.WriteLine("The error is that you have not entered numbers in the
input line");
          }
        Console.WriteLine("------------------------------------------------------------
---------");
      }

    public void DisplayErrors_type4()
      {
    Console.WriteLine("This section will show customised messages using catch and
finally \nand display them in a proper way.");

         try
```

```
            {
             Console.Write("Please enter a number: ");
             val1 = Int32.Parse(Console.ReadLine());
            }
           catch
            {
             Console.WriteLine("The error is that you have not entered numbers in the
input line.");
            }
           finally
            {
             if (val1.Equals(0))
               {
             Console.WriteLine("As you have entered a string, the number cannot be
displayed.");
               }
              else
               {

              Console.WriteLine("You have entered : " + val1);
               }
            }
         Console.WriteLine("-------------------------------------------------------------
---------");
        }

      public void DisplayErrors_type5()
        {
      Console.WriteLine("This section uses multiple catch(s),finally to display
customised error messages \nand display them in a proper way.");

           try
            {
             Console.Write("Please enter a number: ");
             val1 = Int32.Parse(Console.ReadLine());
            }
          catch (FormatException fe)
            {
             Console.WriteLine("The error is, you have not entered any number in the
input line." + fe.Message);
            }
          catch (Exception ee)
            {
             Console.WriteLine("The error is that you have not entered numbers in the
input line." + ee.Message);
            }
```

```
        finally
            {
                if (val1.Equals(0))
                {
                    Console.WriteLine("As you have entered a string, the
number cannot be displayed.");
                }
                else
                {

                    Console.WriteLine("You have entered : " + val1);
                }
            }
            Console.WriteLine("----------------------");
        }



    }

    class Program
    {
        static void Main(string[] args)
        {
            ExceptionHandlers eh = new ExceptionHandlers();
            eh.DisplayErrors_type1();
            eh.DisplayErrors_type2();
            eh.DisplayErrors_type3();
            eh.DisplayErrors_type4();
            eh.DisplayErrors_type5();
        }
    }
}
```

Remarks

In C#, **try** and **catch** block can be used to handle exceptions in an application. The **try** block contains a group of statements that are expected to generate an exception. The **catch** block contains the message that should be displayed when an exception occurs. In C#, each try block contains a corresponding **catch** block. In a C# application, the **finally** keyword can also be used to handle exceptions.

Common Programming Errors

- *Trying to catch the same type of exception in two different catch blocks associated with a particular try block.*
- *Using an incorrect catch argument to catch an exception.*
- *Placing catch (Exception e) before other catch blocks.*

- *Placing a catch that catches a super class object before a catch that catches an object of its subclass.*
- *Placing a general catch handler catch before other catch blocks.*
- *When creating user-defined exception classes, forgetting to implement all the three constructors of System.Exception class.*
- *Forgetting to use the checked operator on operations that are likely to cause overflow.*
- *Misspelling of exception class names in catch handlers.*

Review Questions

18.1 State whether the following statements are true or false.
(a) Syntax errors will be caught and informed by the compiler.
(b) A run-time error will always terminate a program execution.
(c) An exception is caused by a syntax error.
(d) An exception handler is used to correct a run-time error.
(e) A catch block may be placed anywhere in a program.
(f) A try block normally encloses a block of code that contains a run-time error.
(g) A catch block need not necessarily contain any statements.
(h) When using multiple catch statements, C# permits them to be placed in any order.
(i) The parameterless catch statement must be placed at the end of all catch handlers.
(j) A try block must always be followed by a catch block.
(k) A finally block can be used to handle any exception generated by a try block.
(l) The checked operator is used for checking syntax errors.
18.2 What are compile-time errors? Give some examples. How are they detected?
18.3 What are run-time errors? Give some examples. How are they detected?
18.4 What is an exception? Give some examples.
18.5 How do you write code to handle various exceptions differently?
18.6 Why is it necessary to handle an exception by a special code?
18.7 Describe the tasks involved in handling exceptions.
18.8 How do we define the following?
(a) try block
(b) catch block
(c) finally block
18.9 List at least four exceptions that occur commonly in C# programs.
18.10 What happens when a try block is neither followed by a catch block, nor by a finally block?
18.11 Distinguish between the parameterless catch and the catch(Exception e) statements.
18.12 How would you implement a method that is likely to cause an exception the type of which is not known?
18.13 What is the finally statement? What is its use?
18.14 What is the purpose of using a finally block? Give some examples.
18.15 What happens when an exception is caused in an inner try block of a nested try block?
18.16 When do you consider it necessary to define your own exception classes?
18.17 When do we use a parameterless catch handler?
18.18 Explain how exception handling mechanism can be used for debugging a program.
18.19 Why is a proper ordering of catch blocks necessary in C#?

18.20 State what will happen in the following situations:
(a) An exception is thrown outside a try block
(b) No catch handler matches the type of exception thrown
(c) Several handlers match the type exception thrown
(d) A catch handler throws an exception
(e) Catch is the first handler of multiple catch handlers

18.21 Find errors, if any, in the following code segments:

(a)
```
try
{
    byte m = 1000 * 5000;
    go to catch;
}
```
(b)
```
try
{
    byte m = checked (1000 * 2000);
}
catch (FormatException e)
{
. . . .
}
```
(c)
```
try {. . .}
catch (Exception e) {. . .}
catch (ArithmeticException e) {. . .}
finally { }
```
(d)
```
try { . . . }
finally { . . .}
catch {. . .}
```
(e)
```
try {. . .}
catch (DividebyzeroException e)
{. . .}
finally {. . .}
```

Debugging Exercises

18.1 Will the given program compile? If not, why?

```
using System;
using System.Collections.Generic;
using System.Text;

namespace debigAppl_chap18
{
   class Program
    {
```

```
        static void Main(string[] args)
         {
          Console.WriteLine("This display is from the Main function.");
           Program pp = new Program();
            pp.show();
          Console.WriteLine("Exit from the Main function.");
         }

        public void show()
         {
          Console.WriteLine("Showing message from the Show function.");

         }

        public void check()
         {
          Console.WriteLine("Entering Check function.");
            try
             {
              Console.WriteLine("Entering try block of check function.");
              Console.WriteLine("Exiting from try block.");
             }
            catch
             {
              Console.WriteLine("Exception caught and handled!");
             }
              catch (NullReferenceException ex)
             {
              Console.WriteLine(ex.Message);
             }
          Console.WriteLine("Exiting from Check function.");
         }
     }
}
```

18.2 Correct the syntax errors in the given program for demostrating the finally keyword.

```
using System;
using System.Collections.Generic;
using System.Text;

namespace debigAppl_chap18
{
   class Program
```

```
    {
      static void Main(string[] args)
        {
         Console.WriteLine("This display is from the Main function.");
          Program pp = new Program();
           pp.show();
         Console.WriteLine("Exit from the Main function.");
        }

      public void show()
        {
         Console.WriteLine("Showing message from the Show function.");

        }

      public void check()
        {

            try
             {
              Console.WriteLine("Entering try block of check function.");
             }
            finally()
             {
              Console.WriteLine("This is finally block.");
             }
            catch
             {
              Console.WriteLine("Exception caught and handled!");
             }


        }
    }
}
```

18.3 Identify bugs in the following program.

```
using System;
using System.Collections.Generic;
using System.Text;

namespace debugApp3_chap18
{
```

```
    class Program
    {
      static void Main(string[] args)
        {
          string name = "";
          string address = "";
          int age = 0;
         Console.WriteLine("This is another example of try catch block.");
         Console.Write("Please enter your name : ");
         name = Console.ReadLine();
         Console.Write("Please enter your address : ");
         address = Console.ReadLine();
         Console.Write("Please enter your age : ");
           try
           {
            age = Console.ReadLine();
           Console.WriteLine("Hello {0}. You live in {1} and you are {2} years old.",
+name, +address, +age);

         catch (FormatException e)
           {
           Console.WriteLine("Age is a number." + e.Message());
           }
          catch
           {
            Console.WriteLine("error");
           }
           }
          finally
           {
           Console.WriteLine("Press any key to exit.");
           }

        }
    }
}
```

Programming Exercises

18.1 Develop a program containing a possible exception. Use a try block to throw it and a catch block to handle it properly.

18.2 Write a program that illustrates the application of multiple catch handlers.

18.3 Write a program that throws an exception of type "array index out of bounds" and then handles it appropriately.

18.4 Write a program that handles an exception of type ArgumentException effectively.

18.5 Develop a program that is likely to throw multiple exceptions that are handled using catch and finally blocks.

18.6 Define a method that would sort an array of integers. Incorporate exception handling mechanism for "index out of bounds" situations.
Develop a Main program that employs this method to sort a given set of integers.

18.7 Modify the program of Exercise 18.6 incorporating the exception handling mechanism into the Main program that calls the method.

18.8 Write a program that calls a deeply nested method containing an exception. Incorporate necessary exception handling mechanisms.

18.9 Define an exception called "NoMatchException" that is thrown when a string is not equal to "India". Write a program that uses this exception.

18.10 Write a program that will read a name from the keyboad and display it on the screen. The program should throw an exception when the length of the name is more than 15 characters. Design your own exception mechanism.

Chapter

19

Multithreading in C#

19.1 INTRODUCTION

C# supports the concept of multithreading which enables us to execute two or more "parts" of a program concurrently. Each part is known as a *thread.* A thread is basically a separate sequence of instructions designed for performing a "specific task" in the program. A task may represent an operation such as reading data, sending a file over the Internet, printing some results or performing certain calculations. Multithreading, therefore, means performing multiple tasks at the same time during the execution of a program.

The execution of a C# program starts with a single thread called the *main thread* that is automatically run by the Common Language Runtime (CLR) and the operating system. From the main thread, we can create other threads for performing desired tasks in the program. The process of developing a program for execution with multiple threads is called *multithreaded programming* and the process of execution is called *multithreading.*

The main advantage of multithreading is that it enables us to develop efficient programs that could optimize the use of computer resources such as memory, I/O devices and time.

C# defines a namespace known as **System.Threading** containing classes and interfaces that are required for developing and running multithreaded programs. We must, therefore, include the statement

```
using System.Threading;
at the start of any multithreaded program.
```

19.2 UNDERSTANDING THE SYSTEM.THREADING NAMESPACE

The classes and interfaces contained in the **System.Threading** namespace allow us to perform multithreading in C#. We can use the methods and properties contained in these classes to perform tasks such as synchronising the activities of a thread and creating a thread. The following are the important classes contained in the **System.Threading** namespace:

- **Thread**
- **ThreadPool**
- **Monitor**
- **Mutex**

19.2.1 Thread Class

The **Thread** class helps us to perform tasks such as creating and setting the priority of a thread. We can use this class to control a thread and obtain its status.

The **Thread** class provides various properties that allow us to perform tasks such as obtaining the status of a thread and specifying a name for the thread. Table 19.1 shows some of the important properties contained in the **Thread** class.

Table 19.1 *Properties of the Thread class*

Property	Task
CurrentThread	To retrieve the name of the thread, which is currently running.
IsAlive	To retrieve a value to indicate the current state of thread execution. The value of the **IsAlive** property is true if the thread has been started or has not terminated, otherwise the value is false.
IsThreadPoolThread	To retrieve a value to indicate whether a thread is part of a thread pool or not. The value of **IsThreadPoolThread** property is true if the thread is part of a thread pool, otherwise it is false.
Name	To specify a name for a thread.
Priority	To obtain a value, which indicates the scheduling priority of a thread. By default, the value of the **Priority** property is **Normal**. We can assign either **Highest, AboveNormal**, **Normal**, **BelowNormal** or **Lowest** value to the **Priority** property. The **Highest** value indicates that the thread must be scheduled for running before any other thread. The **AboveNormal** value indicates that the thread with this value must be scheduled after the **Highest** priority threads but before the **Normal** priority thread. The **Normal** value indicates that threads with this value must be scheduled after the threads with **AboveNormal** priority but before the **BelowNormal** priority threads. The **BelowNormal** value indicates that the thread with this value must be scheduled after a **Normal** priority value thread but before the **Lowest** priority thread. The **Lowest** value indicates that the thread with this value must be scheduled for running after the threads with other priorities such as **Normal** and **Highest** have executed.
ThreadState	To retrieve a value, which indicates the state of a thread. By default, the value of the **ThreadState** property is **Unstarted**. The **ThreadState** property can have a value indicating the state of the current thread. These values can be **Running, Stopped, Suspended, Unstarted** or **WaitSleepJoin**. The **Running** value indicates that the current thread is running. The **Stopped** value indicates that the current thread is stopped. The **Suspended** value indicates that the current thread is suspended. The **Unstarted** value indicates that the thread has not been started yet using the **Start** method of the **Thread** class. The **WaitSleepJoin** value indicates that the current thread is blocked because of call to **Wait, Sleep** or **Join** method.

The Thread class also provides certain methods that can be used to manage operations such as starting a thread and resuming a suspended thread. Table 19.2 shows some of the important methods of the **Thread** class.

Table 19.2 *Methods of the Thread class*

Method	Task
Interrupt	To interrupt the thread, which is in the **WaitSleepJoin** state.
Join	To block a thread until another thread has terminated.
Resume	To resume a thread, which has been suspended earlier.
Sleep	To block the current thread for a specified time period.
SpinWait	To make a thread wait the number of times specified in **Iterations** parameter.
Start	To start a thread.
Suspend	To suspend a thread.

19.2.2 ThreadPool Class

The **ThreadPool** class provides a pool of threads that helps us to perform tasks such as processing of asynchronous I/O and waiting on behalf of another thread.

The **ThreadPool** class provides various methods, which we can use to perform tasks such as determining whether two thread pools are equal or not. Table 19.3 shows some of the important methods of the **ThreadPool** class.

Table 19.3 *Methods of the ThreadPool class*

Method	Task
Equals	To determine whether two thread pools are equal or not.
GetType	To obtain the type for the current thread pool.
QueueUserWorkItem	To allow a method to be queued for execution. The method, which has been queued, executes when a thread pool thread is made available.
SetMaxThreads	To specify the number of requests to the thread pool that can be concurrently active.
SetMinThreads	To specify the number of idle threads that can be maintained by a thread pool for new requests.

19.2.3 Monitor Class

The **Monitor** class provides control access to an object by granting a lock for the object to a single thread. When an object is locked for a thread then access to a specific program code is restricted. The **Monitor** class defines a few methods that allow us to perform tasks such as acquiring and releasing a lock for an object. Table 19.4 shows some of the important methods of the **Monitor** class.

Table 19.4 *Methods of the Monitor class*

Method	Task
Enter	Allows a thread to obtain a lock on a specified object.
Exit	Allows a thread to release a lock on a specified object.
TryEnter	Allows a thread to try and obtain a lock on a specified object.
Wait	Allows a thread to release the lock on an object and block the thread for the time period until which it again acquires the lock.
GetType	Allows us to obtain the type for the current instance of the **Monitor** class.

19.2.4 Mutex Class

A **mutex** is a synchronisation primitive that helps to perform interprocess synchronisation in C#. **Mutex** allows a thread to have exclusive access to shared resources. When a thread acquires a mutex, another thread, which wants to obtain the mutex, is suspended until the first thread releases the mutex.

The **Mutex** class provides the **Handle** and **SafeWaitHandle** properties that can be used to retrieve the handle for the operating system. This class also provides the various methods, which allow us to perform tasks such as setting access control security for a mutex and opening a named mutex. Table 19.5 shows some of the important methods provided by the **Mutex** class.

Table 19.5 *Methods of the Mutex class*

Method	Task
Close	To release the resources held by an object of the **WaitHandle** class. The **WaitHandle** class encapsulates objects specific to operating system that are waiting for access to shared resources.
Equals	To determine whether two mutex are equal or not.
OpenExisting	To open an existing mutex.
ReleaseMutex	To release a mutex once.
SetAccessControl	To set the access control security for a specified mutex.

19.3 CREATING AND STARTING A THREAD

A thread can be created in C# using the constructor of the **Thread** class. We need to pass the **ThreadStart delegate** to the **Thread** class constructor along with the name of the method from which the execution should start. The **ThreadStart delegate** is defined as:

```
public delegate void ThreadStart();
```

To create a new thread in a C# program, we can use the following statement:

```
Thread threadname = new Thread(new ThreadStart(methodname);
```

Here, *threadname* represents the name of the new thread and *methodname* is the name of the method from which the execution starts.

For example, to create a thread **t1**, we can use the following code:

```
Thread t1 = new Thread(new ThreadStart(First);
```

If the method from which execution needs to start is defined in a class other than the class in which the thread is created then we must use an object of that class to access the method. **The Start()** method of the **Thread** class starts a new thread. Program 19.1 illustrates how a thread is created and executed.

Program 19.1 Creating and starting threads

```
using System;
using System.Threading;
public class AOne
{
        public void First()
        {
        Console.WriteLine("First method of AOne class is running on T1 thread.");
        Thread.Sleep(1000);
        Console.WriteLine("The First method called by T1 thread has ended.");
    }
    public static void Second()
    {
        Console.WriteLine("Second method of AOne class is running on T2 thread.");
        Thread.Sleep(2000);
        Console.WriteLine("The Second method called by T2 thread has ended.");
    }
}
public class ThreadExp
{
    public static int Main(String[] args)
    {
        Console.WriteLine("Example of Threading");
```

```
        AOne a = new AOne();
        Thread T1 = new Thread(new ThreadStart(a.First)); // Creating thread T1
        T1.Start(); // Starting thread T1
        Console.WriteLine("T1 thread started.");
        Thread T2 = new Thread(new ThreadStart(AOne.Second)); Creating T2
        T2.Start(); Starting thread T2
        Console.WriteLine("T2 thread started.");
        return 0;
    }
}
```

In the above C# program, two threads **T1** and **T2** have been created and started using the **Start()** method of the **Thread** class. In case of **T1** thread, the execution starts from **First method** defined in the **AOne** class while in case of **T2** thread the execution starts from **Second method**, which is a static method. The output of the above program is:

```
Example of Threading
First method of AOne class is running on T1 thread.
T1 thread started.
Second method of AOne class is running on T2 thread.
T2 thread started.
The First method called by T1 thread has ended.
The Second method called by T2 thread has ended.
Press any key to continue . . .
```

We can also create a thread and attach a timer to it as illustrated by Program 19.2.

Program 19.2 starts a thread and displays the thread information after every second. When the **Enter** key is pressed, the thread stops. On pressing the **Enter** key again, the thread resumes. On pressing the **Enter** key again, the thread is destroyed.

Program 19.2 Using timer

```
using System;
using System.Collections.Generic;
using System.Text;
using System.Timers;

namespace ThreadTimers
{
    class Program
    {
        static void Main()
        {
            Timer time = new Timer();            // Doesn't require any args
            time.Interval = 1000;
            time.Elapsed += time_Elapsed;        // Uses an event
            time.Start();                        // Start the timer
            Console.ReadLine();
```

```
            time.Stop();
            Console.WriteLine("Timer Thread stopped. Enter again to start.\nTwice entering
would destroy the Thread."); // Pause the timer
            Console.ReadLine();
            time.Start();                    // Resume the timer
            Console.ReadLine();
            Console.WriteLine("Thread destroyed!");
            time.Dispose(); //Destroy the Timer Thread
        }
        static void Display(object ObjTime)
        {
            // This runs on a pooled thread
            Console.WriteLine(ObjTime);      // Writes "tick..."
        }

        static void time_Elapsed (object sender, EventArgs e)
        {
            Console.Write("The Time is passing by : ");
            Console.WriteLine(DateTime.Now);
        }
    }
}
```

In the above program, a thread with a timer is created. The **Start** and **Stop** methods of the timer is used to start and stop the timer and the thread. The output of Program 19.2 is:

```
The Time is passing by : 7/25/2007 4:03:24 PM
The Time is passing by : 7/25/2007 4:03:25 PM
The Time is passing by : 7/25/2007 4:03:26 PM
The Time is passing by : 7/25/2007 4:03:27 PM
The Time is passing by : 7/25/2007 4:03:28 PM
The Time is passing by : 7/25/2007 4:03:29 PM
The Time is passing by : 7/25/2007 4:03:30 PM
The Time is passing by : 7/25/2007 4:03:31 PM
The Time is passing by : 7/25/2007 4:03:32 PM
The Time is passing by : 7/25/2007 4:03:33 PM
The Time is passing by : 7/25/2007 4:03:34 PM
The Time is passing by : 7/25/2007 4:03:35 PM
The Time is passing by : 7/25/2007 4:03:36 PM
The Time is passing by : 7/25/2007 4:03:37 PM
The Time is passing by : 7/25/2007 4:03:38 PM
The Time is passing by : 7/25/2007 4:03:39 PM
The Time is passing by : 7/25/2007 4:03:40 PM
The Time is passing by : 7/25/2007 4:03:41 PM
```

19.4 SCHEDULING A THREAD

In C#, threads can be scheduled by setting the priority of the thread using the **Priority** property of the **Thread** class. We can provide the following values for the **Priority** property of a specific thread:

- **Highest**
- **AboveNormal**
- **Normal**
- **BelowNormal**
- **Lowest**

For example, to set the priority of **tr1** thread as Lowest, we can use the code:

```
trl.Priority = ThreadPriority.Lowest;
```

After a priority is set for a thread, the thread runs according to its priority. This means that a **BelowNormal** priority thread runs after the **Normal** priority thread but before the Lowest priority thread. Program 19.3 illustrated the use of **Priority** property.

Program 19.3 Setting thread priority

```
using System;
using System.Threading;
public class ABC
    {
    public void A()
    {
        Console.WriteLine("Thread A");
        Console.WriteLine("Priority of Thread A is: "
    +Thread.CurrentThread.Priority.ToString());
        Console.WriteLine(" ");
        Thread.Sleep(5000);
        Console.WriteLine("Thread with "+Thread.CurrentThread.Priority.ToString()+" has
run");
        Console.WriteLine(" ");
    }

    public void B()
    {
        Console.WriteLine("Thread B");
        Console.WriteLine("Priority of Thread B is: " +
    Thread.CurrentThread.Priority.ToString());
        Console.WriteLine(" ");
        Thread.Sleep(5000);
        Console.WriteLine("Thread with " + Thread.CurrentThread.Priority.ToString() + " has
run");
        Console.WriteLine(" ");
    }
    public static void C()
    {
    Console.WriteLine("Thread C");
    Console.WriteLine("Priority of Thread C is: " +
Thread.CurrentThread.Priority.ToString());
```

```
    Console.WriteLine(" ");
    Thread.Sleep(3000);
    Console.WriteLine("Thread with " +
Thread.CurrentThread.Priority.ToString() + " has run");
    Console.WriteLine(" ");
    }
    }

    public class ScheduleExp
    {
    public static int Main(String[] args)
    {
        Console.WriteLine("............Scheduling Threads");
        Console.WriteLine(" ");
        ABC a1 = new ABC();
        Thread tr1 = new Thread(new ThreadStart(a1.A));
        Thread tr2 = new Thread(new ThreadStart(a1.B));
        Thread tr3 = new Thread(new ThreadStart(ABC.C));
        Console.WriteLine("Priority of Main method is: "
    +Thread.CurrentThread.Priority.ToString());
        Console.WriteLine(" ");
        tr1.Name = " Thread A";
        tr2.Name = " Thread B";
        tr3.Name = " Thread C";
        tr1.Priority = ThreadPriority.AboveNormal;
        tr2.Priority = ThreadPriority.BelowNormal;
        tr3.Priority = ThreadPriority.Lowest;
        tr1.Start();
        tr2. Start();
        tr3.Start();
        return 0;
    }
}
```

In the above C# program, the priorities of the three threads **tr1**, **tr2**, and **tr3** have been set to **AboveNormal**, **BelowNormal** and **Lowest**. The priority of **Main** thread is also displayed using the **Console.WriteLine()** method. The output of the above program is:

```
............Scheduling Threads

Priority of Main method is: Normal

Thread A
Priority of Thread A is: AboveNormal

Thread B
Priority of Thread B is: BelowNormal
```

```
Thread C
Priority of Thread C is: Lowest

Thread with Lowest has run

Thread with AboveNormal has run

Thread with BelowNormal has run

Press any key to continue . . .
```

19.5 SYNCHRONISING THREADS

In C#, threads can be synchronised by using methods such as **Sleep** and **Join** of the **Thread** class. We can use these methods to block a thread for a specific time period until another thread completes its processing. In a multithreading environment multiple threads are run for the execution of a program. These threads may run asynchronously. As a result, one thread can access the resources such as network connection and file handles that are being used by another thread. This may corrupt the data unpredictably. Therefore it is necessary to synchronise the access to resources by different threads.

Synchronisation of threads means coordinating the access to resources for different threads running for execution. To synchronise the threads, we can either block a thread till the time another thread has completed its tasks or lock access to resource. The **Thread** class provides methods such as **Spinwait** and **Suspend** that help block a thread for a specific time period. In C#, we can use the **lock** keyword to lock access to a resource for a specific thread. The lock keyword marks a block of C# program statement as a critical section by first obtaining a mutually exclusive lock for an object. After obtaining the lock, the program statements are executed. When the program statements have been executed, the mutually exclusive lock is released. The syntax for using the lock keyword is:

```
lock(obj) // obj is the object being synchronised
{
   //set of statements to be synchronised
}
```

Program 19.4 shows how to use the **lock** keyword and **Sleep** method for thread synchronisation.

Program 19.4 Synchronising threads

```
using System;
using System.Threading;

class ABC
{
    private object thislock = new object();
    public void A()
    {
        Console.WriteLine("Starting " + Thread.CurrentThread.Name.ToString());
        int x = 10, z = 30;
```

```
        int sum=0;

        lock (thislock)
        {

            sum = x + z;
            Console.WriteLine(sum);
        }
        Thread.Sleep(5000);
        Console.WriteLine(Thread.CurrentThread.Name.ToString() + " is stopped");
    }
}
public class SyncExp
{
public static void Main()
{
    ABC a1 = new ABC();
    Thread t1 = new Thread(new ThreadStart(a1.A));
    Thread t2 = new Thread(new ThreadStart(a1.A));
    t1.Name = "T1";
    t2.Name = "T2";
    t1.Start();
    t2.Start();
}
```

In the above C# program, the **Sleep** method is used to block the **t1** and **t2** threads. The lock keyword is used to lock specific lines of code in the above C# program. The output of the above C# program is:

```
Starting T1
40
Starting T2
40
T1 is stopped
T2 is stopped
Press any key to continue . . .
```

19.6 THREAD POOLING

In thread pooling, a thread pool is created to perform multiple tasks simultaneously. A thread pool is basically a group of threads that can be run simultaneously to perform a number of tasks in the background. This feature of C# is mainly used in server applications. In server applications, a main thread receives the requests from the client computers and passes it to a thread in the thread pool for processing of the request. In this manner, the main thread functions asynchronously and is free to receive the requests from client computers. The main thread does not have to process the request. Instead, the request is transferred to a thread in the thread pool for processing. A delay in receiving the requests from the client computer does not occur because of implementation of thread pooling in server applications. After the thread in the thread pool completes the task of processing the client request, it waits in a queue for performing another task. In this way, the thread in the thread pool can be reused for performing different tasks. The reusability of the thread because of thread pooling enables a server application to avoid creating a new thread for every task.

A thread pool may contain a number of threads, each performing a specific task. If all the threads in a thread pool are occupied in performing their tasks then a new task, which needs to be processed, waits in a queue until a thread becomes free. The .**NET** framework provides a thread pool through the **ThreadPool** class. We can either implement a custom thread pool in a C# program or use the thread pool provided through the **ThreadPool** class. It is easy to implement a thread pool through the **ThreadPool** class. Program 19.5 shows how to use a thread pool.

This example executes 20 different threads at different intervals of time and subsequently stops all threads one by one after a period of 3 seconds, where each thread is made to wait for 3 seconds.

Program 19.5 Using thread pool

```
using System;
using System.Collections.Generic;
using System.Text;
using System.Threading;

namespace ThreadPooling
{
    class ThreadPoolTest
    {
        static object showThread = new object();
        static int runThreads = 20;

        public static void Main()
        {
            for (int i = 0; i < runThreads; i++)
            {
                ThreadPool.QueueUserWorkItem(Display, i);
            }
            Console.WriteLine("Running 20 threads to one by one and stopping them after 3
seconds.\n");
            lock (showThread)
            {
                while (runThreads > 0) Monitor.Wait(showThread);
            }
            Console.WriteLine("All Threads stopped successfully!");
            Console.ReadLine();
        }
        public static void Display(object threadObj)
        {
            Console.WriteLine("Started Thread : " + threadObj);
            Thread.Sleep(3000);
            Console.WriteLine("Ended Thread : " + threadObj);
```

```
            lock (showThread)
            {
                runThreads–;
                Monitor.Pulse(showThread);
            }
        }
    }
}
```

In the above code, a thread pool with the name **runThreads** is created. In addition the **QueueUserWorkItem** method of **ThreadPool** class and **Wait** method of **Monitor** class are used. The output of the above program is:

```
Running 20 threads to one by one and stopping them after 3 seconds.

Started Thread : 0
Started Thread : 1
Started Thread : 2
Started Thread : 3
Started Thread : 4
Ended Thread : 0
Started Thread : 5
Started Thread : 6
Started Thread : 7
Ended Thread : 1
Started Thread : 8
Started Thread : 9
Ended Thread : 2
Started Thread : 10
Started Thread : 11
Ended Thread : 3
Started Thread : 12
Started Thread : 13
Ended Thread : 4
Started Thread : 14
Started Thread : 15
Ended Thread : 5
Started Thread : 16
Ended Thread : 6
Started Thread : 17
Started Thread : 18
Ended Thread : 7
Started Thread : 19
Ended Thread : 8
Ended Thread : 9
Ended Thread : 10
Ended Thread : 11
Ended Thread : 12
Ended Thread : 13
Ended Thread : 14
Ended Thread : 15
Ended Thread : 16
Ended Thread : 17
Ended Thread : 18
Ended Thread : 19
All Threads stopped successfully!
```

Case Study

Problem Statement

Nishant is working as a programmer for Best IT Solutions Private company, which is involved in developing high-end software applications using technologies and programming languages, such as ASP.NET and C#. Nishant is asked by his Project Manager to create a server-based application for BM Enterprises. The server-based application has to run on a server for processing of specific data. The Project Manager has asked Nishant to take care that the load on the server must be minimum when the application is run. What functionality should Nishant implement in his application to decrease the load on the server?

Solution

Nishant must implement the functionality of multithreading, which is supported in C# programming language to decrease the load on the server. In multithreading, multiple threads are run simultaneously during the execution of a program or application. Nishant first creates the following sample C# application to understand the concept of multithreading.

```
using System;
using System.Collections.Generic;
using System.Text;
using System.Threading;

namespace Threading
{
   class ThreadedApplications
    {
       static void Main(string[] args)
        {
            ApplicationCounter acObj = new ApplicationCounter();
            ThreadStart execute = new
ThreadStart(acObj.application_stopstart);
            ThirdPartyApplications acObj2 =
new ThirdPartyApplications();
            ThreadStart execute2 = new
ThreadStart(acObj2.thirdapplication_stopstart);
            Thread applications1 = new Thread(execute);
            Console.WriteLine("Starting the Primary Application.");
            Thread applications2 = new Thread(execute2);
          //starting the primary application
          applications1.Start();
          applications2.Start();

          //starting the secondary applications
```

```
        for (int m = 1; m <= 5; m++)
         {
             Console.WriteLine("Executing the Primary Application: p{0}", m);
             Thread.Sleep(1000);

         }
          Console.WriteLine("Primary Application ended");
      }
   }

   public class ApplicationCounter
    {

      public void application_stopstart()
       {
        Console.WriteLine("Starting Secondary Applications now.");
        for (int m = 1; m <= 10; m++)
         {
          Console.WriteLine("Executing Secondary Application : s{0}", m);

          Thread.Sleep(500);

         }
       Console.WriteLine("Secondary Application ended");
       }
    }

   public class ThirdPartyApplications
    {

      public void thirdapplication_stopstart()
       {
        Console.WriteLine("Starting Third Party Applications now.");
        for (int m = 1; m <= 15; m++)
         {
          Console.WriteLine("Executing Third Party Application : t{0}", m);

          Thread.Sleep(2000);

            }
         Console.WriteLine("Third Party Applications ended");
      }
    }

}
```

Remarks

In C#, multithreading helps in optimising the resources related to a specific application. It also helps in faster execution of an application. Multithreading can be used in C# to create different threads for performing operations such as checking the user input and performing background tasks.

Common Programming Errors

- *Thread declared but not started using the* ***Start*** *method.*
- *The execution started with a static method, which is called using the object of the class and not the class name.*
- *The* ***System.Threading*** *namespace not used in the C# program using the USING keyword.*
- *In a C# program, a thread is being restarted.*
- *A thread is started before the* ***try/catch*** *or* ***finally*** *block.*
- *The methods of the predefined classes such as* ***Thread*** *and* ***Monitor*** *are called without using the class name.*

Review Questions

19.1 What is a thread? What does it do?
19.2 What is the use of threading in C#?
19.3 How a **Thread** is created? And how is it started?
19.4 What is the function of **Sleep()** and **Join()** methods of the Thread class?
19.5 How is a thread named?
19.6 Can the Threads priority be set?
19.7 What is a thread pool?

Programming Exercises

19.1 Given below is a C# program that creates and starts a thread **t1**. A **write** method is also defined in the program that is used to display a string in the console window when the thread starts. Specify whether the given program will compile or not. If not, then given the reason.

```
using System;
using System.Collections.Generic;
using System.Text;
using System.Threading;



namespace debugAppl_chap19
{
    class ThreadName
    {
```

```
        static void Main(string[] args)
        {
            Thread.CurrentThread.Name = "Thread1";
            Thread t1 = new Thread(write);
            t1.Name = "Thread2";
            t1.Start();
            write();
        }
        public void write()
        {
            Console.WriteLine("Hello from " + Thread.CurrentThread.Name);
        }
    }
}
```

19.2 In the C# program given below, a **show** method is defined for displaying a string in the console window. A thread with the name **th** is also created in the program using the constructor of the **Thread** class. Identify the error in the program and specify the reason for the error.

```
using System;
using System.Collections.Generic;
using System.Text;



namespace debugApp2_chap19
{
    class Program
    {
        static void Main(string[] args)
        {
            Thread th = new Thread(new ThreadStart(show));
            th.Start(); // Run Go() on the new thread.
            th.show(); // Simultaneously run Go() in the main thread.
        }
        static void show() {
            Console.WriteLine("This is a thread start program!");
    }
}
```

19.3 Identify the error in the following C# program and state reason.

```
using System;
using System.Collections.Generic;
using System.Text;
```

```
using System.Threading;

namespace debugApp3_chap19
{
    class Program
    {
        string bar = "This is a string value.";


        public static void Main()
        {
            string text =
            string.Format("{0} contains string value of: {1}",
Thread.CurrentThread.GetHashCode(), bar);
            Console.WriteLine(text);
        }
    }
}
```

Chapter 20

Windows and Web-based Application Development on .NET

20.1 INTRODUCTION

.NET is a software programming architecture provided by Microsoft for developing applications, which can be run on the Web through a network connection such as Internet or Intranet. The applications, which are developed for the Web using .NET are called Web based applications. .NET supports the ASP.NET technology with C# programming for allowing software developers to create Web based applications, which can be used to perform tasks such as validating data received by the server from client computers and doing business on the Internet. The Web-based application can be created for allowing users who are surfing the Internet to do online shopping. .NET also allows software developers to develop Windows based applications for performing tasks such as accessing data from a database and manipulating that data. We can use *Microsoft Visual Studio 2005* to develop Web based and Windows based .NET applications.

20.2 UNDERSTANDING MICROSOFT VISUAL STUDIO 2005

Microsoft Visual studio 2005 is basically a set of software development tools that helps in the rapid development of applications such as Windows applications, Web-based applications and mobile applications. It provides a development environment for creating different applications that can run on platforms such as Windows servers, Smartphones and World Wide Web (WWW) browsers, supported by .NET framework. Microsoft Visual Studio 2005 allows you to create the user interfaces of a Windows and Web-based application using Windows Forms and Web Forms. In order to be able to create .NET Windows and Web based applications using Microsoft Visual Studio 2005, it is essential to first understand its *Integrated Development Environment (IDE)*. Languages such as C#, C++ and Visual Basic.NET can be used in Microsoft Visual Studio 2005 to specify a specific functionality such as connecting to a database and displaying that data.

20.2.1 Overview of Microsoft Visual Studio 2005 IDE

The Microsoft Visual Studio 2005 IDE is a window, which contains components, such as **Toolbox, Solution Explorer, menubars** and **toolbars** that help in the rapid development of Windows and Web-based applications. To access the Microsoft Visual Studio 2005 IDE, we must select Start->Programs->Microsoft Visual Studio 2005->Microsoft Visual Studio 2005. Figure 20.1 shows the Microsoft Visual Studio IDE.

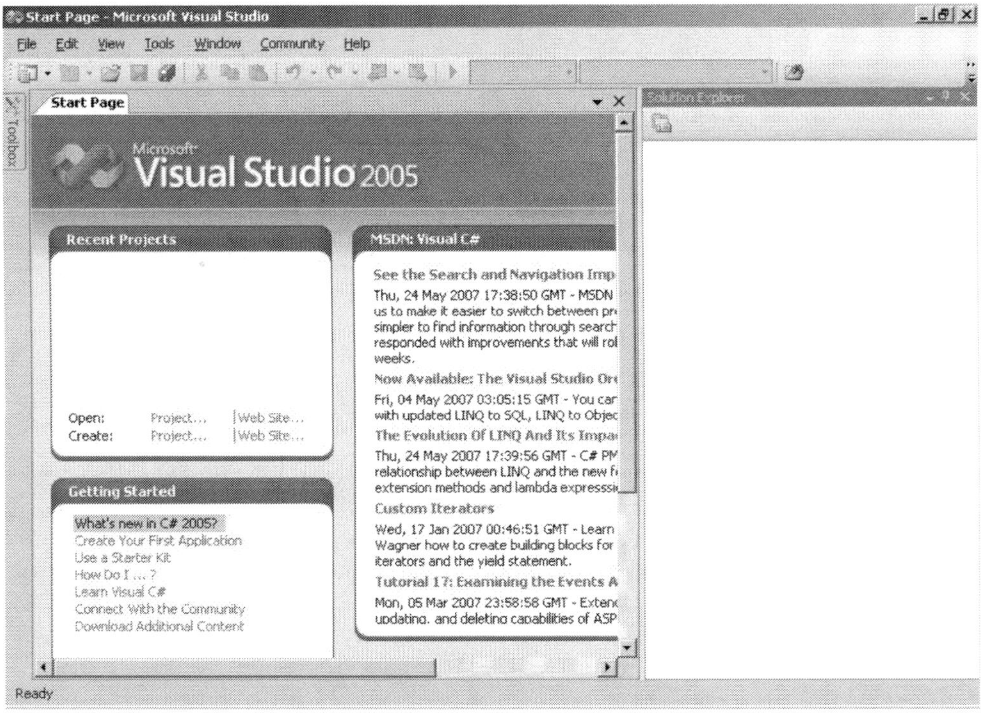

Figure 20.1 *The microsoft visual studio IDE*

After the Microsoft Visual Studio IDE appears, select File->New->Project to display the New Project dialog box as shown in Fig. 20.2.

Select the Windows Application template, which appears in the **Templates** section when we choose Visual C# in the **Project Type** pane, to specify a template for a new project. After this, enter a name for the new project in the **Name** text box and click OK. The components such as **Solution Explorer** and **Properties** window of the Microsoft Visual Studio 2005 IDE appear with a Windows form. Figure 20.3 shows the different components of Microsoft Visual Studio 2005 IDE.

The various components of the Microsoft Visual Studio IDE are the following:

- **Menubar:** This contains various menus such as **File** and **View** that help perform tasks such as creating a new project and displaying the **Toolbox** if it is hidden.
- **Toolbar:** This contains command buttons that we can use to quickly perform tasks, which include saving a project and running a Windows or Web-based application.
- **Toolbox:** This contains various controls such as **TextBox** and **Label** that help create the user interface of a Windows or Web based application. If the **Toolbox** does not appear in the Microsoft

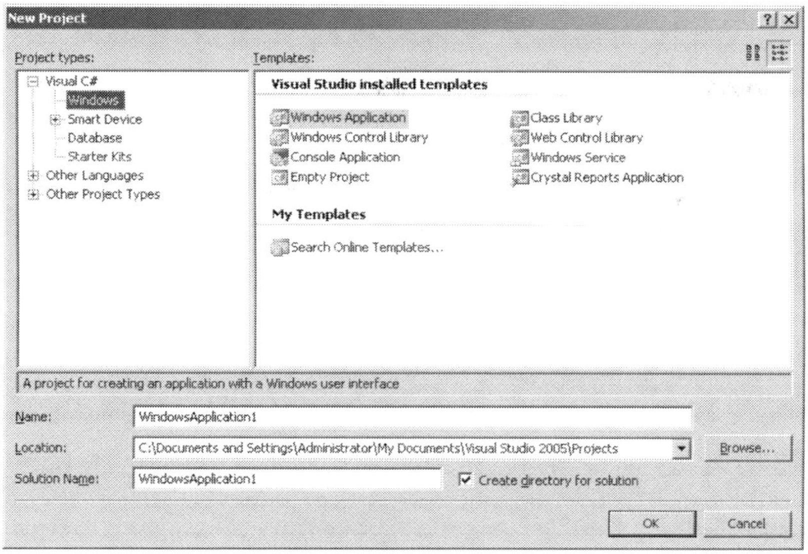

Figure 20.2 *The new project dialog box*

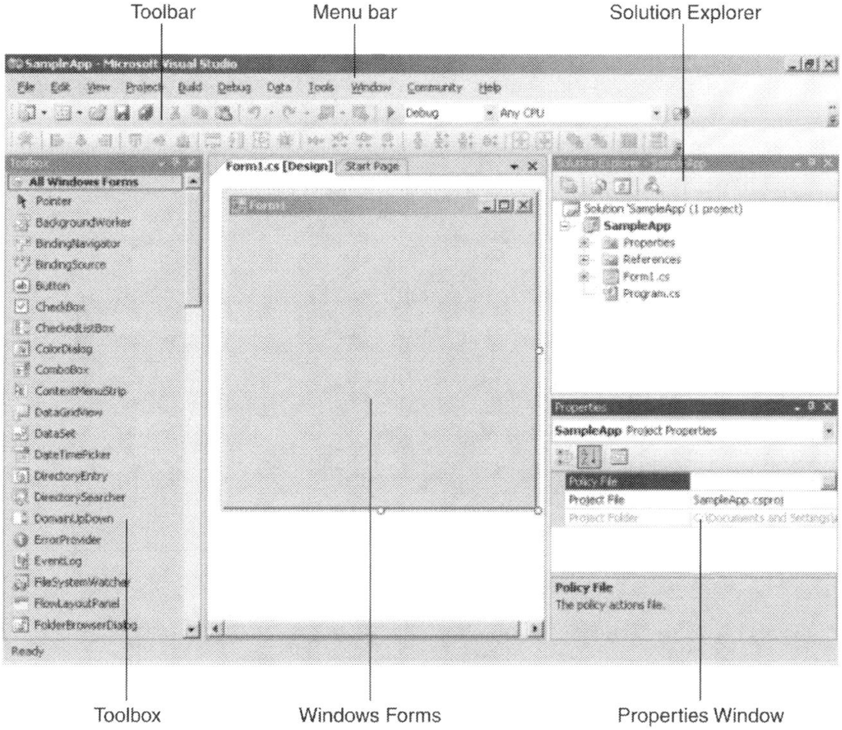

Figure 20.3 *Components of microsoft visual studio IDE*

Visual Studio IDE then we can view it by selecting View->Toolbox option in the Microsoft Visual Studio IDE.

- **Solution Explorer:** This provides an organised view of files related to a project. In the **Solution Explorer**, we can access commands to perform specific tasks such as adding an item, file or form to a project and viewing the C# code for a specific form. If the **Solution Explorer** does not appear in the Microsoft Visual Studio IDE then you can view it by selecting View->Solution Explorer option.
- **Properties window:** This window displays the properties of the Windows or Web forms and controls, which have been added to the project. It also displays the values of the properties and allows us to edit them. If the **Properties** window does not appear in the Microsoft Visual Studio 2005 IDE then we can view it by selecting View->Properties Window option.

20.2.2 Controls in Microsoft Visual Studio

Controls in Microsoft Visual Studio are the elements that can be used to design the user interface of applications such as Windows applications and Web based applications. Microsoft Visual Studio 2005 provides a set of built-in controls in the **Toolbox** that can be easily dragged and dropped on to a Windows or Web form for designing of the user interface of an application. These help in the fast development of an application as you do not have to create these controls but need to simply add them to a form from the **Toolbox**. Table 20.1 lists some of the important controls available in the **Toolbox** component of Microsoft Visual Studio IDE.

Table 20.1 *Controls in microsoft visual studio*

Name	Purpose
Button	To perform a specific task by causing an event to occur on clicking.
CheckBox	To select or deselect an option.
CheckBoxListBox	To display a list of items with a corresponding check box appearing on the left for each item.
ComboBox	To display an editable text box with a drop-down list, which contains a specified set of values.
DateTimePicker	To select a date and time and display it in a specific format.
Label	To display a text on a Windows or Web form.
LinkLabel	To display a text on a Windows or Web form with hyperlink, formatting or tracking capability.
ListBox	To display a list from which a user can select a specific item.
ListView	To display a list of items with an icon and a label associated with each item.

(Contd.)

Name	Purpose
MonthCalendar	To display a month calendar from which the user of an application can select a specific date.
NotifyIcon	To display an icon during run time in the notification area, which appears on the right in the Windows taskbar.
PictureBox	To display an image in a Windows or Web form.
ProgressBar	To display a bar, which indicates the progress of an operation.
RadioButton	To select an option from a group of options, specified through other **RadioButton** controls.
TextBox	To accept text from the user of an application.
ToolTip	To display text when the user of an application moves the mouse pointer to the associated control.
TreeView	To display a collection of items with corresponding labels in a hierarchical manner.

There are various properties such as **Text** and **BackColor** associated with these controls. These properties can be used to perform tasks such as specifying the text for a control and changing the background colour of a control. The **Properties** window can be used to change the value of properties associated with a control, Windows form or Web form.

20.3 CREATING AND RUNNING A SAMPLEWINAPP WINDOWS APPLICATION

SampleWinApp is a Windows application that allows a user to click on a button named *Click Me*. When the button is clicked, the **Click** event occurs through which the *Click Me* text of the button changes to **Hello** and vice-versa. To create the SampleWinApp Windows application on .NET using Microsoft Visual studio, we must:

- Create the Form1 form
- Add code to Form1 form

We can use **F5** key or a combination of **Ctrl** and **F5** that is **Ctrl+F5** keys to run the SampleWinApp Windows application.

Creating Form1 Form

The **Form1 form** allows a user to click a button to change the text *Click Me* to *Hello*. To create the **Form1 form**:

1. Select *Start->Programs->Microsoft Visual Studio 2005-> Microsoft Visual Studio 2005* to display the Microsoft Visual Studio IDE.
2. Select *File->New->Project* to display the New Project dialog box.

3. Select the Windows Application template, which appears in the **Templates** section when you choose Visual C# in the Project Type pane, to specify a template for a new project.
4. Enter a name, such as SampleWinApp in the **Name** text box to specify a name for the Windows application. Figure 20.4 shows the **New Project** dialog box with the template and name specified for the SampleWinApp Windows application.

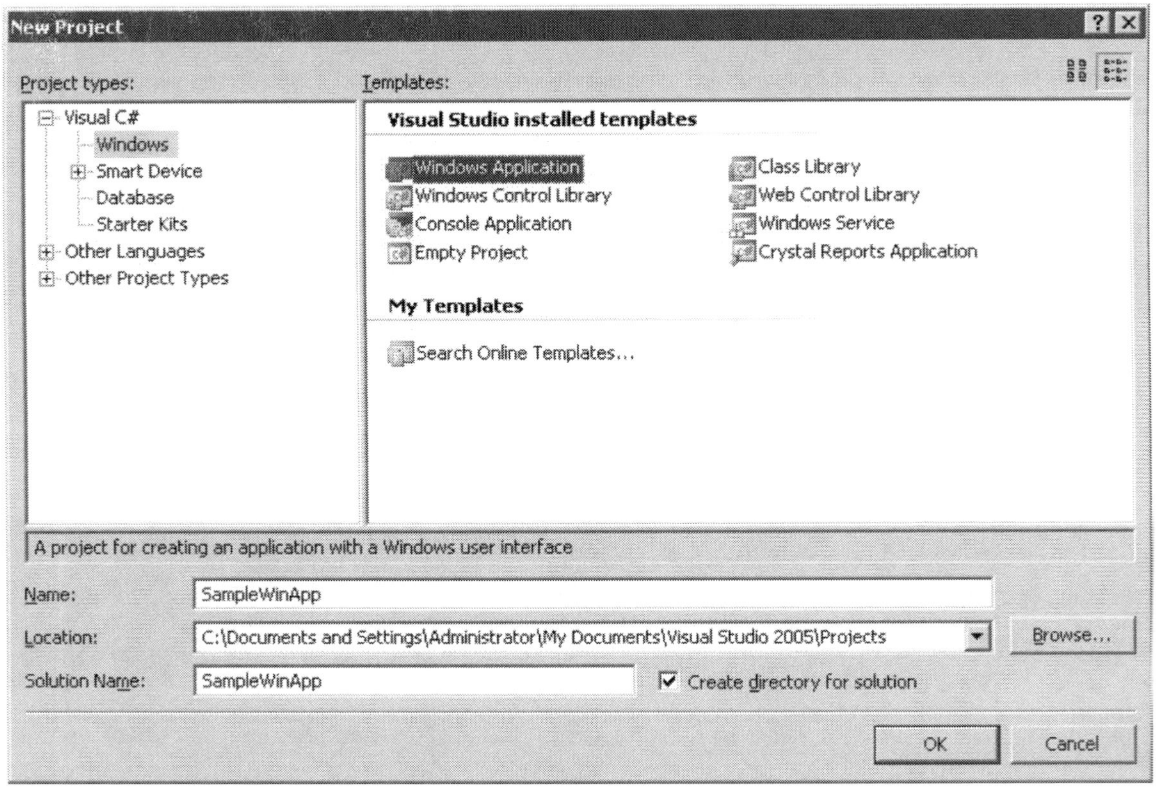

Figure 20.4 *The new project dialog box for C# windows application*

5. Click OK to close the **New Project** dialog box. The Microsoft Visual Studio IDE appears with a Windows form and **Solution Explorer** as shown in Fig. 20.5.
6. Change the value of **Text Property** for **Form1 form** from **Form1** to **Button Test** in the **Properties** window as shown in Fig. 20.6.
7. Drag a **Button** control from the **Toolbox** on to the **Form1 form** to add the **Button** control to **Form1 form**.
8. Change the value of the **Text property** for the **Button** control to *Click Me* in the Properties window.
9. Change the **BackColor** and **ForeColor** properties of the **Button** control as required in the **Properties** window.
10. Drag a **Label** control from the **Toolbox** on to the **Form1 form** to add the control to the form.

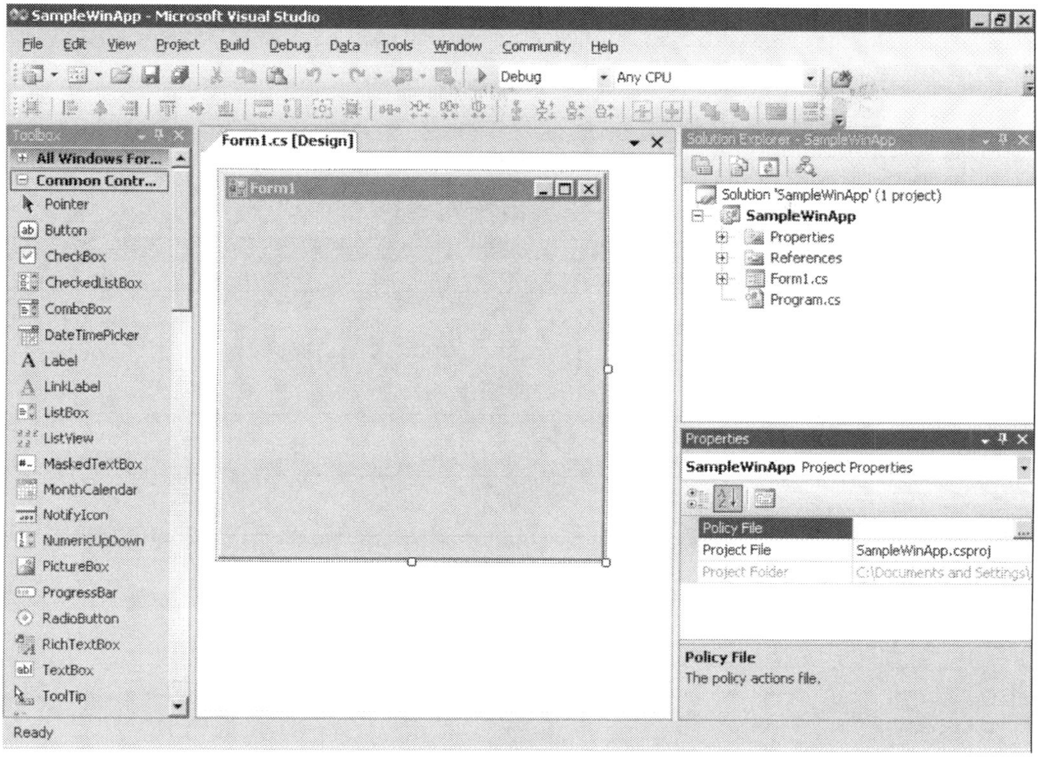

Figure 20.5 *The Form1 form*

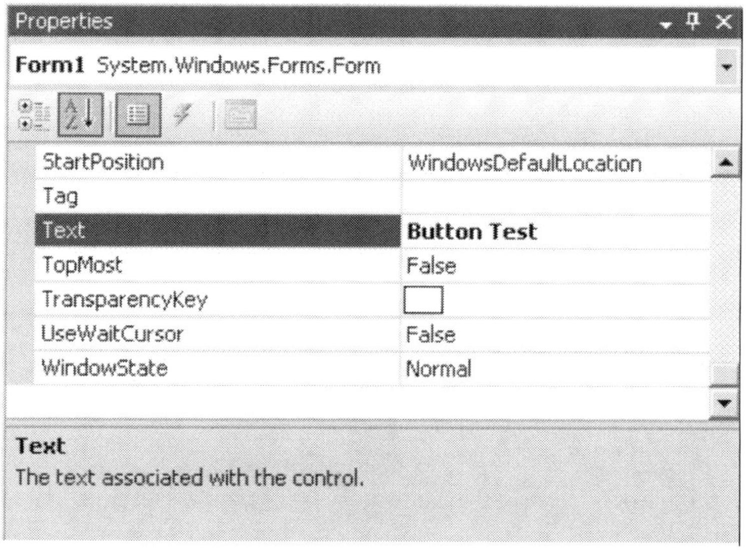

Figure 20.6 *Properties window with changed text*

11. Change the value of the **Text property** for the **Label** control from *label1* to *This is a Label, TextBox* and *Button Control Application.*
12. Change the **BackColor** and **ForeColor** properties of the **Label** control as required in the **Properties** window.
13. Drag a **TextBox** control from the **Toolbox** on to the **Form1 form** to add the control to the form. Figure 20.7 shows the **Form1 form** with controls added to it.

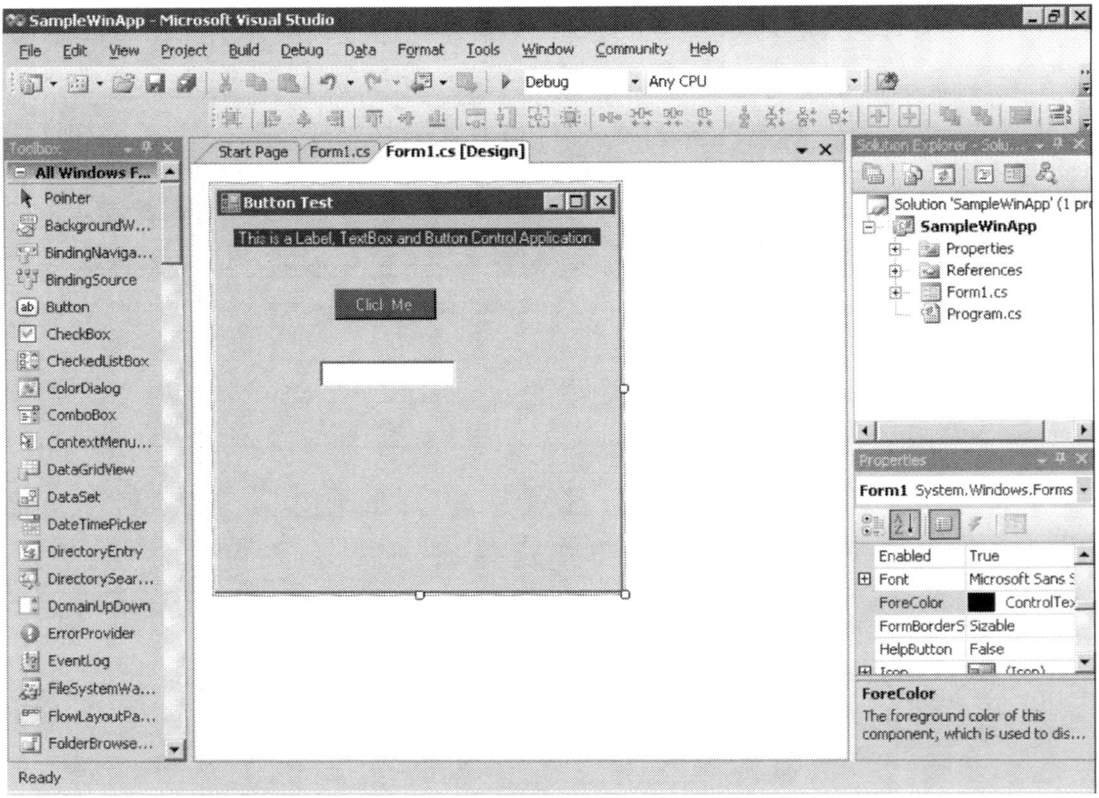

Figure 20.7 *Form1 form with added controls*

This completes the task of creating the **Form1 form**.

Adding Code to Form1 Form

You need to add code to Form1 form to specify the functionality for changing the text *Click Me* that appears in the **Button** control to *Hello*. To add code to **Form1 form**:

1. Right click the **Form1.cs** node in the **Solution Explorer** to display a shortcut menu.
2. Select the **View Code** option to display the **Code Editor** with **Form1.cs** file associated to **Form1 form** as shown in Fig. 20.8:

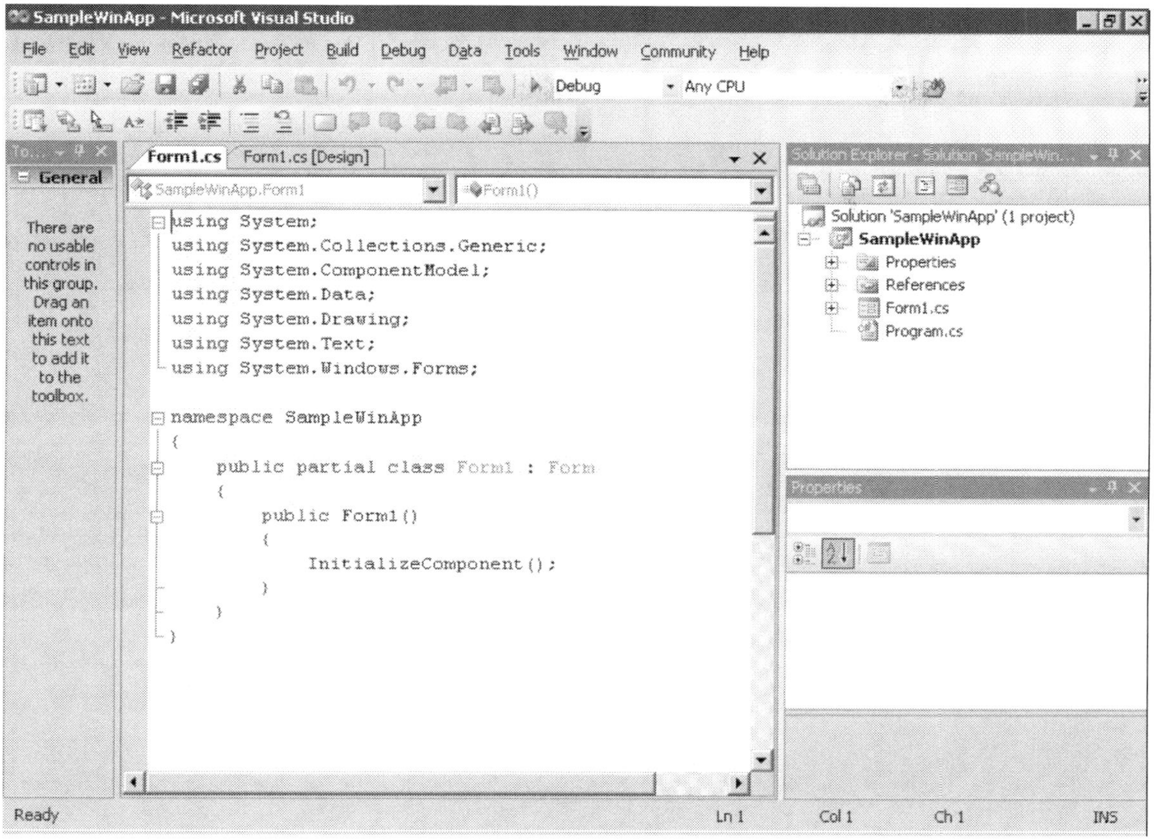

Figure 20.8 *The code editor in microsoft visual studio*

3. Enter the following code before **public Form1()** to declare a variable of integer type:

```
int i = 0;
```

4. Click the **Close** button represented by a cross that appears at the top in the **Code Editor** window to close the **Editor**. The **Form1.cs [Design]** view appears.
5. Double click anywhere on the **Form1 form** to display the **Code Editor** window with the **Form1.cs** file containing the **Form1_Load** event handler as shown in Fig. 20.9.

   Note: To handle events generated by a control, an event handler needs to be defined for the event. An event handler is a Visual C# method containing a code that is executed when an event occurs on a control. These event handlers help perform specific tasks, such as displaying a button or modifying the colour of a control at runtime.
6. Enter the following code in the **Form1_Load** event handler to specify the functionality for hiding the **TextBox** control, which you have added to **Form1** form.

   // Hiding the **TextBox** on start of the application

```
textBox1.Hide();
```

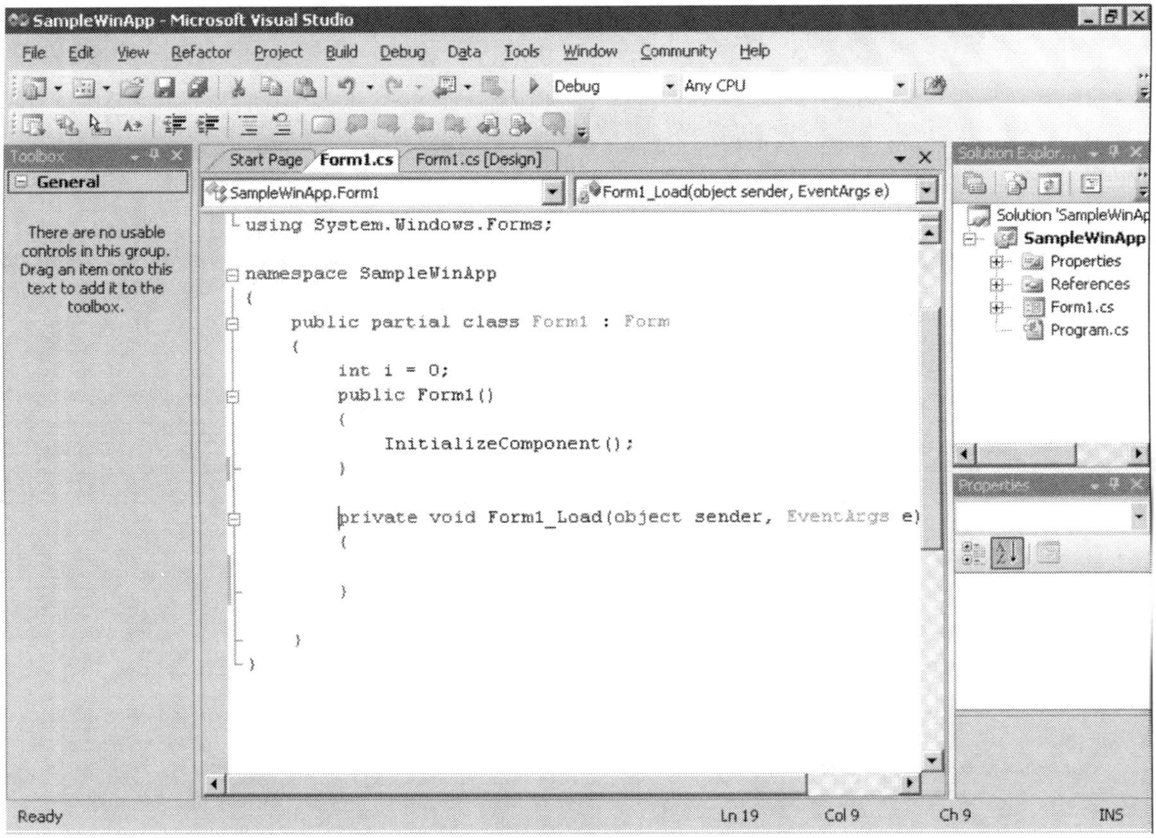

Figure 20.9 *The code editor with form1_load event handler*

7. Click the **Close** button represented by cross that appears at the top in the **Code Editor** window to close the **Editor**. The **Form1.cs [Design]** view appears.
8. Double click the Button control that appears in the **Form1 form** to display the **Code Editor** window with **Form1.cs** file containing the **button1_Click** event handler.
9. Enter the following code to the **button1_Click** event handler to specify the functionality for changing the text of the **Button** control:

Program 20.1 shows a form and a button. When the button is clicked the text of the button changes and a read-only textbox appears with the message '*click the button again*'. Clicking the button again sets the button text to its original text. This can be repeated as required by the user.

Program 20.1 Changing text of a button control

```
i++;
        if (i % 2 != 0)
        {
```

```
        button1.Text = "Hello";
        MessageBox.Show("The button has been clicked and the button text changed to
'Hello'.");
    }
    else
    {
        button1.Text = "Click Me";
        MessageBox.Show("The button has been clicked and the button text changed to 'Click
Me'.");
    }

    //Unhiding the text box here
    textbox1.Show();
    textbox1.Text = "Click the button again and the button text changes.";
    textbox1.ReadOnly = true;
```

In the above program, the **button1_Click** event handler for the **Click** event contains the code for changing the *Click Me* text, which appears in the **Button** control. This completes the task of adding code to **Form1 form**.

Running the SampleWinApp Windows Application

We can also click the **Start Debugging** button to run the Windows application, which we have created. The **Button Test** window then appears as shown in Fig. 20.10.

In the **Button Test** window, click the *Click Me* button. This causes the Click event of the Button control to occur. A message box appears to confirm that the *Click Me* button has been clicked as shown in Fig. 20.11.

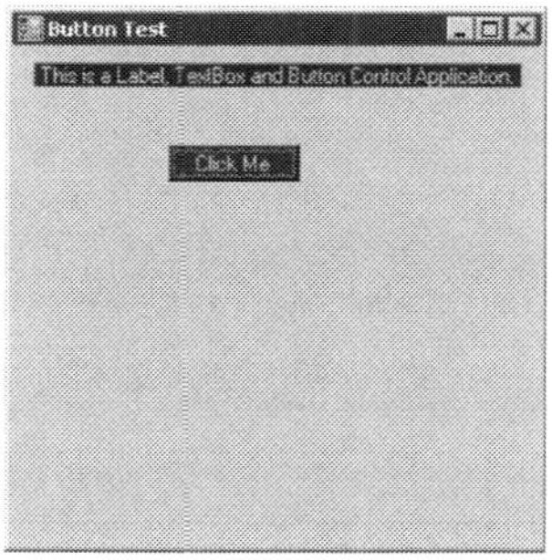

Figure 20.10 *The button test window*

Figure 20.11 *The message box confirming button click*

We can click the OK button in the message box to display the **Button Test** window with changed text of the **Button Control** as shown in Fig. 20.12.

We can click the button that appears in the **ButtonTest** window again to change the text from *Hello* to *Click Me* and vice versa.

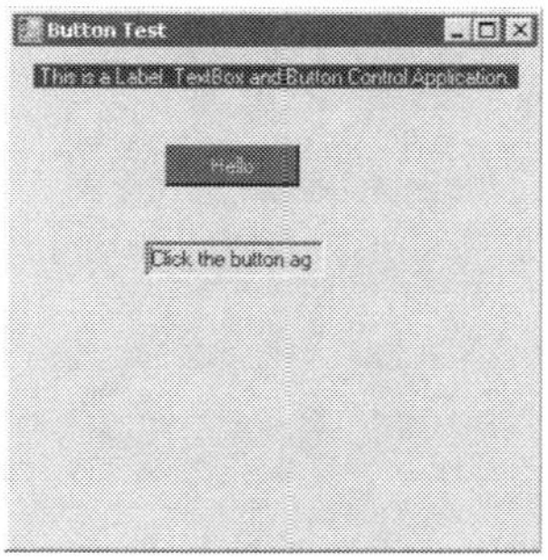

Figure 20.12 *The button test window with changed text*

20.4 CREATING AND RUNNING A SAMPLEWINAPP2 WINDOWS APPLICATION

SampleWinApp2 Windows application is an Single Document Interface (SDI) application that allows us to make a selection, calculate factorial and check whether a number is a prime number or not. An SDI application is an application that may contain multiple Windows Forms and each Windows Form opens in a separate application window. Each Windows form of an SDI application contains a separate menu bar, status bar, and toolbar. An example of an SDI application is the **WordPad** application available in Windows. The WordPad application allows us to open only one document at a time. In WordPad, we need to close a document before opening another document. Microsoft Visual Studio 2005 provides the **Start Debugging** button in the Toolbar, which appears at the top in the IDE, to run a application. We can also press **F5** or a combination of **Ctrl** and **F5** keys to run the SampleWinApp2 Windows application.

Creating a SampleWinApp2 Windows Application

A SampleWinApp2 Windows application contains three forms: **FirstForm**, **factorial** and **PrimeNo.** In the **FirstForm** form, we can make a choice for calculating factorial or checking for prime number. In the factorial form, a number is entered by a user and when a button is clicked, the factorial of the entered number is displayed. In the **PrimeNo** form, a number entered by a user is checked to determine whether it is a prime number or not. To create the SampleWinApp2 Windows application, we must:

- Create the **FirstForm** form.
- Add code to **FirstForm** form.
- Create the **factorial** form.
- Add code to the **Factorial** form.
- Create the **PrimeNo** form.
- Add the code to **PrimeNo** form.

Creating the FirstForm Form

We need to create the **FirstForm** form in order to make a choice for calculating factorial or checking for prime number. To create the **FirstForm** form:

1. Select *Start->Programs->Microsoft Visual Studio 2005->Microsoft Visual Studio 2005* to display the Microsoft Visual Studio IDE.
2. Select *File->New->Project* to display the **New Project** dialog box.
3. Select the Windows Application template, which appears in the **Templates** section when you choose Visual C# in the **Project Type** pane, to specify a template for the new Windows application.
4. Enter a name, such as SampleWinApp2 in the **Name** text box to specify a name for the new Windows application.
5. Click OK to close the **New Project** dialog box. The Microsoft Visual Studio IDE appears with the **Form1 form** and **Solution Explorer**.
6. Right click the **Form1.cs** node in the **Solution Explorer** to display a shortcut menu.
7. Select the **Rename** option to display the editable text box, which allows us to enter the new name for **Form1.cs** file.
8. Enter **FirstForm.cs** in the editable text box to specify a new name for **Form1.cs** file.
9. Change the value of **Text property** for the **FirstForm** form from **Form1** to **SDI Application in C#** in the **Properties** window.

10. Change the **BackColor** property of the **FirstForm** form in the **Properties** window as per requirement.
11. Drag a **Label** control from the **Toolbox** on to the **FirstForm** form to add the control to the form.
12. Change the **Text** property of the **Label** control from **Label 1** to **This is an SDI application**, in the **Properties** window.
13. Change the **BackColor** of the **Label Control** in the **Properties** window as per requirement.
14. Expand the **Font** node that appears in the **Properties** window to display the **Bold** option.
15. Click the editable text box that appears next to the **Bold** option to display an arrow.
16. Click the arrow to display a list containing **True** and **False** values as shown in Fig. 20.13.
17. Select **True** from the list to make the text appearing in the **Label** control bold.

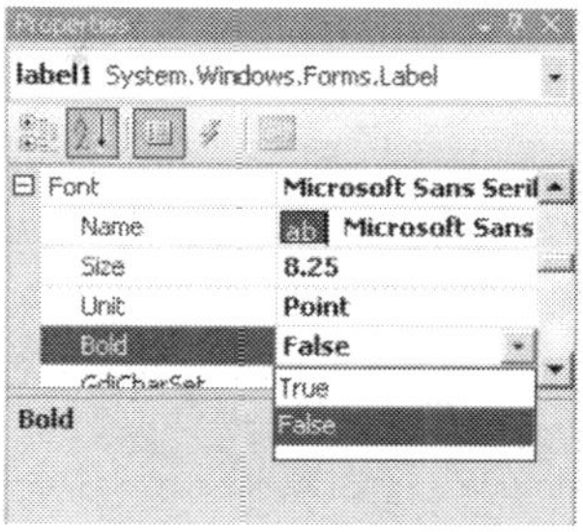

Figure 20.13 *The properties window with bold option*

18. Drag a **RadioButton** control from the **Toolbox** on to the **FirstForm** form to add the **RadioButton** control.
19. Similarly, add another **RadioButton** control to the **FirstForm** form.
20. Change the value of **Text property** of the first **RadioButton** control, which you have added, to **Check** factorial in the **Properties** window.
21. Expand the **Font** node in the **Properties** window to display the **Bold** option.
22. Click the editable text box, which appears next to the **Bold** option to display an arrow.
23. Click the arrow to display a list containing **True** and **False** values.
24. Select **True** from the list to make the text appearing with the **RadioButton** control bold.
25. Similarly, change the value of **Text property** for the second **RadioButton** control to **Check Prime Number**.
26. Make the text, **Check Prime Number**, which appears with the second **RadioButton** control, bold using the **Font** node in the Properties window.
27. Drag a **Button** control from the **Toolbox** on to the **FirstForm** form to add the control to the form.
28. Change the value of the **Text property** of the **Button** control to **Start** in the Properties window.
29. Expand the Font node in the **Properties** window to display the **Bold** option.
30. Click the editable text box, which appears next to the **Bold** option to display an arrow.
31. Click the arrow to display a list containing **True** and **False** values.
32. Select **True** from the list to make the text appearing in the **Button** control bold.
33. Add another **Button** control from the **Toolbox** to the **FirstForm** form.
34. Change the value of the **Text property** for the **Button** control to **Exit** in the **Properties** window.
35. Change the **BackColor** properties of the two **Button** controls, which you have added using the **Properties** window as per requirement.
36. Select the first **Button** control, which was added to the **FirstForm** form to display the **ForeColor** option in the **Properties window** as shown in Fig. 20.14.

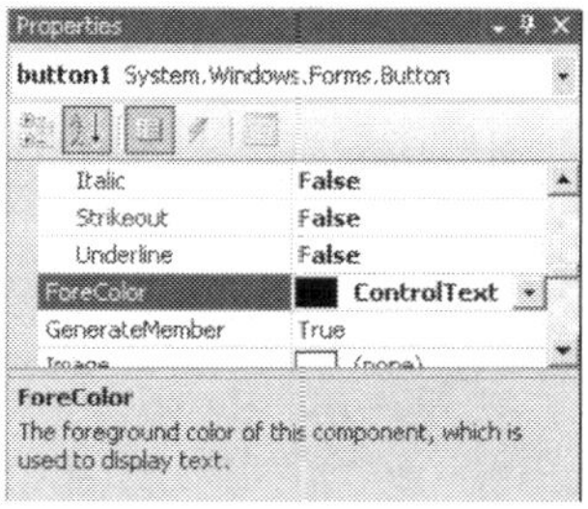

Figure 20.14 *The properties window with ForeColor option*

37. Click the arrow that appears in the editable text box next to the **ForeColor** option to display a colour palette.
38. Select a colour from palette to change the colour of the text that appears in the first **Button** control.
39. Change the colour of the text appearing in the second **Button** control in the same way as you have changed the colour of the text in first **Button** control. This completes the task of creating the **FirstForm** form. Figure 20.15 shows the **FirstForm** form with added controls:

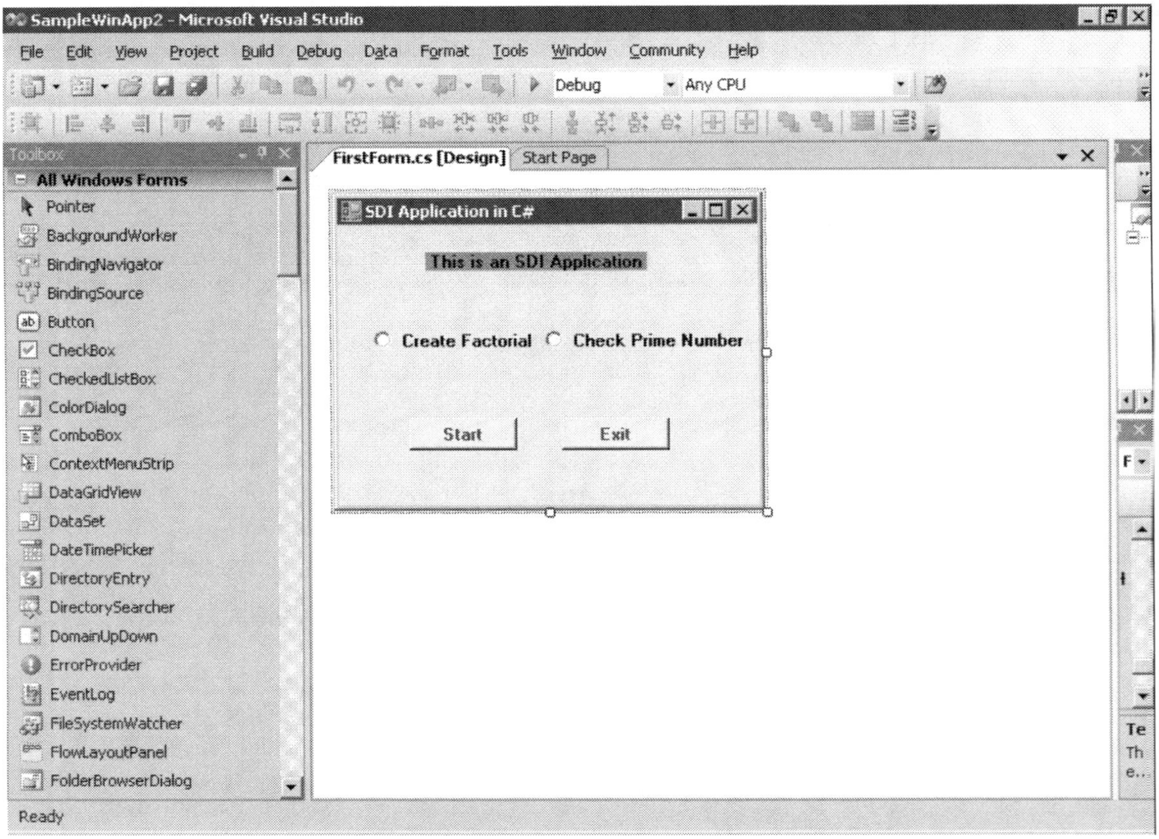

Figure 20.15 *The FirstForm Form in design view*

Adding Code to FirstForm Form

In the **FirstForm** form, we need to add code to specify the functionality for navigating either to **factorial** form or **PrimeNo** form. To add code to the **FirstForm** form:

1. Double click the **Button** control with **Start** text to display the **Code Editor** with **FirstForm.cs** file associated to the **FirstForm** form as shown in Fig. 20.16.
2. Add the following code to the **button1_Click** event handler, which appears in the **Code Editor** to specify the functionality for displaying the forms to calculate the factorial and check the prime number.

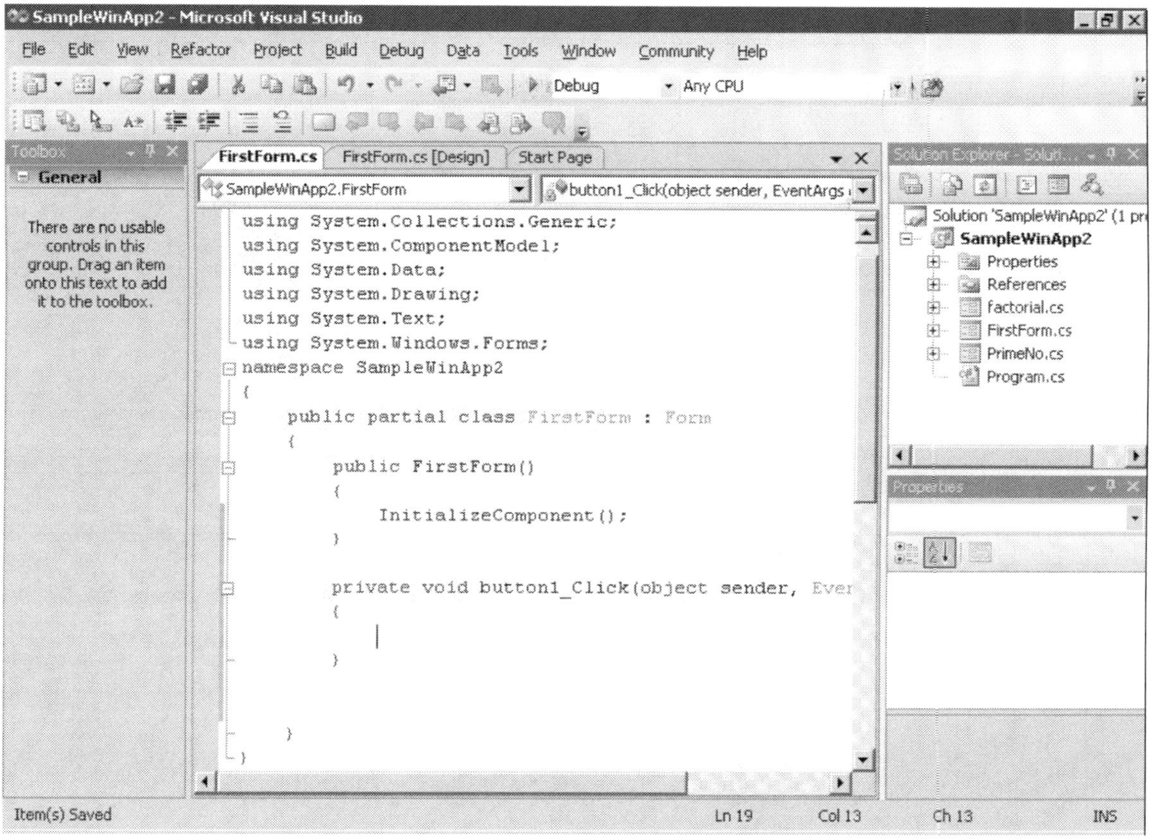

Figure 20.16 *The code editor window for FirstForm Form*

Program 20.2 Displaying forms

```
if (radioButton1.Checked == true)
    {
        factorial f1 = new factorial();
        f1.Show();
    }
    else if (radioButton2.Checked == true)
    {
        PrimeNo f2 = new PrimeNo();
        f2.Show();
    }
```

3. Click the **Close** button represented by a cross, which appears at the top in the **Code Editor** to close the **Editor**. The **FirstForm.cs [Design]** view for the **FirstForm** form appears.

4. Double click the **Button** control with the text **Exit** to open the **Code Editor** with **FirstForm.cs** file and **button2_Click** event handler.
5. Add the following code in the **button2_Click** event handler to specify the functionality for closing the **FirstForm** form:

```
this.Close();
```

This completes the process of adding code to **FirstForm** form.

Creating the Factorial Form

We need to create a factorial form to enable us to enter a number and calculate its factorial. To create the **Factorial** form:

1. Right click the SampleWinApp2 node in the **Solution Explorer** to display a shortcut menu.
2. Select **Add->New Item** to display the **Add New Item** dialog box as shown in Fig. 20.17.

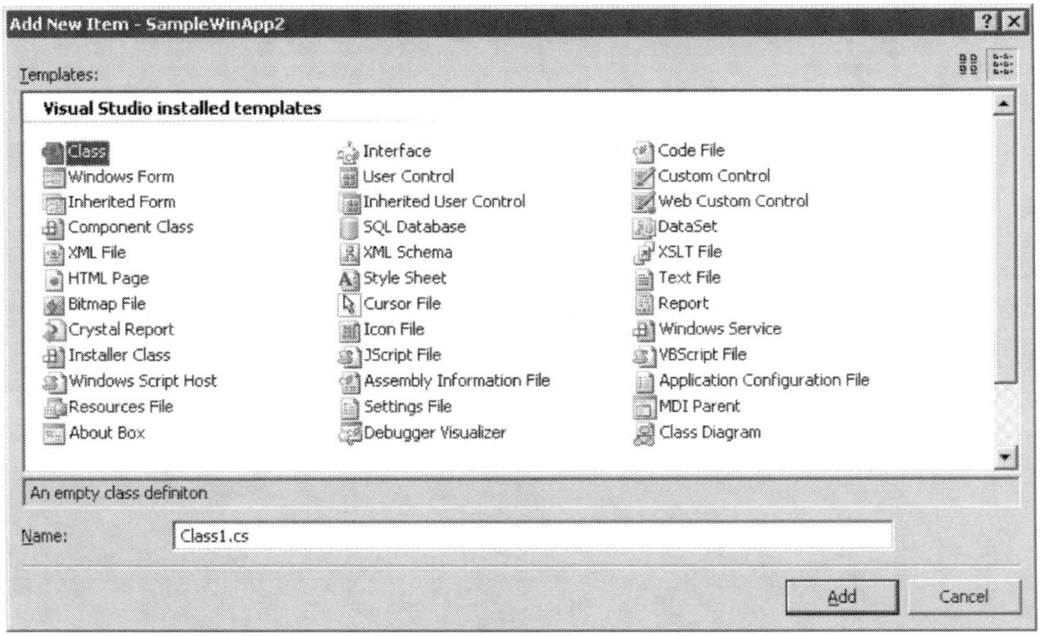

Figure 20.17 *The add new item dialog box*

3. Select Windows Form template and enter **factorial.cs** in the **Name** text box to specify the name for the new form.
4. Click **Add** to add the factorial form to the SampelWinApp2 application. The Microsoft Visual Studio IDE appears with **Factorial** form.
5. Change the value of the **Text property** for the factorial form to **Factorial Form**.
6. Change the **BackColor** property of the factorial form using the **Properties** window as per requirement.
7. Add a **Label** control to the factorial form and change its **BackColor** property using the **Properties** window.

8. Change the value of **Text property** of the **Label** control to *Please enter values up to 25* in the **Properties** window.
9. Add a **Label** control to the factorial form and change its **BackColor** property using the **Properties** window.
10. Change the value of **Text property** of the **Label** control to *Enter a number*: in the **Properties** window.
11. Make the text appearing in the **Label** control bold using the **Font** node in the **Properties** window.
12. Add a **TextBox** control to the factorial form using the **Toolbox**.
13. Add two **Button** controls to the factorial form using the **Toolbox**.
14. Change the **BackColor** properties of the **Button** controls using the **Properties** window.
15. Select the first **Button** control and change the value of its **Text property** to *Calculate Factorial* in the **Properties** window.
16. Similarly, change the value of the **Text property** for the second **Button** control to *Back to FirstForm*.
17. Make the text that appears in the **Button** controls, bold using the **Font** node in the **Properties** window.
18. Change the **ForeColor** property of the **Button** controls to modify the colour of the text that appears in the controls.
19. Add a **Label** control and a **TextBox** control to the factorial form.
20. Change the value of **Text** property of the **Label** control to *The Factorial* is in the **Properties** window.
21. Make the text appearing in the **Label** control bold using the **Font** node appearing in the **Properties** window.
22. Change the **BackColor** properties of the **Label** and **TextBox** controls using the **Properties** window. This completes the process of creating the factorial form. Figure 20.18 shows the factorial form with added controls.

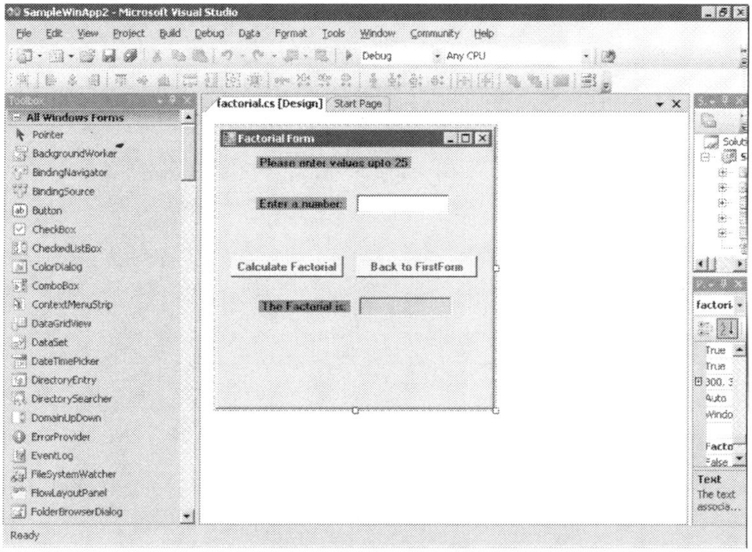

Figure 20.18 *The Factorial form in design view*

Adding Code to a Factorial Form

We need to add code to the **factorial** form to specify the functionality for calculating the factorial of a number and closing the factorial form. To add the code to factorial form:

1. Double click the **Button** control with the text **Calculate Factorial** to display the **Code Editor** with **factorial.cs** file and **button1_Click** event handler for the **Button** control.
2. Add the following code in the **button1_Click** event handler to specify the functionality for calculating factorial:

Program 20.3 Calculating factorial

```
int num;
        long fact;
        if (textBox1.Text.Length > 2)
        {
        MessageBox.Show("Numbers upto 25 are allowed.");
    }
    num = Int32.Parse(textBox1.Text);

    fact = 1;
    while (num >= 1 && num <= 25)
    {
        fact *= num;
        num–;
    }
    textBox2.Text = fact + "";
```

3. Click the **Close** button represented by a cross, which appears at the top in the **Code Editor** to close the **Editor**. The **factorial.cs [Design]** view for the factorial form appears.
4. Double click the **Button** control with the text *Back to FirstForm* to open the **Code Editor** with **factorial.cs** file and **button2_Click** event handler.
5. Add the following code in the **button2_Click** event handler to specify the functionality for closing the factorial form:

```
this.Close();
```

This completes the process of adding code to the **factorial** form.

Creating PrimeNo Form

We need to create the **PrimeNo** form to enable us to enter a number and check whether that number is a prime number or not. To create the **PrimeNo** form:

1. Add a new form with the name **PrimeNo** to the SampleWinApp application using the **Solution Explorer**.
2. Change the value of the **Text property** for the **PrimeNo** form to *Prime Number Form* using the **Properties** window.

3. Change the **BackColor** property of the **PrimeNo** form using the **Properties** window.
4. Add a **Label** control and a **TextBox** control to the **PrimeNo** form using **Toolbox**.
5. Change the **BackColor** property of the **Label** control using the **Properties** window.
6. Change the value of the **Text** property for the **Label** control to **Enter** a number: using the **Properties** window.
7. Make the text appearing in the **Label** control bold using the **Font** node of the **Properties** window.
8. Add two **Button** controls to **PrimeNo** form using **Toolbox**.
9. Change the value of the **Text property** for the first **Button** control, which you have added, to *Check Prime No* using the **Properties** window.
10. Change the value of the **Text property** for the second **Button** control to *Back to FirstForm* using the **Properties** window.
11. Make the text that appears in the **Button** controls, bold using the **Font** node of the **Properties** window.
12. Change the **BackColor** and **ForeColor** properties of the **Button** controls using the **Properties** window. This completes the task of creating the **PrimeNo** form. Figure 20.19 shows the **PrimeNo** form with added controls.

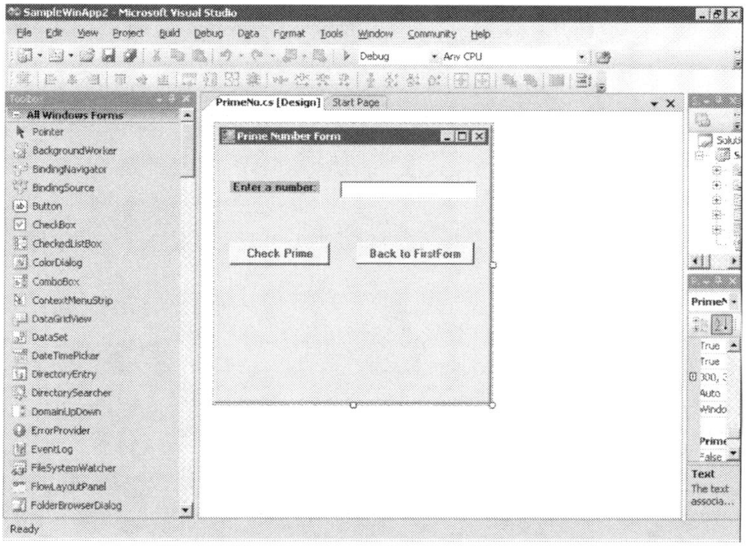

Figure 20.19 *The PrimeNo form in design view*

Adding Code to PrimeNo Form

We need to add code to the **PrimeNo** form to specify the functionality for checking whether a number is a prime number or not and closing the **PrimeNo** form. To add the code to **PrimeNo** form:

1. Double click the **Button** control with the text *Check Prime* to display the **Code Editor** with **PrimeNo.cs** file and **button1_Click** event handler for the **Button** control.
2. Add the following code in the **button1_Click** event handler to specify the functionality to check for a prime number:

Program 20.4 Checking for prime number

```
int num;
double prime, squr_root;
prime = Double.Parse(textBox1.Text);
squr_root = Math.Sqrt(prime);

for (num = 2; num <= squr_root; num++)
{
    if (prime % num == 0)
    {
        MessageBox.Show(prime + " is not a prime number");
        break;
    }
}
if (num > squr_root)
    MessageBox.Show(prime + " is a prime number");
```

3. Click the **Close** button represented by a cross, which appears at the top in the **Code Editor** to close the **Editor**. The **PrimeNo.cs [Design]** view for the **PrimeNo** form appears.
4. Double click the Button control with the text **Back to FirstForm** to open the **Code Editor** with *PrimeNo.cs* file and **button2_Click** event handler.
5. Add the following code in the **button2_Click** event handler to specify the functionality for closing the **PrimeNo** form:

```
this.Close();
```

This completes the process of adding code to the **PrimeNo** form.

Running SampleWinApp2 Windows Application

A SampleWin App2 Windows application can be run to test whether or not it is generating the desired output. This windows example opens a window form and displays two options '*factorial*' and '*prime number*'. Two different windows open on clicking the different options available on the window. The *factorial* calculates the factorial of a number entered and the *prime number* option checks the whether the number is a prime number or not and displays the result in a pop-up window. To run the SampleWinApp2 Windows application:

1. Press **Ctrl + F5** to initiate the process of running SampleWinApp2 Windows application. The SDI Application in C# window appears as shown in Fig. 20.20.
2. Click the **Start** button to display the **Factorial Form** window as shown in Fig. 20.21.

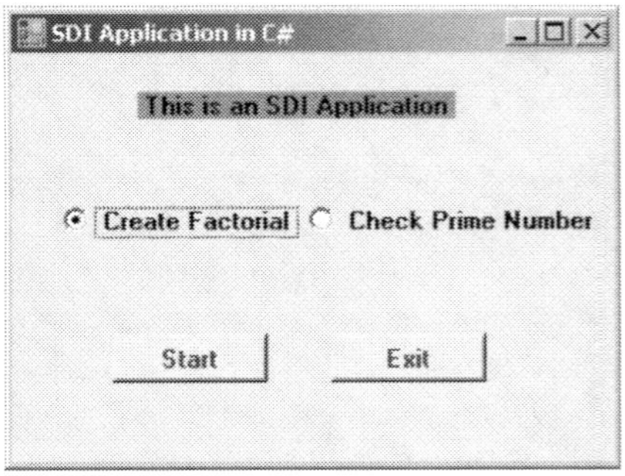

Figure 20.20 *The SDI application in C# window*

3. Enter a number such as 23 in the *Enter a number* text box and click the **Calculate Factorial** button. The factorial of the entered number appears in the *The Factorial is* text box as shown in Fig. 20.22.
4. Click the *Back to FirstForm* button to display the SDI Application in C# window.
5. Select the *Check Prime Number* radio button and click **Start** to display the **Prime Number Form** window as shown in Fig. 20.23.
6. Enter a number, such as 3 and click the *Check Prime button* to display a message box as shown in Fig. 20.24.

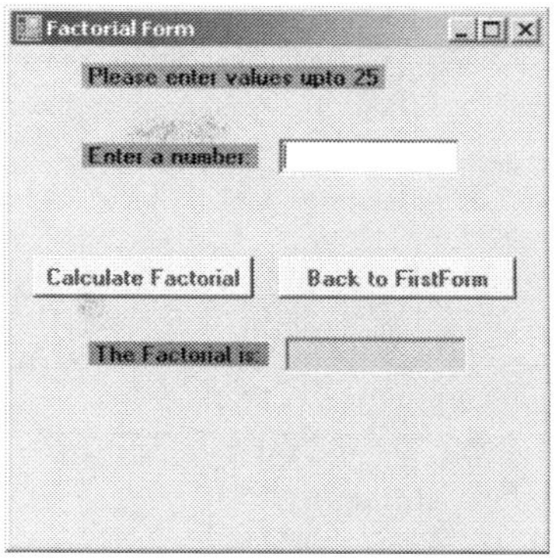

Figure 20.21 *The Factorial form window*

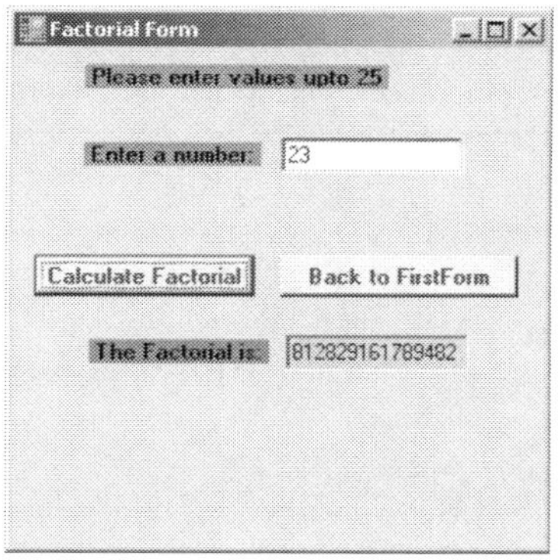

Figure 20.22 *Finding factorial of a number*

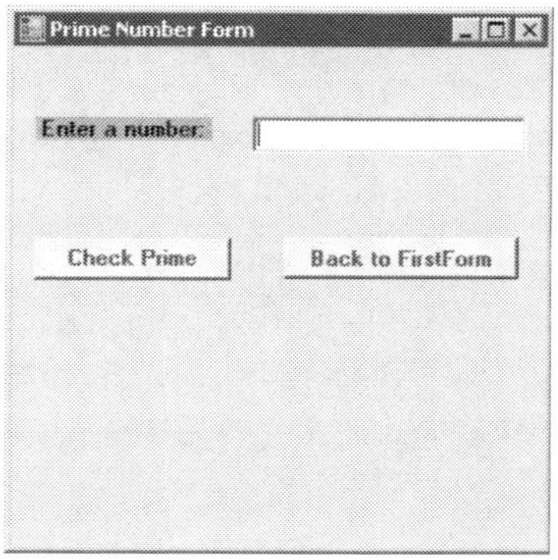

Figure 20.23 *The Prime Number form window*

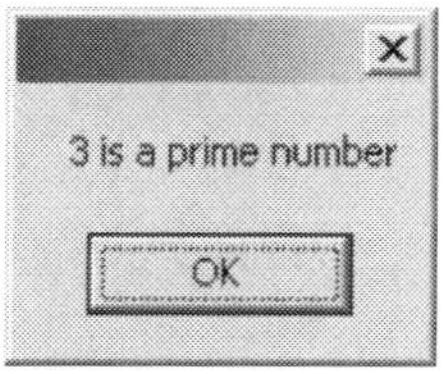

Figure 20.24 *The message to confirm prime number*

7. Click OK to close the message box. The **Prime Number Form** appears again.
8. Click the *Back to FirstForm* button to display the SDI Application in C# window.
9. Click **Exit** to exit from the SampleWinApp2 Windows application.

This completes the task of running SampleWinApp2 Windows application.

20.5 WEB-BASED APPLICATION ON .NET

In Microsoft Visual Studio, we can use ASP.NET technology and a programming language such as C# to create a Web based application. The controls available in the **Toolbox** of the Microsoft Visual Studio

IDE help us to create the user interface of a Web based application. For example, we can create a Web based application to accept name and city from the user and when the user clicks a button, the entered name and city can be displayed.

20.5.1 Creating and Running a Website1 Web-based Application

Website1 Web based application allows users to enter their information, such as name and address. When the user clicks a button on the **Default.aspx** page of the Website1 Web-based application, another page appears displaying the entered information. You can run the Website1 Web-based application by pressing the **F5** key or the **Start Debugging** button of the toolbar, which is present in the Microsoft Visual Studio IDE.

20.5.2 Creating a Website1 Web-based Application

A Website1 Web based application is an ASP.NET website containing pages such as **Default.aspx** and **show.aspx**. This is a web application in C# using ASP.NET. This website makes uses of **label, textbox, check box, radio button, dropdownlist, button** and **hyperlink** controls. The form checks that all the values are inserted and displays the data in another page. To create the Website1 Web-based application, we must:

- Create the **Default.aspx** page
- Add code for the **Default.aspx** page
- Create the **show.aspx** page
- Add code to the **show.aspx** page
- Create the **terms.aspx** page

Creating the Default.aspx Page

The **Default. aspx** page of the Website1 Web-based application allows a user to enter details such as name and address and click a button to submit the details. To create the **Default.aspx page**:

1. Open Microsoft Visual Studio IDE.
2. Select *File-> New->Web Site* to display the **New Web Site** dialog box as shown in Fig. 20.25.

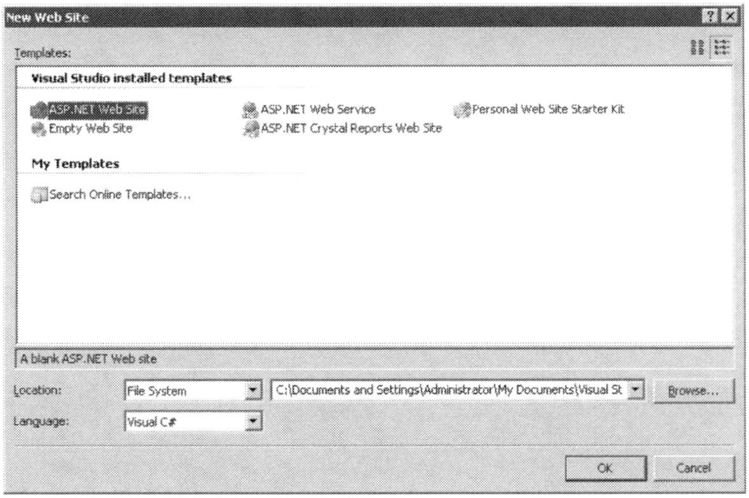

Figure 20.25 *The new web site dialog box*

Note: By default, the ASP.NET Web Site template appears selected and the Name text box contains *C:\Documents and Settings\Administrator\My Documents\Visual Studio 2005\WebSites\WebSite1* to specify the location and name for the Web based application. We can change the name of the Web-based application as per requirement. By default, Visual Basic appears in the Language list. When you select Visual C# for a specific Web-based application then after that Visual C# appears automatically in the Language list.

3. Accept the default settings and click OK to close the **New Web Site** dialog box. The Microsoft Visual Studio IDE appears with the Source view of the **Default.aspx** page, which is the default page of the Web Site1 Web based application as shown in Fig. 20.26.

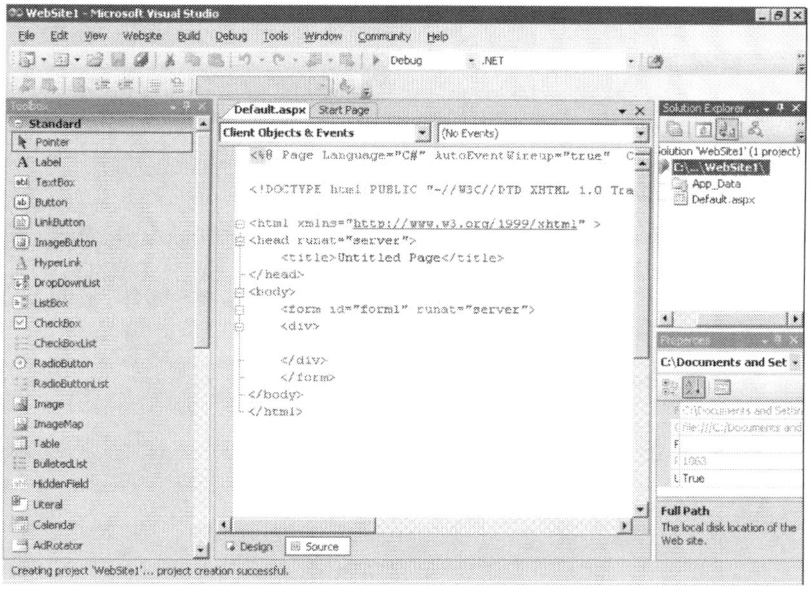

Figure 20.26 *The source view of the Default.aspx page*

4. Change the value of the title element that appears in the **Source View** to *Using Controls in* ASP.NET and *C#.*
5. Click the **Design** tab that appears at the bottom to display the **Design view** of the **Default.aspx** page as shown in Fig. 20.27.

Note: When we open the Design view of the **Default.aspx** page, the **Toolbox** containing controls such as **Button** and **Label** appears automatically.

6. Drag a **Label Control** from **Toolbox** on to the **Default.aspx** page to add the control to the page.
7. Change the value of the **Text** property for the **Label** control to **Using Controls** in ASP.NET and C# using the **Properties** window.
8. Change the **BackColor** property of the **Label** control using the **Properties** window as required.
9. Similarly, add four more **Label** controls to the **Default.aspx** page.
10. Change the **Text Property** of the first **Label** control that you have added to *Name* using the **Properties** dialog box.
11. Change the **Text Property** of the second **Label** control that you have added to *Address* using the **Properties** dialog box.

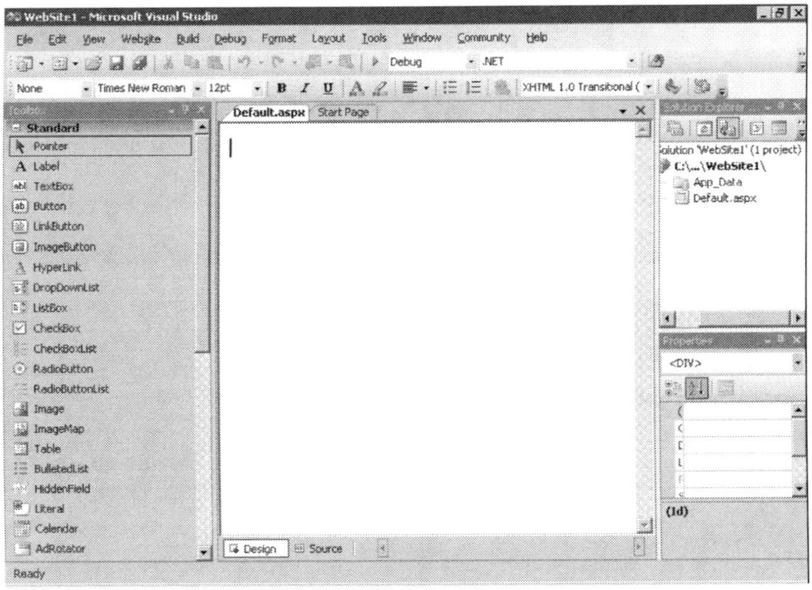

Figure 20.27 *The design view of the Default.aspx page.*

12. Change the **Text Property** of the third **Label** control that you have added to *City* using the **Properties** dialog box.
13. Change the **Text Property** of the first **Label** control that you have added to *Gender* using the **Properties** dialog box.
14. Drag a **TextBox** control from the **Toolbox** on to **Default.aspx** page to add the control to the page.
15. Similarly add another **TextBox** control to the **Default.aspx** page.
16. Drag a **DropDownList** control on to the **Default.aspx** page to add the control to the page.
17. Click the **Items Property** in the **Properties** window to display an ellipse button as shown in Fig. 20.28.

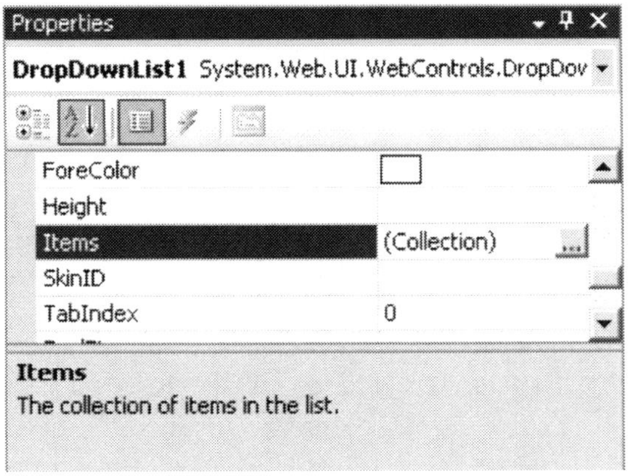

Figure 20.28 *The properties window for DropDownList control*

18. Click the ellipse button to display the **ListItem Collection Editor** as shown in Fig. 20.29.

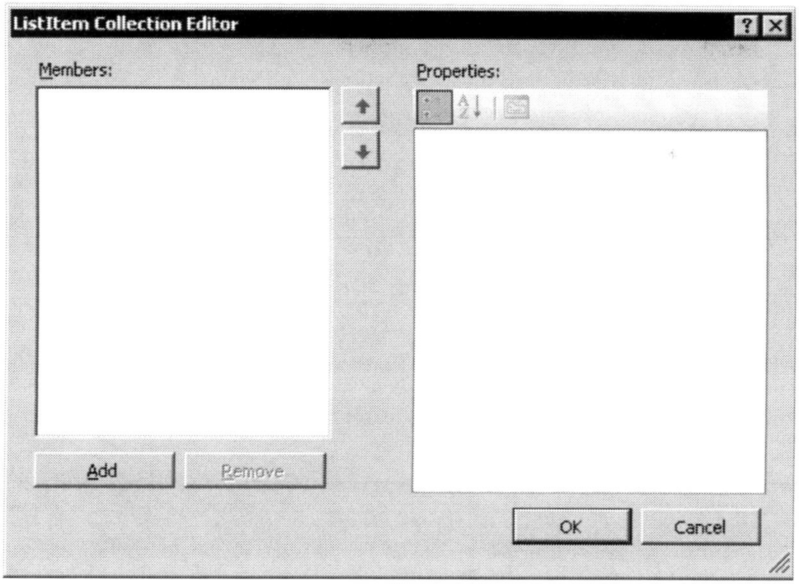

Figure 20.29 *The ListItem Collection Editor*

19. Click the **Add button** to add an item to the **DropDownList** control. Fig. 20.30 shows the **List Collection Editor** with an item added for the **DropDownList** control.

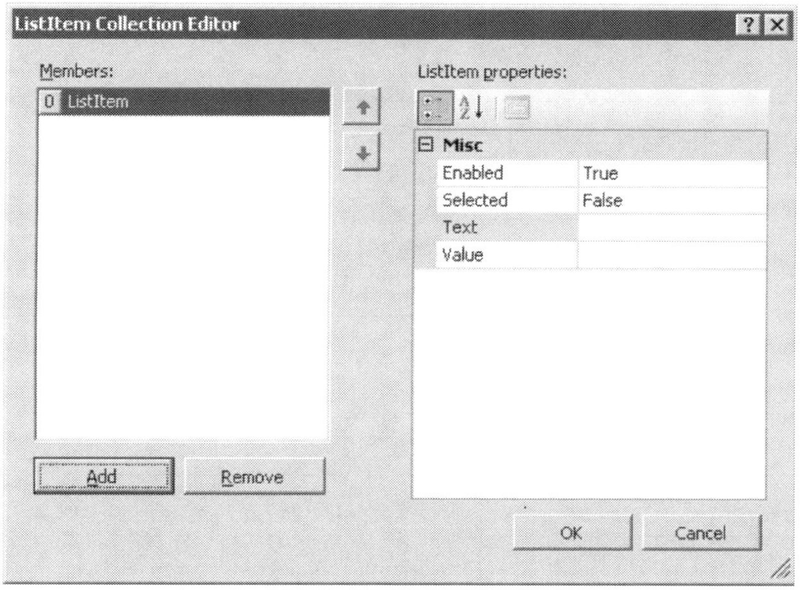

Figure 20.30 *The ListItem Collection Editor with added item*

20. Enter a text such as *Select* in the editable text box next to the **Text property** to specify the text for the new item.
21. Similarly, add items such as *Bangalore*, *Delhi*, *Calcutta* and *Chennai* to the **DropDownList** control.
22. Click the OK button to close the **ListItem Collection Editor**.
23. Drag a **RadioButton** control from the **Toolbox** on to the **Default.aspx** page to add the control to the page.
24. Similarly, add another **RadioButton** control to the **Default.aspx** page.
25. Change the value of the **Text property** of the first **RadioButton** control that you have added to *Male*.
26. Change the value of the **Text property** for the second **RadioButton** control that you have added to *Female*.
27. Drag a **Hyperlink** control, which is available in the **Toolbox** on to the **Default.aspx** page to add the control to the page.
28. Change the value of the **Text property** for the **Hyperlink** control to *Terms & Conditions* using the Properties window.
29. Drag a **CheckBox** control from the **Toolbox** on to the **Default.aspx** page to add the control to the page.
30. Change the value of **Text property** for the **CheckBox** control to *Read Terms*.
31. Drag a **Button** control from the **Toolbox** on to the **Default.aspx** page to add the control to the page.
32. Change the value of the **Text property** for the **Button control** to *Submit Form* using the **Properties** window.
33. Change the **BackColor property** for the **Button** control using the *Properties* window.
34. Add a **Label** control to the **Default.aspx** page and change its **ForeColor** property using the **Properties** window as per requirement. This completes the task of creating the **Default.aspx** page. Figure 20.31 shows the **Default.aspx** page with added controls:

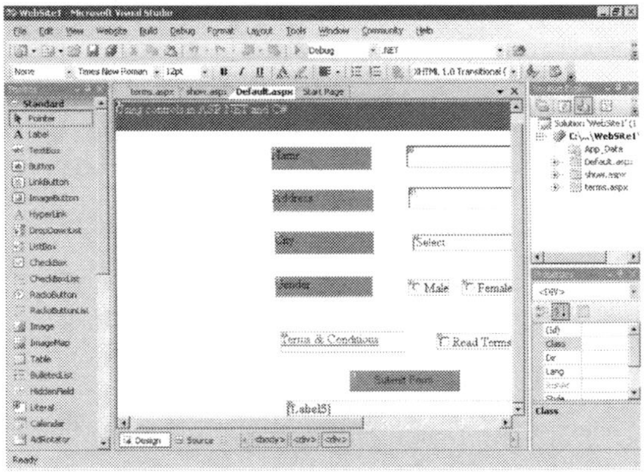

Figure 20.31 *The design view Default.aspx page for Website1 application*

Note: The **Hyperlink** control is available in Microsoft Visual Studio 2005 when we create an ASP.NET Web site. This control allows us to create a hyperlink in a Web-based application to navigate between the pages included in the application.

Adding Code to the Default.aspx Page

We need to add code to the **Default.aspx** page to specify the functionality such as hiding a label and initialising values for controls added to the **Default.aspx** page. To add code to the **Default.aspx** page:

1. Right click the **Default.aspx** node that appears in the **Solution Explorer** to display a shortcut menu.
2. Select the **View Code** option to display the **Code Editor** with the **Default.aspx.cs** file containing the **Page_Load** event handler as shown in Fig. 20.32.

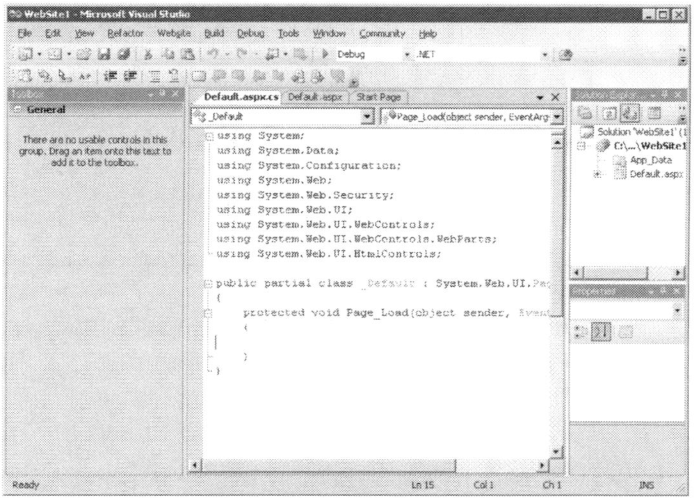

Figure 20.32 *The code editor window with Default.aspx.cs file*

3. Enter the following code in the **Page_Load** event handler to specify the functionality for hiding the label that appears at the end in the **Default.aspx** page and initialising the **TextBox** and **DropDownList** controls.

```
Label5.Visible = false;
   if (!IsPostBack)
   {
      TextBox1.Text = “ ”;
      TextBox2.Text = “ ”;
      DropDownList1.SelectedValue = “Select”;
   }
```

4. Click the **Close** button that appears at the top in the **Code Editor** to close the **Code Editor**. The **Default.aspx** page in **Design** view appears.

Note: If the **Default.aspx** page appears in the **Source** view after closing the **Code Editor** then we can click the **Design** tab to display the **Design** view of the **Default.aspx** page.

5. Double click the **Button Control** with the text *Submit Form* to display the **Code Editor** with **Default.aspx.cs** file containing the **Button1_Click** event handler, as shown in Fig. 20.33.

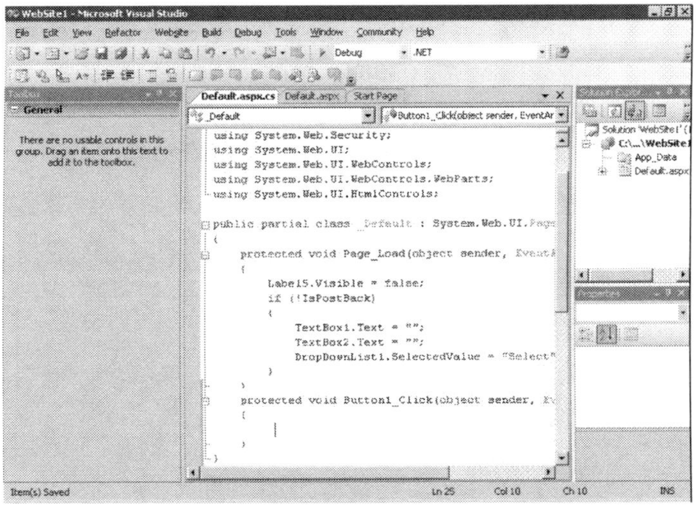

Figure 20.33 *The code editor with Button_Click handler*

6. Enter the following code in the **Button_Click** event handler to specify the functionality for displaying the label, which appears at the end and checking the value of the controls such as **RadioButton** and **TextBox** to determine whether they are empty or not.

Program 20.5 Code for Button-Click handler

```
Label5.Visible = true;
   if (TextBox1.Text == "" || TextBox2.Text == "" ||
DropDownList1.SelectedValue == "Select" || (RadioButton1.Checked == false &&
RadioButton2.Checked == false) || CheckBox1.Checked == false)
   {

   Label5.Text = "All values are required";
}
else
{
   if (RadioButton1.Checked==true)
      Response.Redirect("show.aspx?Name=" + TextBox1.Text+ "&Address="
+TextBox2.Text+"&City="+DropDownList1.SelectedValue+"&Gender="+RadioButton1.Text);
      else
      Response.Redirect("show.aspx?Name=" + TextBox1.Text + "&Address=" +
TextBox2.Text + "&City=" + DropDownList1.SelectedValue + "&Gender=" +
RadioButton2.Text);
      }
```

This completes the process of adding code to the **Default.aspx** page.

Creating the show.aspx Page

The **show.aspx** page displays the values entered by a user in the **Default.aspx** page of the Website1 Web-based application. To create the **show.aspx** page:

1. Right click the *C:\...\Website1\ node* that appears in the **Solution Explorer** to display a shortcut menu.
2. Select the **Add New Item** option to display **Add New Item** dialog box as shown in Fig. 20.34.

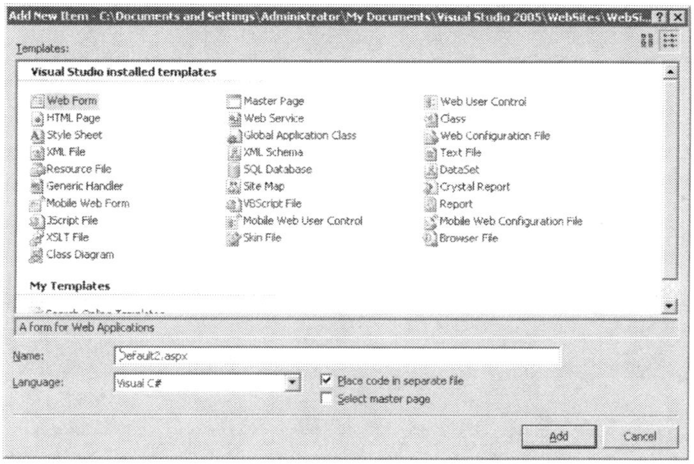

Figure 20.34 *The add new item dialog box for adding show.aspx page*

3. Enter a name such as **show.aspx** in the **Name** text box to specify a name for new page, which you need to add to Website1 Web-based application.
4. Click **Add** to add a new page with the name **show.aspx** to the Website1 Web-based application. The **Source View** for the **show.aspx** page appears in the Microsoft Visual Studio IDE.
5. Change the value of the title HTML element that appears in the **Source** view to Show the details.
6. Click the **Design** tab that appears at the bottom in the Microsoft Visual Studio 2005 IDE to display the **Design** view for the show.aspx page.
7. Drag a **Label** control from the **Toolbox** on to the **show.aspx** page to add the control to the page.
8. Change the **ForeColor** property of the **Label** control using the **Properties** window.
9. Change the value of the **Text Property** for the **Label** control to *Values entered in the previous form are*:
10. Drag a **Button** control from the **Toolbox** on to the **show.aspx** page to add the control to the page.
11. Change the value of the **Text property** for the **Button** control to *Back*. This completes the task of creating the **show.aspx** page.

Adding Code to the show.aspx Page

We need to add code to the **show.aspx** page to specify the functionality for displaying the values entered by a user in **Default.aspx** page. To add code to the **show.aspx** page:

1. Click the Source tab that appears at the bottom in the Microsoft Visual Studio 2005 IDE to display the **Source** view of the **show.aspx** page.
2. Add the following code to the **Default.aspx** page to specify the functionality of displaying the values entered in the **Default.aspx** page:

```
<font color="maroon"><strong>
        <%string name = ""; string address = ""; string city = ""; string
gender = "";
        name = Request.QueryString.Get("Name");
        address = Request.QueryString.Get("Address");
        city = Request.QueryString.Get("City");
        gender = Request.QueryString.Get("Gender");
        Response.Write("Name is : " +name+"<br>");

        Response.Write("Address is : " +address+"<br>");
        Response.Write("City is : " + city + "<br>");
        Response.Write("Gender is : " + gender + "<br>");

        %></strong></font><br />
```

3. Modify the ASP code for the **Button** control by replacing it with following code to specify functionality for navigating back to **Default.aspx** page.

```
<asp:Button ID="Button1" runat="server" PostBackUrl="~/Default.aspx"
Text="Back"
        Width="114px" OnClick="Button1_Click" />
```

This completes the process of adding code to the **show.aspx** page. Figure 20.35 shows the **Source** view of the **show.aspx** page with added code.

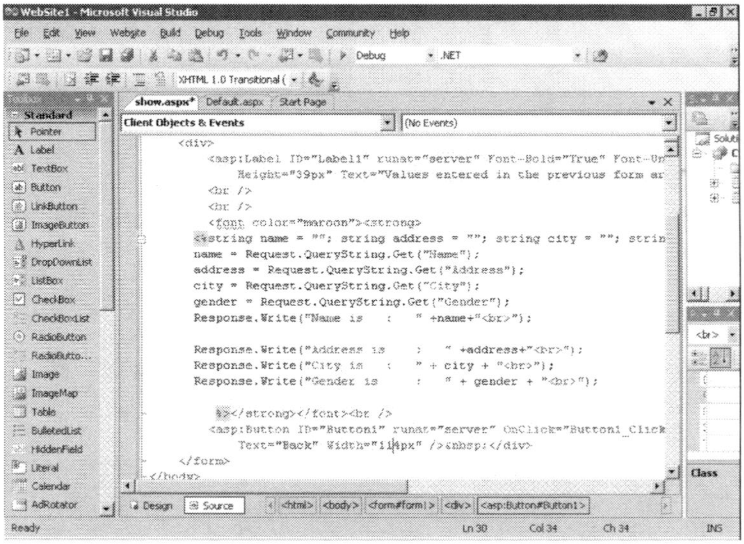

Figure 20.35 *Source view with code added for show.aspx page*

Creating the terms.aspx Page

The **terms.aspx** page allows you to view the terms and conditions related to the ASP.NET Web Site created as Web-based application. To create the **terms.aspx** page:

1. Add a new page with the name **terms.aspx** to the Website1 Web based application. The **Source** view of the **terms.aspx** page appears.
2. Change the value of title HTML element that appears in the Source view to *Terms and Conditions Page*.
3. Add the following code below the div HTML element to display a text string in the **terms.aspx** page:

```
<font color="blue" face="Verdana"><strong>The terms and conditions
of the company is described here.</strong></font>
```

This completes the process of creating the **terms.aspx** page. Figure 20.36 shows the **Source** view of the **terms.aspx** page with added code.

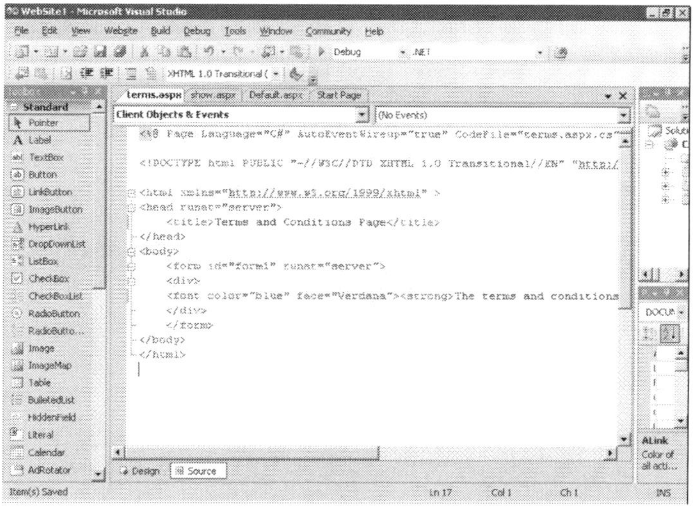

Figure 20.36 *The source view with code added for terms.aspx page*

Running the Website1 Web-based Application

After creating the Website1 Web-based application, you need to run the application to test whether or not it is generating the desired output. To run the Website1 Web-based application:

1. Press **Ctrl+F5** to initiate the process of running the Website1 Web-based application. The **Default.aspx** page appears as shown in Fig. 20.37.
2. Enter a name such as *Nikita* in the **Name** text box and address, such as *23, Anjali Apts. Gandhi Colony* in the **Address** text box.

Note: When a user does not enter any value in the **Default.aspx** page of the Website1 Web-based application then an error message appears at the bottom.

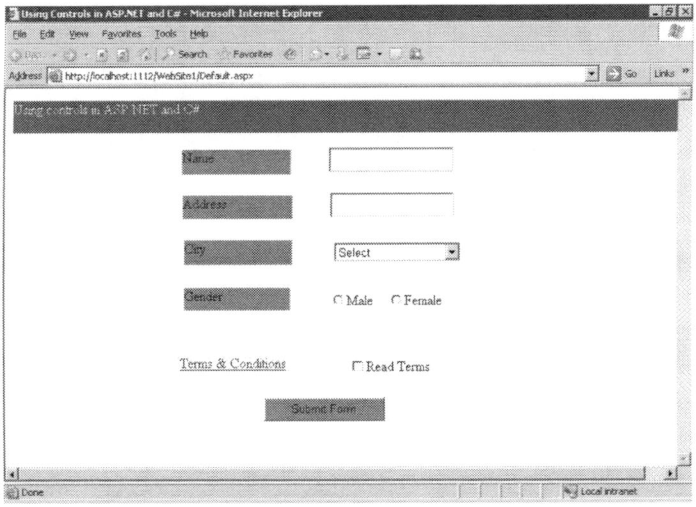

Figure 20.37 *The Default.aspx Page*

3. Click the arrow that appears in the **City drop down list** to display a list of cities.
4. Select a city such as *Calcutta* to specify a city.
5. Select the *Read Terms* check box to specify that *Terms* must be displayed for reading.
6. Click the *Submit Form* button to display the **show.aspx** page with the values entered in the **Default.aspx** page as shown in Fig. 20.38.

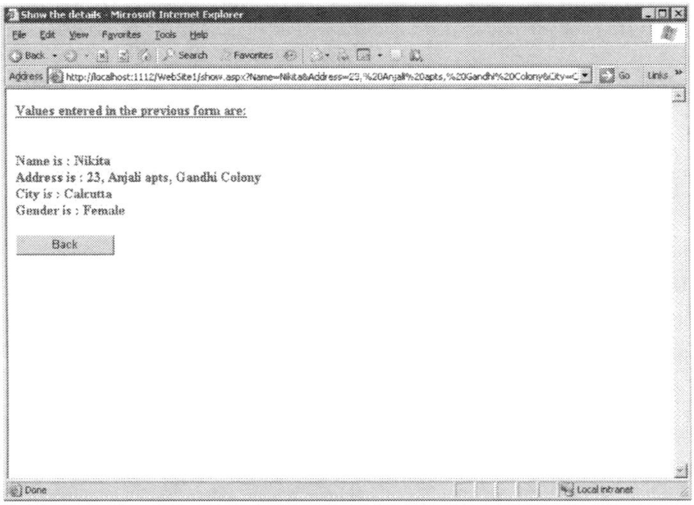

Figure 20.38 *The show.aspx page*

7. Click the **Back** button to navigate back to the **Default.aspx** page.

Note: In the **Default.aspx** page, you can click the *Terms & Conditions* hyperlink to display the **terms.aspx** page.

Creating and Running the Website2 Web-based Application

The Website2 application allows a user to enter a name and select a city. When the user clicks a button, the entered values are displayed. We can run the Website2 Web-based application in the same way as we have run the Website1 Web based application using the **F5** key.

Creating the Website2 Web-based Application

WebSite2 Web-based application is also an ASP.NET Web site that contains a single page **Default.aspx** page. This is a web application in C# using ASP.NET and uses **label, button, textbox, dropdownlist** controls. The form validates if the name and the city are inputted. If the user has entered the required information, the program displays it on the same page.To create the Website2 Web-based application:

- Create the **Default.aspx** page.
- Add code to **Default.aspx** page.

Creating the Default.aspx Page

The **Default.aspx** page allows a user to enter a name and select a city. After the user clicks a button, the entered values are displayed in the **Default.aspx** page. To create the **Default.aspx** page:

1. Open the Microsoft Visual Studio IDE.
2. Select *File->New->Website* to display the New Web Site to display the New Web Site dialog box.

Note: By default, the ASP.NET Web Site template appears selected and the **Name** text box contains *C:\Documents and Settings\Administrator\My Documents\Visual Studio 2005\WebSites\WebSite2* to specify the location and name for the Web based application. You can change the name of the Web-based application as per requirement.

3. Accept the default settings and click OK to close the **New Web Site** dialog box. The Microsoft Visual Studio IDE appears with the **Source** view of the **Default.aspx** page for the Website2 Web-based application.
4. Change the value of the title HTML element to ASP.NET with C# (Form and Controls).
5. Click the **Design** tab to display the **Design** view for the **Default.aspx** page.
6. Add three **Label** controls to the **Default.aspx** page.
7. Select the first Label control, which you have added to the **Default.aspx** page and change the **ForeColor** property of the control, using the **Properties** window.
8. Change the value of **Text property** for the first **Label** control to *This is an example of simple form validation in* ASP.NET and *C#* using the **Properties** window.
9. Expand the **Font** node in the **Properties** window to display the **Bold** option.
10. Click the editable text box that appears next to the **Bold** option to display a list with two values **True** and **False**.
11. Select the **True** value from the list to make the text appearing in the first **Label** control as bold.
12. Select the second **Label** control added to the **Default.aspx** page and change its **Text** property to *Enter your Name*.
13. Select the third **Label** control added to the **Default.aspx** page and change its **Text** property to *Select your City*.

14. Add a **TextBox** control to the **Default.aspx** page using the **Toolbox**.
15. Add a **DropDownList** control to the **Default.aspx** page using the **Toolbox.**
16. Add items such as *Select, Bangalore, Calcutta* and *Chennai* to the **DropDownList** control using the **Items** property available in the **Properties** window.
17. Add a **Button** control to the **Default.aspx** page.
18. Change the value of the **Text property** for the **Button** control to *Submit Values* using the **Properties** window.
19. Change the **BackColor** and **BorderColor** properties of the **Button** control using the **Properties** window.
20. Add a **Label** control to the **Default.aspx** page and change its **Text property** to **Label** using the **Properties** window.
21. Change the **ForeColor** property of the **Label** control using the **Properties** window. This completes the process of creating the **Default.aspx** page. Figure 20.39 shows the **Default.aspx** page with added controls.

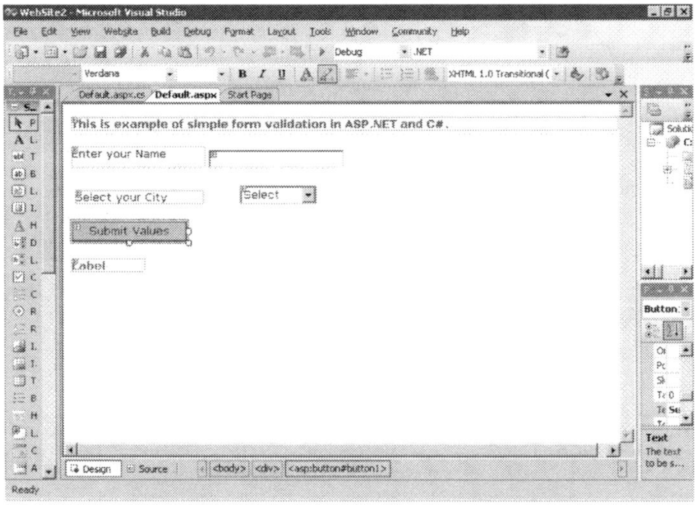

Figure 20.39 *The design view of website2 web-based application*

Adding Code to Default.aspx Page of the Website2 Web-based Application

We need to add code to **Default.aspx** page to specify the functionality for displaying the name and city entered by a user. To add code to the Website2 Web-based application:

1. Right click the **Default.aspx** node that appears in the **Solution Explorer** to display a shortcut menu.
2. Select the **View Code** option to display the **Code Editor** with the **Default.aspx.cs** file containing the **Page_Load** event handler.
3. Enter the following code in the **Page_Load** event handler to specify the functionality for hiding the label, which appears in the end in the **Default.aspx** page.

```
Label3.Visible = false;
```

4. Click the **Close** button that appears at the top in the **Code Editor** to close the **Editor**. The **Default.aspx** page in **Design** view appears.

Note: If the **Default.aspx** page appears in the **Source** view after closing the **Code Editor** then you can click the **Design** tab to display the **Design** view of the **Default.aspx** page.

5. Double click the **Button** control with the text *Submit Values* to display the **Code Editor** with Button1_Click event handler.
6. Enter the following code in the **Button1_Click** event handler to specify the functionality for displaying the values entered by a user.

```
Label3.Visible = true;
    if (TextBox1.Text == "" || DropDownList1.SelectedValue.Contains("0"))

        Label3.Text = "Please enter all the values.";

    else

        Label3.Text = "You have entered your Name as " + TextBox1.Text + "
and City as " + DropDownList1.SelectedValue.ToString();
```

This completes the process of adding code to the **Default.aspx** page of the Website2 Web-based application.

Running the Website2 Web-based Application

After completing the task of creating the Website2 Web-based application, we can run the application to test whether or not it is producing the desired result. To run the Website2 Web-based application:

1. Press **Ctrl+F5** to initiate the process of running the Website2 Web-based application. The **Default.aspx** page for the Website2 Web-based application appears as shown in Fig. 20.40.

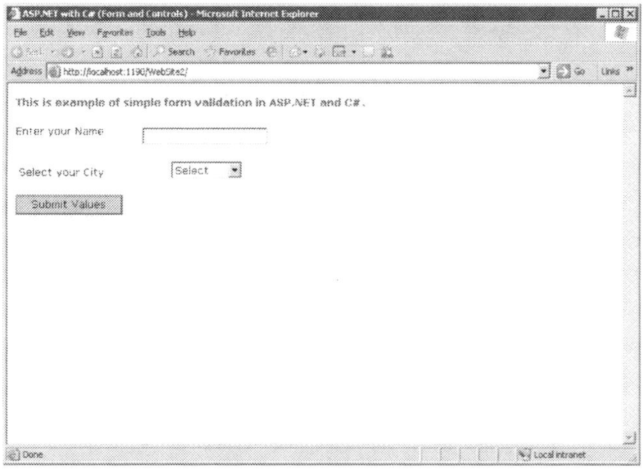

Figure 20.40 *The Default.aspx page for the website2 web-based application*

2. Enter a name such as *Shivani* in the *Enter your Name* text box.
3. Click the arrow that appears in the *Select your City* drop down list to display a list of cities.
4. Select a city such as *Mumbai* from the list and click the *Submit Values* button to display text string with entered values as shown in Fig. 20.41.

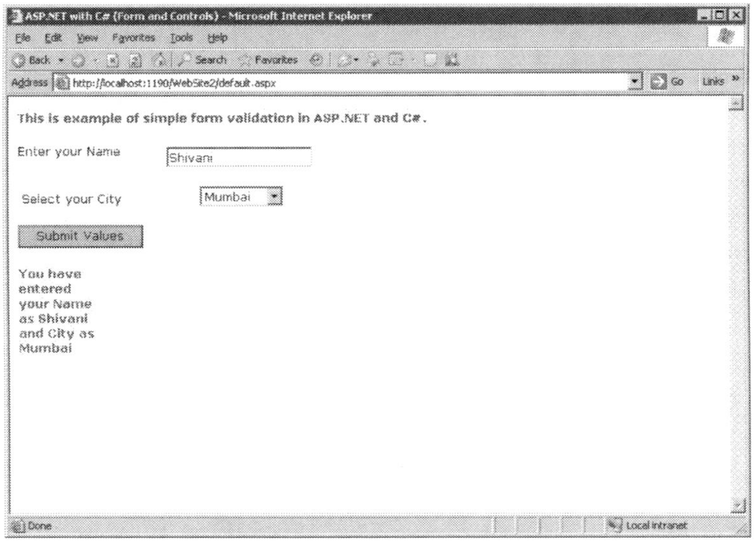

Figure 20.41 *Displaying values entered by a user*

This completes the process of running the Website 2 Web-based application.

Case Study

Problem Statement

E-Store is an online store available on the Internet that is involved in the sale of household products such as jams and sauces. Users surfing the Internet are currently able to only view the products sold by E-Store. The management of the organisation, which has hosted the E-Store online store on the Internet, has now decided that the users must have the facility of selecting a product from the E-Store and placing it in a shopping chart. It is important that the users have the facility of a shopping chart while surfing the E-Store online store because it will encourage more customers to purchase products from E-Store. How can the management add the functionality of a shopping chart to the E-Store?

Solution

In order to add the facility of a shopping chart to the E-Store online store, the organisation which has hosted E-Store on the Internet, contacts Web Solutions Private Limited. Web Solutions Private Limited is involved in creating Web based applications, which can be hosted on the

Internet. The programmers at Web Solutions Private Limited first design a page **Default.aspx** to allow users to enter their details such as name and address and select products for purchase using Microsoft Visual Studio 2005 as shown below.

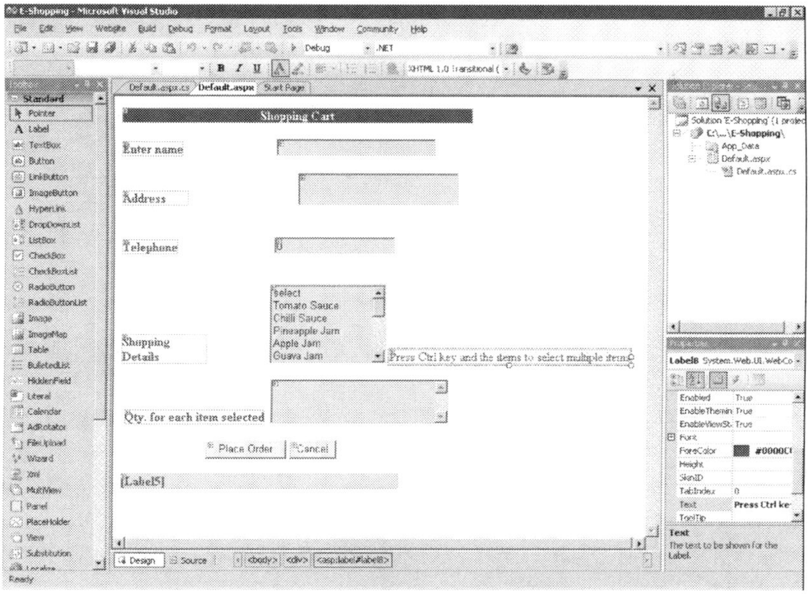

Figure 20.42 *The online shopping chart of E-store*

In addition, programmers at Web Solutions Private Limited added the following code to the Web page:

```
using System;
using System.Data;
using System.Configuration;
using System.Web;
using System.Web.Security;
using System.Web.UI;
using System.Web.UI.WebControls;
using System.Web.UI.WebControls.WebParts;
using System.Web.UI.HtmlControls;

public partial class _Default : System.Web.UI.Page
{
   protected void Page_Load(object sender, EventArgs e)
    {
      Label5.Visible = false;
    }
   protected void Button2_Click(object sender, EventArgs e)
```

```
{
    TextBox1.Text = "";
    TextBox2.Text = "";
    TextBox3.Text = "0";
    ListBox1.SelectedItem.Value = "select";
    TextBox4.Text = "";
}
protected void Button1_Click(object sender, EventArgs e)
{
    double telephone=0;
    string msg = "";
    try
    {
        telephone = Int32.Parse(TextBox3.Text);
    }
    catch(Exception ee)
    {
        Label5.Text ="Error is " + ee.Message;
    }

    Label5.Visible = true;

    if (TextBox1.Text == "")
    {
        Label5.Text = "Please enter your Full Name.";
    }
    else if (TextBox2.Text == "")
    {
        Label5.Text = "Please enter your Address.";
    }
    else if (telephone.Equals("") || telephone.Equals(0))
    {
        Label5.Text = "Please enter your resdence telephone or mobile no.";
    }
    else if (ListBox1.Text == "" || ListBox1.Text == "select")
    {
        Label5.Text = "Shopping item detail is required";
    }
    else
    {
        telephone = Int64.Parse(TextBox3.Text);
        if (ListBox1.Items[1].Selected)
        {
```

```
    msg = ListBox1.Items[1].Text + "<br />";

   }

if (ListBox1.Items[2].Selected)
   {

    msg = msg + ListBox1.Items[2].Text + "<br />";

   }

if (ListBox1.Items[3].Selected)
   {

    msg = msg + ListBox1.Items[3].Text + "<br />";

   }

if (ListBox1.Items[4].Selected)
   {

    msg = msg + ListBox1.Items[4].Text + "<br />";

   }

if (ListBox1.Items[5].Selected)
   {

    msg = msg + ListBox1.Items[5].Text + "<br />";

   }

if (ListBox1.Items[6].Selected)
   {

    msg = msg + ListBox1.Items[6].Text + "<br />";

   }
if (ListBox1.Items[7].Selected)
   {

    msg = msg + ListBox1.Items[7].Text + "<br />";

   }
if (ListBox1.Items[8].Selected)
```

```
    {

      msg = msg + ListBox1.Items[8].Text + "<br />";

    }
 if (ListBox1.Items[9].Selected)
    {

      msg = msg + ListBox1.Items[9].Text + "<br />";

    }
 if (ListBox1.Items[10].Selected)
    {

      msg = msg + ListBox1.Items[10].Text + "<br />";

    }
 if (ListBox1.Items[11].Selected)
    {

      msg = msg + ListBox1.Items[11].Text + "<br />";

    }
 if (ListBox1.Items[12].Selected)
    {

      msg = msg + ListBox1.Items[12].Text + "<br />";

    }
 if (ListBox1.Items[13].Selected)
    {

      msg = msg + ListBox1.Items[13].Text + "<br />";

    }
 if (ListBox1.Items[14].Selected)
    {

      msg = msg + ListBox1.Items[14].Text + "<br />";

    }
 if (ListBox1.Items[15].Selected)
    {
```

```
              msg = msg + ListBox1.Items[15].Text + "<br />";

             }
          if (ListBox1.Items[16].Selected)
             {

              msg = msg + ListBox1.Items[16].Text + "<br />";

             }

          Label5.Text = "Thank you for entering your details.<br/>You have entered the
following details :<br/>Name " + TextBox1.Text + "<br />Address : " + TextBox2.Text + "<br/
>Telephone : " + telephone + "<br/>Shopping Items to deliver: <br/>" + msg + "<br/><br/>";
           if (TextBox4.Text == "")
             {
             Label5.Text = Label5.Text + "<br/>As you have not entered any quantity so
each Item to be delivered will be of 1 quantity.<br/>";
             }
            else
             {
             Label5.Text = Label5.Text + "<br/> Quantities of items entered are " +
TextBox4.Text +"<br/>";
             }
          Label5.Text = Label5.Text + "Your shopping items will be delivered within 6
hours.";

        }



        }
```

The code added to the **Default.aspx** page helps specify the functionality of displaying details such as name and address entered by a user. A message is also displayed to indicate that the product ordered by the user will be delivered within six hours.

Common Programming Errors

- *Typecasting is a common programming error, which may occur when running an ASP.NET/C# program.*
- *Error occurs when a programmer deletes the event handler such as **button1_click** for a control from the .cs or **.aspx.cs** file but does not delete the reference of the event handler from the **.Designer.cs** file or **.aspx file.***

- *The runat="server" syntax not added while changing code in the **.aspx** file.*
- *A programmer may copy and paste a C# 2003 file in a C# 2005 Windows **.cs** file without knowing that the following statements need to be declared in the **.Designer.cs class:***

```
private System.ComponentModel.IContainer components = null;.
protected override void Dispose(bool disposing)
private void InitializeComponent() {}
```

- *A programmer may be trying to execute a program without calling the classes, which are required for program execution.*
- *A programmer may be trying to run a program without using exceptions, which are required for the processes in a program. This can result in crashing of the program.*

Review Questions

20.1 What is Microsoft Visual Studio IDE?
20.2 What are controls?
20.3 What is the use of creating a Windows based application?
20.4 Why do you need to create a Web-based application?
20.5 Explain two differences between a Windows based application and a Web-based application.
20.6 What is the use of the **Properties** window?
20.7 Explain in two three lines the use of **Solution Explorer**.
20.8 How can you create a new project in Microsoft Visual Studio window?

Programming Exercises

20.1 Write a windows application in C#, having two textbox controls and a button control, validate that the textbox values are not blank. Clicking the button will interchange the textbox values and the button disappears.

20.2 Write an ASP.NET/C# application with two textbox controls (Name and Telephone) and a button control. After submitting the form, the application redirects to another page and displays the Name and Telephone in it in Blue Color.

*20.3 Write an ASP.NET/C# application with a textbox username, textbox password and a button. Entering "admin" as username and password as "aspnetc#" will take you to another page with a welcome message written in blue-green. Entering the wrong password will take you to another page called **error.aspx** and show you the message wrong password.*

Minor Project

RANDOM NUMBER APPLICATION

The **Random Number** application is a console-based application that generates a set of random numbers which lie in the range from 0 to any number inputted by the user. It also asks for user input about the number of random numbers to be generated in the given range. The program utilizes the various loops, such as **while** and **for**. It also demonstrates the use of Random in-built class of C# for generating random numbers. In addition, the application checks for various exceptions that may occur during program execution.

```
using System;
using System.Collections;
using System.Collections.Generic;
using System.Text;

namespace RandomNoApplication
{

    interface Values
    {
        void show(Stack s);
    }
    interface GenerateRandomValues
    {
        void Display();
    }
```

```
    public class Class1 : Values
    {
        public static int num, valuerange;

        public void show(Stack s)
        {
            for (int i = 1; i <= num; i++)
            {
                System.Console.WriteLine(s.Pop());
            }
        }
    }
public class Class2 : Class1, GenerateRandomValues
{
    //Method overloading from the Interface GenerateRandomValues
    public void Display()
    {
        valuerange = 0;
        num = 1;
        try
        {
            while (valuerange < num)
            {
                System.Console.WriteLine("============================");
                System.Console.WriteLine("Random number generator");
                System.Console.WriteLine("============================");
                System.Console.Write("Enter the number of random numbers to be generated :
");
                num = Convert.ToInt32(System.Console.ReadLine());
                System.Console.Write("Enter the range within which the numbers to be
generated: ");
                valuerange = Convert.ToInt32(System.Console.ReadLine());
            }
        }
        catch (FormatException fe)
        {
        Console.WriteLine("Cannot enter null values " + fe.Message);
        Console.WriteLine("Press key to exit or enter some value.");
        Environment.Exit(0);
```

```
    }
    catch (ArgumentNullException ane)
    {
        Console.WriteLine(ane.Message);
        Console.WriteLine("Press key to exit or enter some value.");
        Environment.Exit(0);
    }
    catch (ArithmeticException ae)
    {
        Console.WriteLine(ae.Message);
        Console.WriteLine("Press key to exit or enter some value.");
        Environment.Exit(0);
    }
    catch (Exception ee)
    {
        Console.WriteLine(ee.Message);
        Console.WriteLine("Press key to exit or enter some value.");
        Environment.Exit(0);
    }
}

//Method to generate the random numbers from the RanDomNo
public void GenerateNumbers(Stack s, RandomiseValues rn)
{
    bool exists = true;
    int m;
    int val;

    for (m = 1; m <= num; m++)
        {
            val = rn.Throw(valuerange); //Roll the Dice
            exists = s.Contains(val);
            while (exists == true) //Reroll the Dice to remove duplicate numbers
            {
                val = rn.Throw(valuerange);
                exists = s.Contains(val);
            }
            s.Push(val);
        }
    }
```

```
}
//The Main class calling the all the classes
class RandomNoClass
{
    static int Main(string[] args)
    {
        Stack s = new Stack();
        RandomiseValues rv = new RandomiseValues();

        Class2 c2 = new Class2();

            //Display the Menu
            c2.Display();

            //Generate the random numbers
            c2.GenerateNumbers(s, rv);

            //Display the random numbers generated.
            Console.WriteLine(Class1.num + " Random Numbers Generated are in the range of "
+ Class1.valuerange);
            c2.show(s);
            return 0;
        }
    }

    //Calss to generate the Random Numbers
    public class RandomiseValues
    {
        public int Throw(int val)
        {
            Random r = new Random();
            int i = 0;
            while (i == 0)
            {
                i = r.Next(val + 1);//Get another random number but
            } //discard zero values
            return i;
        }
    }
}
```

C:\WINNT\system32\cmd.exe

```
============================
Enter the number of random numbers to be generated : 20
Enter the range within which the numbers to be generated: 2000
20 Random Numbers Generated are in the range of 2000
145
1428
1755
1037
1365
647
1930
256
1539
821
1148
431
758
40
1323
1650
932
214
542
1825
Press any key to continue . . .
```

Major Project

BASICIO EXAMPLE

The given application presents the ordering system of a garment shop. The user can select the following options from three sets of menus available as follows:

- Type of Item
- Color Choice
- Type of Delivery

The user also inputs the quantity and item price, based on which the system generates a summary sheet containing the delivery details. The summary includes the information entered by the user and the total price calculated based on the quantity and price information previously entered.

The application is console-based, which uses various concepts of C# to achieve the functionality. There are 3 classes in the project, namely, **Items, InputOutputExample** and **ProcessOrder** which are contained in BasicIOExample namespace. The project uses enumerations to record the user-entered data. The application also applies exception handling to the code and uses methods for processing.

```
using System;
using System.Collections.Generic;
using System.Text;

namespace BasicIOExample
{
    class InputOutputExample
    {
```

```
        static void Main(string[] args)
        {
            ProcessOrder order = new ProcessOrder();

            order.OrderPlaced();
            Console.WriteLine();

            order.OrderDetails();
            Console.WriteLine();

        }
    }
}



using System;
using System.Collections.Generic;
using System.Text;

namespace BasicIOExample
{
    public enum ItemType
    {
        Shirts = 1,
        Trousers,
        TShirts,
        SafariSuits,
        Jeans,
        Polos,
        Jackets
    }

    public enum ItemColor
    {
        Red = 1,
        White,
        Yellow,
        Maroon,
        Orange,
        Blue,
        Lavender,
```

```
        Black
    }

    public enum ItemDelivery
    {
        Packet = 1,
        SoftCase,
        Box,
        Other
    }

    class Items
    {
        public ItemType typeofitem;
        public ItemColor itemcolor;
        public ItemDelivery itemdeliver;
        public decimal itemprice;

        public Items()
        {
            typeofitem = ItemType.Jackets;
            itemcolor = ItemColor.Black;
            itemdeliver = ItemDelivery.Packet;
            itemprice = 0.00M;
        }
        public Items(ItemType type)
        {
            typeofitem = type;
            itemcolor = ItemColor.Black;
            itemdeliver = ItemDelivery.Packet;
            itemprice = 0.00M;
        }
        public Items(ItemType type, ItemColor color,
            ItemDelivery deliver, decimal price)
        {
            typeofitem = type;
            itemcolor = color;
            itemdeliver = deliver;
            itemprice = price;
        }
    }
}
```

```
C:\WINNT\system32\cmd.exe
Enter the Type of Item to Order :
1. Shirts
2. Trousers
3. TShirts
4. SafariSuits
5. Jeans
6. Polos
7. Jackets
Please enter your Choice: 5
Enter the Color
1. Red
2. White
3. Yellow
4. Maroon
5. Orange
6. Blue
7. Lavender
8. Black
Your Choice: 6
Enter the Type of Delivery :
1. Packet
2. Soft Case
3. Box
4. Other
Your Choice: 1
Enter the Item Price:
200
Enter Item Quantity:
3

======================
======Garment Getup Shop======
The Delivery Details are :
----------------------
Item Type:              Jeans
Item Color:             Blue
Delivery Type:          Packet
Price(per item):        $200.00
Quantity:               3
Shipping Charges:        $24.00
-----------------------------------------
Total Price:            $624.00
==========================================

Press any key to continue . . .
```

BASICIOEXAMPLE APPLICATION

Introduction

BasicIOExample is a C# application that allows us to select a product for purchase and calculate the total price. We can create the BasicIOExample using Microsoft Visual Studio 2005, which is a rapid development tool for creating different types of applications such as Windows applications, Console applications and Web based applications. Microsoft Visual Studio 2005 supports the C# programming language to specify the functionality for an application. The BasicIOExample C# application contains three .cs files namely; **InputOutputExample.cs**, **ProcessOrder.cs** and **Items.cs**.

The **InputOutExample.cs** contains the code for calling the **OrderPlaced** and **OrderDetails** methods defined in the **ProcessOrder.cs** file using the order object of the **ProcessOrder** class. The **ProcessOrder** class is defined in the **ProcessOrder.cs** file. The **ProcessOrder.cs** file contains the code to specify the functionality for displaying the products and obtaining information such as quantity and price from the users. The **ProcessOrder.cs** file also contains the code for showing the details of the order placed by a user as a customer. The **Items.cs** file contains the code for defining the enumerations for **item type**, **item colour** and **item delivery**.

Creating the InputOutputExample.cs File

The **InputOutputExample.cs** file is created using the Microsoft Visual Studio 2005 so that OrderPlaced and OrderDetails methods of the ProcessOrder class can be called. To create the **InputOutputExample.cs** file:

1. Select *Start->Programs->Microsoft Visual Studio 2005->Microsoft Visual Studio 2005* to display the Microsoft Visual Studio IDE.
2. Select *File->New->Project* to display the **New Project** dialog box.
3. Select **Visual C#** from **Project types** section and **Console Application** from the Templates section to specify that you want to create a new console application.
4. Enter a name such as **BasicIOExample** in the **Name** text box to specify a name for the new console application.
5. Click OK to close the **New Project** dialog box. The **Code Editor** with **Program.cs** file and **Solution Explorer** appear in the Microsoft Visual Studio IDE.
6. Right click the **Program.cs** node that appears in the **Solution Explorer** to display a shortcut menu.
7. Select the **Rename** option from the shortcut menu to display an editable text box.
8. Enter a name such as **InputOutputExample.cs** in the editable text box to modify the name of the **Program.cs** file.
9. Add the following code in the **Main( )** method that appears in the **InputOutputExample.cs** file to specify the code for calling the **OrderPlaced** and **OrderDetails** methods of the **ProcessOrder** class:

```
ProcessOrder order = new ProcessOrder();

 order.OrderPlaced();
 Console.WriteLine();

 order.OrderDetails();
 Console.WriteLine();
```

This completes the process of creating the **InputOutputExample.cs** file.

Creating the ProcessOrder.cs File

The **ProcessOrder.cs** file contains the functionality for allowing users to select a product, colour and delivery type. It also contains the functionality for displaying the delivery details for the product selected by a user. To create the **ProcessOrder.cs** file:

1. Right click the **BasicIOExample** node that appears in the **Solution Explorer** to display a shortcut menu.
2. Select *Add->New* Item to display the **Add New** dialog box as shown in the following figure:

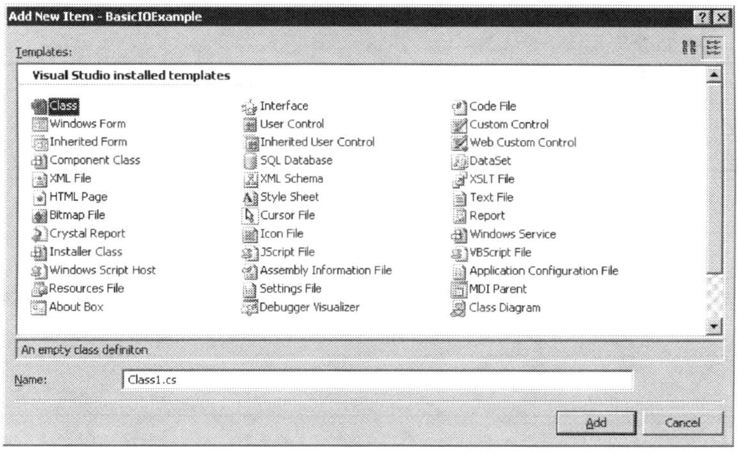

The Add New Item Dialog Box

Note: By default, the **Class** option appears selected in the **Add New Item** dialog box.

3. Enter a name such as **ProcessOrder.cs** in the **Name** text box to specify a name for the new **.cs** file, which you want to add to the **BasicIOExample** application.
4. Click **Add** to add a new **.cs** file to the **BasicIOExample** application. The Code **Editor** appears containing the **ProcessOrder.cs** file, as shown in the following figure:

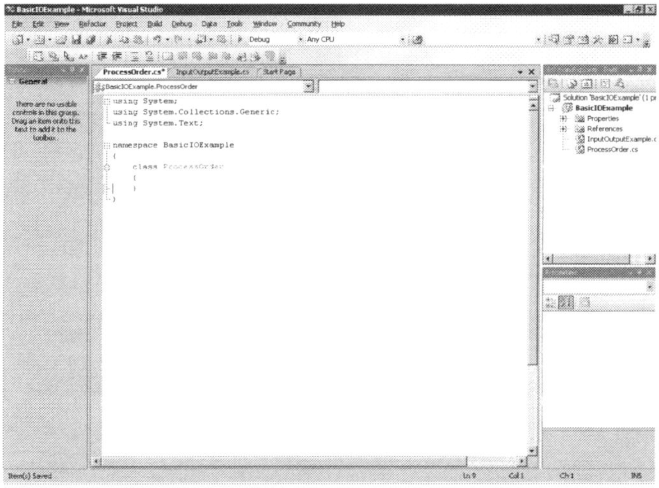

The Code Editor with ProcessOrder.cs File

5. Add the following code in the **ProcessOrder** class to specify the functionality to display a menu to the user for selecting a product and showing the delivery details.

```
public Items Order;
    public int Quantity;
    public decimal ShippingCharges;

    public ProcessOrder()
    {
        Order = new Items();
    }

    public decimal GetTotalPrice()
    {
        return Quantity * Order.itemprice + ShippingCharges;

    }

    public void GetItemType()
    {
        try
        {

            int choice = 0;

            Console.WriteLine("Enter the Type of Item to Order : ");
            Console.WriteLine("1. Shirts");
            Console.WriteLine("2. Trousers");
            Console.WriteLine("3. TShirts");
            Console.WriteLine("4. SafariSuits");
            Console.WriteLine("5. Jeans");
            Console.WriteLine("6. Polos");
            Console.WriteLine("7. Jackets");
            Console.Write("Please enter your Choice: ");
            choice = int.Parse(Console.ReadLine());

            switch (choice)
            {
                case 1:
                    Order.typeofitem = ItemType.Shirts;
                    break;
                case 2:
                    Order.typeofitem = ItemType.Trousers;
                    break;
                case 3:
                    Order.typeofitem = ItemType.TShirts;
                    break;
                case 4:
```

```
                    Order.typeofitem = ItemType.SafariSuits;
                    break;
                case 5:
                    Order.typeofitem = ItemType.Jeans;
                    break;
                case 6:
                    Order.typeofitem = ItemType.Polos;
                    break;
                default:
                    Order.typeofitem = ItemType.Jackets;
                    break;
            }
        }
        catch (FormatException fe)
        {
            Console.WriteLine("The error is " + fe.Message);
            Environment.Exit(0);
        }
        catch (Exception e)
        {
            Console.WriteLine(e.Message);
            Environment.Exit(0);
        }
    }

    public void GetItemColor()
    {
        try
        {
            int choice = 0;

            Console.WriteLine("Enter the Color");
            Console.WriteLine("1. Red");
            Console.WriteLine("2. White");
            Console.WriteLine("3. Yellow");
            Console.WriteLine("4. Maroon");
            Console.WriteLine("5. Orange");
            Console.WriteLine("6. Blue");
            Console.WriteLine("7. Lavender");
            Console.WriteLine("8. Black");
            Console.Write("Your Choice: ");
            choice = int.Parse(Console.ReadLine());

            switch (choice)
            {
                case 1:
```

```
                Order.itemcolor = ItemColor.Red;
                break;
            case 2:
                Order.itemcolor = ItemColor.White;
                break;
            case 3:
                Order.itemcolor = ItemColor.Yellow;
                break;
            case 4:
                Order.itemcolor = ItemColor.Maroon;
                break;
            case 5:
                Order.itemcolor = ItemColor.Yellow;
                break;
            case 6:
                Order.itemcolor = ItemColor.Blue;
                break;
            case 7:
                Order.itemcolor = ItemColor.Lavender;
                break;
            default:
                Order.itemcolor = ItemColor.Black;
                break;
        }
    }
    catch (FormatException fe)
    {
        Console.WriteLine("The error is " + fe.Message);
        Environment.Exit(0);
    }
    catch (Exception e)
    {
        Console.WriteLine(e.Message);
        Environment.Exit(0);
    }
}

public void GetItemDelivery()
{
    try
    {
        int choice = 0;

        Console.WriteLine("Enter the Type of Delivery : ");
        Console.WriteLine("1. Packet");
        Console.WriteLine("2. Soft Case");
```

```
        Console.WriteLine("3. Box");
        Console.WriteLine("4. Other");
        Console.Write("Your Choice: ");
        choice = int.Parse(Console.ReadLine());

        switch (choice)
        {
            case 1:
                Order.itemdeliver = ItemDelivery.Packet;
                break;
            case 2:
                Order.itemdeliver = ItemDelivery.SoftCase;
                break;
            case 3:
                Order.itemdeliver = ItemDelivery.Box;
                break;
            default:
                Order.itemdeliver = ItemDelivery.Other;
                break;
            }
        }

        catch (FormatException fe)
        {
            Console.WriteLine("The error is " + fe.Message);
            Environment.Exit(0);
        }
        catch (Exception e)
        {
            Console.WriteLine(e.Message);
            Environment.Exit(0);
        }
    }

    public void OrderPlaced()
    {
        GetItemType();
        GetItemColor();
        GetItemDelivery();
        try
        {
            Console.WriteLine("Enter the Item Price: ");
            Order.itemprice = decimal.Parse(Console.ReadLine());

            Console.WriteLine("Enter Item Quantity: ");
```

```
                Quantity = int.Parse(Console.ReadLine());

                ShippingCharges = ((Quantity * Order.itemprice * 4) / 100);
            }

            catch (FormatException fe)
            {
                Console.WriteLine("The error is " + fe.Message);
                Environment.Exit(0);
            }
            catch (Exception e)
            {
                Console.WriteLine(e.Message);
                Environment.Exit(0);
            }
        }

        public void OrderDetails()
        {
            Console.WriteLine("======================");
            Console.WriteLine("======Garment Getup Shop======");
            Console.WriteLine("The Delivery Details are : ");
            Console.WriteLine("——————————");
            Console.WriteLine("Item Type: {0}",
    Order.typeofitem);
            Console.WriteLine("Item Color: {0}",
    Order.itemcolor);
            Console.WriteLine("Delivery Type: {0}", Order.itemdeliver);
            Console.WriteLine("Price(per item): {0:C}", Order.itemprice);
            Console.WriteLine("Quantity: {0}", Quantity);
            Console.WriteLine("Shipping Charges: {0:C}", ShippingCharges);
            Console.WriteLine("————————————————");
            Console.WriteLine("Total Price: {0:C}", GetTotalPrice());

    Console.WriteLine("==========================================");
            }
```

This completes the process of creating the **ProcessOrder.cs** file.

Creating the Items.cs File

The **Items.cs** file contains the code for defining the enumerations such as *ItemType*, *ItemColor* and *ItemDelivery*. To create the **Items.cs** file:

1. Right click the **BasicIOExample** node that appears in the **Solution Explorer** to display a shortcut menu.
2. Select *Add->New Item* to display the **Add New Item** dialog box.

3. Enter a name such as **Items.cs** in the **Name** text box and click **Add** to add a new **.cs** file to the **BasicIOExample** application.
4. Add the following code before the **Items** class declaration and below the **BasicIOExample** namespace to define the enumerations; *ItemType*, *ItemColor* and *ItemDelivery*:

```
public enum ItemType
{
    Shirts = 1,
    Trousers,
    TShirts,
    SafariSuits,
    Jeans,
    Polos,
    Jackets
}

public enum ItemColor
{
    Red = 1,
    White,
    Yellow,
    Maroon,
    Orange,
    Blue,
    Lavender,
    Black
}

public enum ItemDelivery
{
    Packet = 1,
    SoftCase,
    Box,
    Other
}
```

Add the following code between the opening and closing braces for the **Items** class to define the **Items** class:

```
public ItemType typeofitem;
    public ItemColor itemcolor;
    public ItemDelivery itemdeliver;
    public decimal itemprice;

    public Items()
    {
        typeofitem = ItemType.Jackets;
```

```
        itemcolor = ItemColor.Black;
        itemdeliver = ItemDelivery.Packet;
        itemprice = 0.00M;
    }
    public Items(ItemType type)
    {
        typeofitem = type;
        itemcolor = ItemColor.Black;
        itemdeliver = ItemDelivery.Packet;
        itemprice = 0.00M;
    }
    public Items(ItemType type, ItemColor color,
            ItemDelivery deliver, decimal price)
    {
        typeofitem = type;
        itemcolor = color;
        itemdeliver = deliver;
        itemprice = price;
    }
```

This completes the task of creating the **Items.cs** file.

Testing the BasicIOExample Application

We need to test the **BasicIOExample** application to verify whether or not it is running properly. To test the **BasicIOExample** application:

1. Press **Ctrl + F5** to initiate the process of running the **BasicIOExample** application. A console window with a menu for selecting a product appears as shown in the following figure:

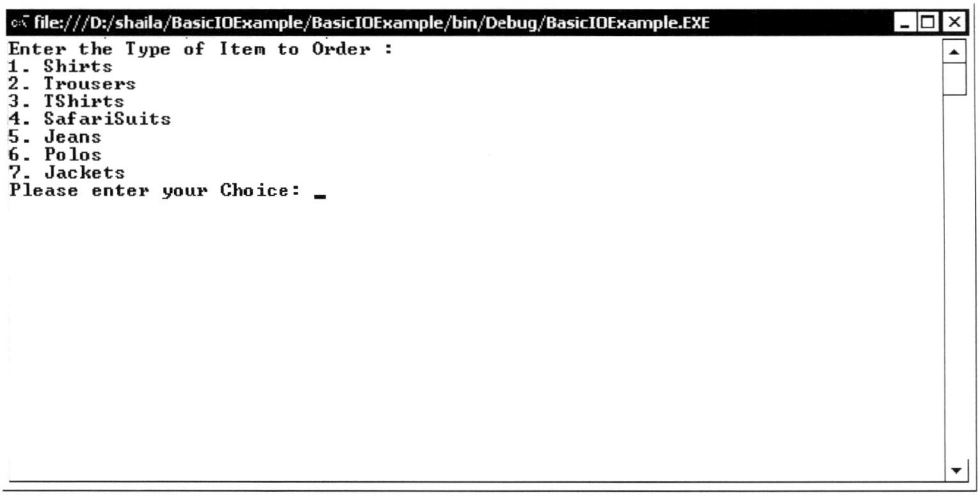

Figure 1 *The console window for selecting a product*

2. Enter a number such as 1 to specify a product to purchase and press **Enter**. A menu for choosing the colour for the selected product appears in the console window as shown in the following figure:

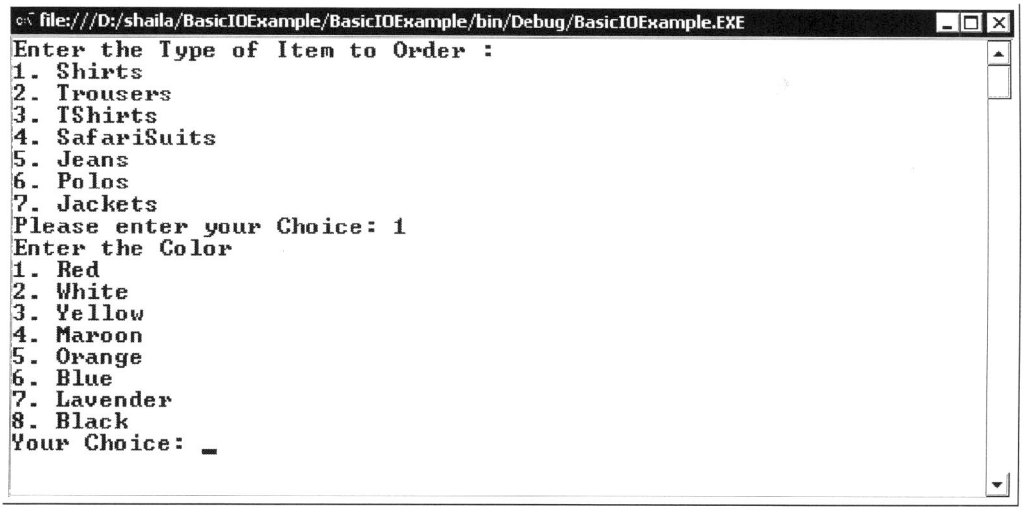

Figure 2 *The console window for selecting colour*

3. Enter a number such as 5 to specify a colour in which you want to purchase the selected product and press **Enter**. A menu appears that allows a user to select a delivery type for the selected product as shown in the following figure:

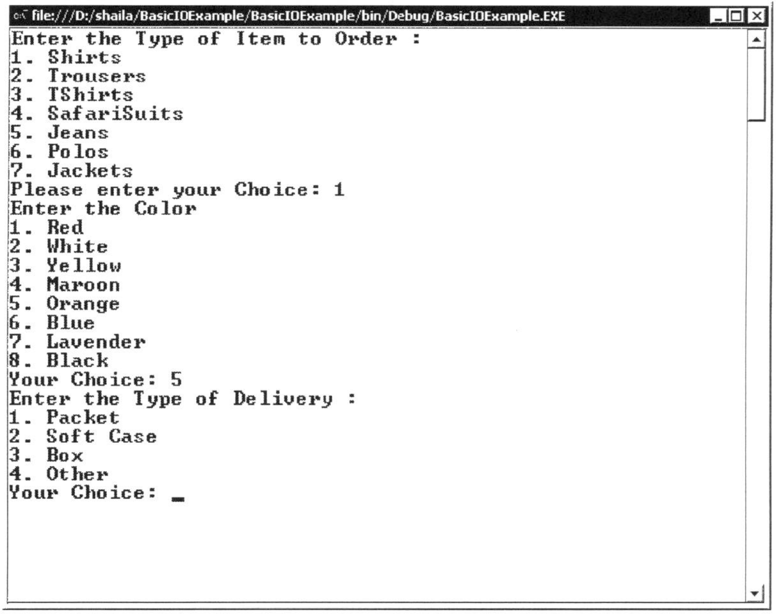

Figure 3 *The console window for selecting a delivery type*

4. Enter a number such as 3 to specify the delivery type through which you want the product to be delivered and press **Enter**. The console window appears prompting the user to enter the price for the selected product.
5. Enter a number such as 450 to specify the price for the selected product and press **Enter**. The console window prompting the user to enter the quantity for the selected product appears.
6. Enter a number such as 10 to specify the quantity that you want to be delivered for the selected product and press **Enter**. The delivery details for the selected product is displayed as shown in the following figure:

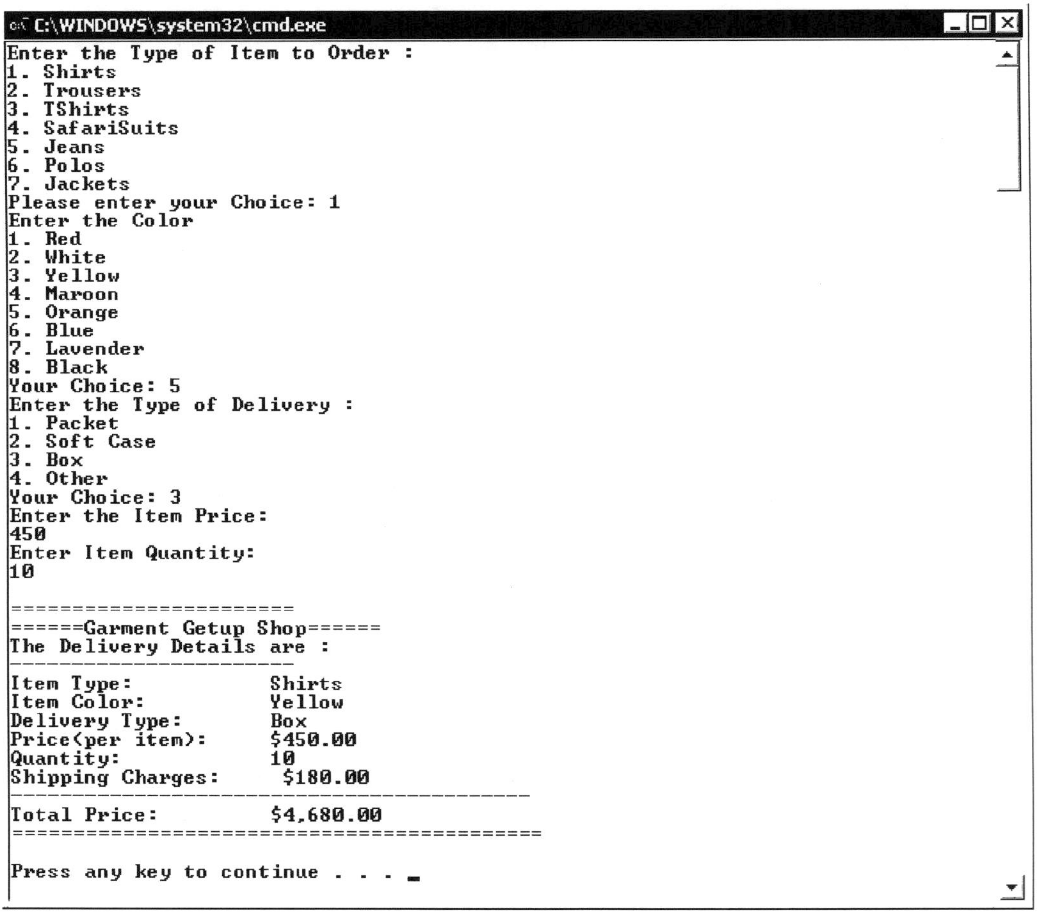

Figure 4 *The console window with delivery details*

This completes the process of testing the BasicIOExample application.

Bibliography

1. Albahari, B., et al., *C# Essentials*, O'Reilly, 1998.
2. Archer, T., *Inside C#*, Microsoft Press, 2001.
3. Conard, J., et al., *Introducing .NET*, Wrox Press, 2000.
4. Gunnerson, E., *A Programmer's Introduction to C#*, Apress, 2000.
5. Harvey, B., et al., *C# Programming*, Wrox Press, 2002.
6. Liberty, J., *Programming C#,* O'Reilly, 2001.
7. Mayo, J., *C# Unleashed*, Sams Publishing, 2002.
8. Michaelis, M. and Spokas, P., *C# Developer's Headstart*, Osborne/McGraw-Hill, 2001.
9. MSDN, *C# Language Specifications*, Microsoft Press, 2001.
10. Oberg, R.J., *C# Using .NET*, Pearson Education, 2002.
11. Robinson, S., et al., *Professional C#,* Wrox Press, 2001.
12. Schildt, H., *C#2.0: The Complete Reference,* McGraw-Hill, 2006.
13. Troelsen, A., *C# and the .NET Platform*, Apress, 2001.
14. Watson, K., *Beginning C#*, Wrox Press, 2001.
15. Wille, C., *Presenting C#*, Sams Publishing, 2000.

Index

D

E

F

G

Made in the USA
Coppell, TX
05 November 2026

11289013R00297